Sustainable Development Goals Series

The **Sustainable Development Goals Series** is Springer Nature's inaugural cross-imprint book series that addresses and supports the United Nations' seventeen Sustainable Development Goals. The series fosters comprehensive research focused on these global targets and endeavours to address some of society's greatest grand challenges. The SDGs are inherently multidisciplinary, and they bring people working across different fields together and working towards a common goal. In this spirit, the Sustainable Development Goals series is the first at Springer Nature to publish books under both the Springer and Palgrave Macmillan imprints, bringing the strengths of our imprints together.

The Sustainable Development Goals Series is organized into eighteen subseries: one subseries based around each of the seventeen respective Sustainable Development Goals, and an eighteenth subseries, "Connecting the Goals", which serves as a home for volumes addressing multiple goals or studying the SDGs as a whole. Each subseries is guided by an expert Subseries Advisor with years or decades of experience studying and addressing core components of their respective Goal.

The SDG Series has a remit as broad as the SDGs themselves, and contributions are welcome from scientists, academics, policymakers, and researchers working in fields related to any of the seventeen goals. If you are interested in contributing a monograph or curated volume to the series, please contact the Publishers: Zachary Romano [Springer; zachary.romano@springer.com] and Rachael Ballard [Palgrave Macmillan; rachael.ballard@palgrave.com].

Mark Matsa
Lazarus Chapungu • Godwell Nhamo
Editors

Climate Change Resilience in Rural Southern Africa

Dynamics, Prospects and Challenges

Springer

Editors
Mark Matsa
Department of Geography
Environmental Sustainability
and Resilience Building
Midlands State University
Gweru, Zimbabwe

Lazarus Chapungu
Institute for Corporate Citizenship
University of South Africa
Pretoria, South Africa

Godwell Nhamo
Institute for Corporate Citizenship
University of South Africa
Pretoria, South Africa

ISSN 2523-3084 ISSN 2523-3092 (electronic)
Sustainable Development Goals Series
ISBN 978-3-031-73599-8 ISBN 978-3-031-73600-1 (eBook)
https://doi.org/10.1007/978-3-031-73600-1

This Springer imprint is published by the registered company Springer Nature Switzerland AG
The registered company address is: Gewerbestrasse 11, 6330 Cham, Switzerland

Preface

Climate change is a critical global issue that is affecting communities worldwide. In rural Southern Africa, the reliance on natural resources and agriculture makes the subregion particularly vulnerable to these impacts. This book, *Climate Change Resilience in Rural Southern Africa: The Dynamics, Prospects and Challenges*, explores the connection between climate change and rural resilience in this highly vulnerable region. Southern Africa's climate vulnerability is expected to worsen, yet the population in most of the countries remains rural and dependent on the sectors and ecosystems that are vulnerable to climate and climate change vicissitudes.

Southern Africa is predicted to become warmer and drier, with variable and erratic rainfall episodes. The intensity of rapid and sudden extreme events, which include droughts, floods, tropical cyclones, diseases, pest infestations, and veld fires is expected to increase. The 2016 drought, for example, caused the loss of lives and economic and livelihood assets, which escalated to food security and health-related crises and population displacements. Three years later, in 2019, severe floods in the region killed more than 1000 people and inundated and destroyed vast areas of crops, food stocks, livestock, and infrastructure. Similar trends are also taking place regarding slow onset events like sea-level rise and ocean acidification. These scenarios always result in increased shocks and stressors for the rural economies of Southern Africa, which are intrinsically sensitive to the direct impacts of climate change. Most of these rural economies are highly dependent on rainfed agriculture and natural ecosystems due to the high levels of poverty and geographic exposure to climate extremes.

It is estimated that 83% of the extremely poor population of Southern Africa live in rural areas, and apart from the flagship subsistence, to semi-intensive agriculture, also depend on other rural economic activities such as pottery, fishery, forestry, and tourism, all of which are vulnerable to climate change. In the face of the threat to climate change and climate variability, and given the importance of the natural ecosystem to rural economies in Southern Africa, sustainability of rural economies should strive to achieve climate resilience, which is the ability of social, economic, and environmental systems to withstand climate change-related impacts, so that they can thrive despite the impact.

The impetus for this work stems from a recognition that the effects of climate change are not uniform worldwide with rural Southern Africa facing unique challenges that require tailored approaches to resilience and adaptation.

This book complements the Southern Africa Development Community (SADC) Regional Resilience Framework 2030-2030, which guides member states' thrust in the design and implementation of the broad spectrum of resilience programmes in support of the regional body's vision, mission, and goals. It advocates for rural inclusion in the region's thrust for resilience building. Rural areas are generally habited by communities who are concerned more about survival than politics and their voices hardly reach the corridors of power to motivate timely, efficient, and transformative action. This book seeks to be the voice of these communities, which incidentally still make up the majority of the population of the regional body.

This book incorporates interdisciplinary research, bringing together perspectives from a diversity of fields such as climatology, agriculture, sociology, and policy studies. The work is the outcome of extensive field research, ongoing dialogue with stakeholders, and systematic review of literature. The aim is to present a holistic view of climate resilience in rural Southern Africa and offer practical recommendations for policymakers, development practitioners, and scholars.

This book is premised on the understanding that the Conceptual Framework for Resilience Building in the SADC Region and member states identifies seven resilience priorities, namely: (1) robust and connected infrastructure; (2) food and nutrition security; (3) social and human development and mobility; (4) natural resources management, protection of biodiversity and conservation; (5) integrated governance and informed decision-making; (6) understanding disaster risks including climate change; and (7) sustainable urban centres. Of these regional priorities, only sustainable urban centres do not directly impact rural development. Ironically, the other six priorities, which have a direct impact on rural development, do not have a strong footprint on resilience building effort on the ground in most of rural Southern Africa, with some SADC communities at the coalface of armed conflict, exacerbating climate shocks and stressors already being experienced.

This book advocates for a multi-scale, transboundary resilience-building approach within the elevated disaster profile of Southern Africa, but with a deliberate bias towards rural communities that are at the tipping edge of droughts, floods, cyclones, climate-induced pests and diseases, as well as chronic food insecurity. In these communities, resilience to climate change is in dire need. These peripheral communities need to develop from within, rather than waiting for development to trickle from regional economic core nodes, as the wait may be eternal. While adopting the Sendai Framework for Disaster Risk Reduction (2015–2030) in March 2015, member states committed to disaster risk reduction and building resilience to disasters to be addressed with a renewed sense of urgency in the context of sustainable development and poverty reduction. The emphasis was on the integration of resilience building into plans, policies, programmes, projects, and budgets at all levels. Rural Southern Africa is in dire need of building resilience and this requires strong commitment and involvement of the region's political leadership in every country and at all levels.

Rural resilience is critical to managing the effects of climate change and in achieving the 2030 Agenda for Sustainable Development and its 17 intertwined Sustainable Development Goals (SDG), especially SDG 13 (Climate Action). Resilient communities are able to manage stressors and recover from them. This calls for Southern African leaders to increase investments in early warnings and early action systems on natural hazards which are accountable to communities and invest in long-term programmes that will help build the people's ability to cope with future crises so as to meet the targets of Africa's Agenda 2063. Promotion of agro-ecological approaches to transition towards more just and sustainable food systems and increase farmers' resilience in the face of climate change is vital. Crucial for rural resilience is a system's ability to adapt to external disturbances; or in other words, the sub-regional ability to transform. For a community to thrive over a long period of time, it is essential that resilience emphasizes the ability to transform along with shifting paradigms. This is what needs to be focused on at all levels of leadership for resilience building against climate change in rural areas of southern Africa.

The pathways to climate resilience in rural southern Africa are fraught with challenges but also filled with opportunities for transformative change. Building resilient communities that can thrive in the face of climate adversity requires utilizing local expertise, strengthening policy frameworks, and encouraging cross-sector collaboration. The findings and recommendations in this book aim to aid policymakers, practitioners, and academics in designing comprehensive policies that not only minimize the risks of climate change but also leverage its opportunities to create sustainable development and social fairness. Building climate resilience in rural Southern Africa requires a holistic approach that integrates environmental sustainability, economic development, and social justice. It demands the collective efforts of governments, communities, and international partners to create a sustainable and resilient future for the region.

Gweru, Zimbabwe — Mark Matsa
Pretoria, South Africa — Lazarus Chapungu
Pretoria, South Africa — Godwell Nhamo

Acknowledgements

The Editors (Prof. Mark Matsa, Dr. Lazarus Chapungu, and Prof Godwell Nhamo) are grateful for the invaluable support and contributions leading to the publication of this book, ***Climate Change Resilience in Rural Southern Africa: The Dynamics, Prospects and Challenges***. First and foremost, we extend our heartfelt thanks to the Exxaro Chair in Climate and Sustainability Transition at the University of South Africa for its generous funding, which enabled a rigorous blind peer review process of the work. We are also immensely appreciative of the contributions made by all the authors whose hard work, dedication, and expertise were instrumental in shaping the chapters of this book. Their collective efforts enriched this publication and provided critical insights into the complex issues of climate resilience in rural Southern Africa. Special thanks also go to the two blind book reviewers whose names remain anonymous in line with global standards. Their meticulous and thoughtful feedback edified the product. We also extend our gratitude to the communities and stakeholders who participated in the various studies presented in this book. Their willingness to share experiences and perspectives provided a foundation for understanding the real-world impacts of climate change and the resilience strategies being employed. Finally, we acknowledge the support and collaboration of our partner institutions, the Midlands State University (Zimbabwe) and the University of South Africa (South Africa). The partnership was vital in bringing this project to fruition.

Contents

1 An Overview of Climate Change Resilience in Rural Southern Africa

Mark Matsa, Lazarus Chapungu, and Godwell Nhamo

1.1 Background

Climate change is the shift in climate patterns primarily caused by greenhouse gas emissions from natural systems and human activities (Ahmed, 2020; Fawzy et al., 2020). Such shifts, of necessity, demand that the affected human and natural systems adjust accordingly to avert death and extinction. According to observed rates and projected values, climate change-induced warming across Southern Africa is occurring at a rate twice that of the rest of the world (Engelbrecht et al., 2015). It is projected that the interior of southern Africa may see temperature increases of more than 4 ° C by 2100 if coordinated international action is not taken to reduce greenhouse gas emissions (Kapwata et al., 2018).

In rural southern Africa, climate change trends have been characterized by increased temperatures, more variable rainfall patterns, and extreme weather events (Thomas et al., 2022). These trends result in several impacts on rural communities. Rural communities, which largely depend on agriculture, face a reduction in crop productivity due to increased temperatures and altered precipitation patterns (Nhemachena et al., 2020). The altered rainfall patterns have exacerbated water scarcity problems in the region with consequent effects on water-dependent economic sectors (Kusangaya et al., 2014; Leal et al., 2022). Southern Africa has also experienced many health risks due to climate change (Wright et al., 2021). Notable climate-induced increases in vector-borne diseases have been observed (Chapungu & Nhamo, 2021).

The region faces a variety of climate-related hazards, including droughts, floods, and cyclones, exacerbating existing vulnerabilities and displacing communities from their homes and livelihoods (Jegede et al., 2022). Rural populations are particularly susceptible to displacement, as changing weather patterns disrupt traditional ways of life (Zuniga et al., 2022). In addition to direct impacts, such as loss of crops and infrastructure, displacement can lead to secondary effects, such as food insecurity, conflict over dwindling resources, and increased migration to already overburdened urban centers.

Almost half the global human population (about 3.4 billion people) live in rural communities, and more than 90% live in developing countries where they depend on natural resources and rain-fed agriculture (Le Roux, 2024). In this book, rural areas refer to open fields and small settlements in the countryside with a greater dependency on agriculture and natural resources,

M. Matsa (✉)
Department of Geography, Midlands State University, Gweru, Zimbabwe
e-mail: matsam@staff.msu.ac.zw

L. Chapungu · G. Nhamo
Institute for Corporate Citizenship, University of South Africa, Pretoria, South Africa
e-mail: tchapul@unisa.ac.za; nhamog@unisa.ac.za

M. Matsa et al. (eds.), *Climate Change Resilience in Rural Southern Africa*, Sustainable Development Goals Series, https://doi.org/10.1007/978-3-031-73600-1_1

which make them uniquely vulnerable to the impacts of climate change. Rural areas are generally marginalized and isolated from centers of mainstream economic activities, have a high prevalence of poverty and low human development, and have been neglected by policymakers. Inherent vulnerabilities, such as poverty, lower levels of education, neglect by national centers of power and influence, and exclusion from decision-making, aggravate the impacts of climate change in rural Southern Africa.

The concept of climate resilience, as defined by the IPCC (2022), refers to the capacity of social, economic, and ecological systems to respond effectively and recover from hazardous events, trends, or disturbances while maintaining their essential functions, identity, and structure. The IPCC identifies three main pillars of resilience: social resilience, which denotes a community's ability to rebound from disruptions and losses with minimal external assistance; economic resilience, which encompasses a community's capacity to withstand, recover from, and mitigate disturbances to its economic foundations; and ecological resilience, which refers to the health status of an ecosystem and its ability to swiftly recuperate following disasters. Climate change resilience, as highlighted by Ntontis et al. (2020), Singh (2021), and Sharifi (2021), involves the ability of natural and human systems to endure, adapt to, and rebound from the impacts of climate change while mitigating adverse effects. It encompasses enhancing the foresight, response, and management capacities of individuals, communities, ecosystems, and institutions to address the challenges posed by climate change.

Given the proliferation of climate change-related disasters and the consequent impacts on rural communities, the need to build resilience cannot be overemphasized. This chapter provides an overview of the resilience to climate change in Southern Africa to set the tone for examining trends, dynamics, prospects, and challenges in the region. The overview includes a discussion on the theoretical perspectives of climate resilience, an overview of trends and impacts, vulnerabilities of rural communities, and challenges and opportunities for resilience.

1.2 Resilience to Climate Change: A Theoretical Perspective

Physical scientists first used the idea of resilience to illustrate how springs function to characterize a material's stability and resistance to external shocks. The idea of resilience was applied to the ecological field in later advances during the 1960s. Adger (2000) established the notion of resilience in the social realm, which refers to a community's capacity to tolerate external disruptions to infrastructural conditions. Adaptation is necessary to mitigate the effects of climate change and direct physical impacts that arise within societal and biological systems. Resilience in socioecological systems is frequently called adaptability. The concepts of resilience and adaptation are sometimes used interchangeably; however, resilience is linked to thought processes, whereas adaptations are connected to actors, policies, and activities. Resilience and "adaptation" are frequently associated in the context of climate change, whereas risk takes the place of "vulnerability reduction" in a disaster.

Climate change induces hazards that directly affect the intrinsic resilience of rural communities. The ability of communities to cope with or adapt to physical impacts determines the degree of resilience. Additionally, the level of exposure to hazards and the transformative capacity determine overall resilience. Figure 1.1 illustrates the interconnected roles of exposure, risk, and internal resilience in shaping overall resilience.

Resilience comprises hazards, intrinsic resilience, and exposure, with intrinsic resilience representing the inherent capacity of a socioecological system to withstand diverse hazards. Intrinsic resilience denotes an intrinsic attribute of the system itself, reflecting its ability to cope with disturbances and is influenced by the community's collective experiences and adaptive capacities in response to climate-related disruptions. Therefore, the system's resilience hinges on its capacity to adapt and transform in the face of evolving climate challenges, thereby sustaining its ability to cope with and rebound from future disturbances.

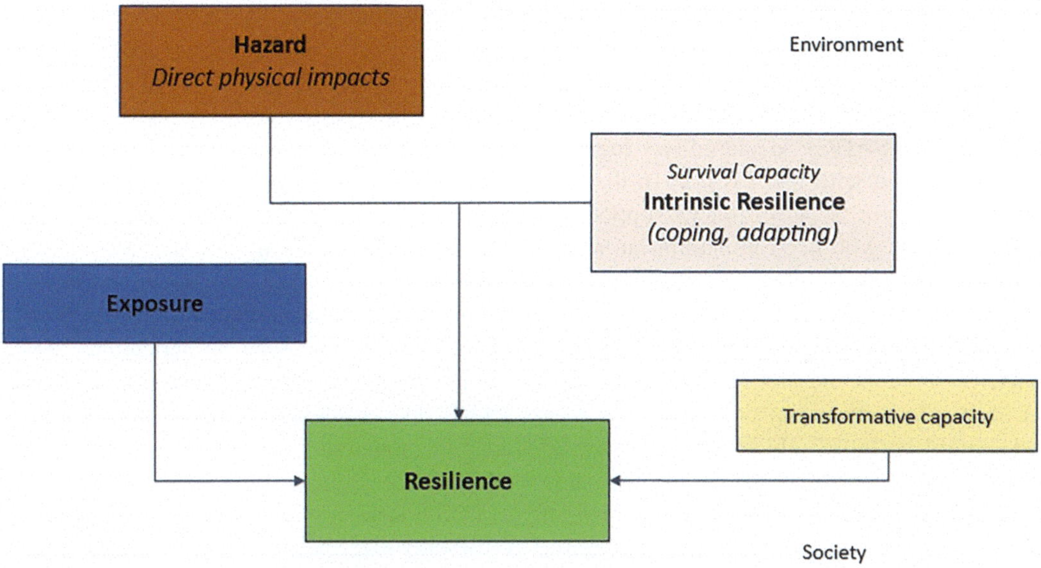

Fig. 1.1 The determinants of climate change (Source: Authors, adapted from Subiyanto et al. (2020))

As postulated by Bogardi and Fekete, resilience is formally defined as the capacity to rebound from hazardous situations and to withstand the impact of unforeseen events. The frequency and severity of climate change-induced occurrences, such as extreme weather events, depend on the societal capacity to adapt and respond to disruptions. Within this framework, the quantification of resilience depends on exposure and intrinsic resilience, assuming a standardized measure of risk. Furthermore, resilience underscores the system's inherent capacity to adapt and evolve, a notion that will be elucidated in the ensuing discourse.

1.3 Rural Southern Africa Disaster Risk Context

Rural Southern Africa is particularly vulnerable to the effects of climate change due to its diverse terrain, abundant biodiversity, and high reliance on natural resources (Trew & Maclean, 2021). Disaster risks are increasing due to the frequency and intensity of hazards, increasing vulnerability, and the limited adaptive capacity of the region. The subregion experiences a variety of climate-related risks such as droughts, floods, heatwaves, shifting rainfall patterns, and tropical cyclones that, as evidenced by the successive major cyclones Idai and Freddy of 2019, occur more frequently and at greater magnitude as a result of climate change (Edokpayi et al., 2020; Defe & Matsa, 2021; Nilsson et al., 2021; Mafongoya et al.,2022). Other hazards that have affected Southern Africa include diseases (pandemics such as COVID-19 and cholera and epidemics such as malaria and measles), pest infestations, fires, and conflict in countries such as Mozambique and the Democratic Republic of Congo.

In Southern Africa, the severity of disasters has been exacerbated by the detrimental impact of climate change and other ills such as poverty, HIV/AIDS, food insecurity, population growth, migration, increased urbanization, food insecurity, and stress on natural resources.

Climate extremes and related disasters have long affected Southern Africa, especially rural communities, undermining the prospects for development. Manatsa et al. noted that the 1992 drought, recorded as the most extreme in the twentieth century, cost the region US $4 billion overall. Thirteen years later, El Nino-induced drought in 2015–2016 was recorded as the worst in 35 years. These drought episodes always have shocks and stressors that are more severe and impact rural communities whose livelihoods are based on nature and climate. The Southern African Development Community (SADC) declared a regional disaster to cope with the 2015–2016 disaster and appealed for $2.4 billion to support 40 million people who need humanitarian assistance (Le Roux, 2024). This declaration exposed the region's institutional and political challenges. To this day, Southern Africa's capacity to reduce disaster risk is limited, and its ability to reduce the occurrence of new risks could be more robust. As noted by Le Roux (2024), there is a disproportionate emphasis on disaster risk management, which emphasizes preparation for relief, response, and recovery, rather than proactive investment in resilience building to reduce disaster risk equitably.

1.4 Resilience-Building Initiatives in Southern Africa

The Southern African region experiences myriad climate change-related shocks and stressors, which has led to the implementation and adoption of various combat strategies with different levels of effectiveness (Udeagha & Muchapondwa, 2022; Musasa & Matsa, 2022; Dube et al., 2022; Chivasa et al., 2022).

The adoption of the Sendai Framework for Disaster Risk Reduction 2015–2030 (SFDRR) introduced the concept of resilience building into the planning blueprints of Southern African countries (Le Roux, 2024). Zimbabwe (2015), Lesotho (2017), Eswatini (2017), and Malawi (2018) have all developed national resilience strategies. The SADC Resilience Framework (2020–2030) was developed in alignment and coherence with these regional initiatives. This integration emphasizes shared governance and informed decision-making for regional resilience building in the context of equitable, sustainable development, reduction of disaster risk, and adaptation to climate change.

Strategies and Plans

The scope of climate change resilience in Southern Africa involves a comprehensive set of strategies and activities that address various aspects of resilience, including adaptation measures (Busayo et al., 2019; Kareem et al., 2020; Matsa et al., 2022; Ndebele-Murisa et al., 2020; Scholz & Methner, 2020). It further encompasses the depth of actions taken to mitigate and adapt to the impacts of climate change and also takes note of different sectors such as agriculture, water resources, energy, transport, urban planning, and social and economic dimensions. Southern African countries are implementing various adaptation measures to reduce their vulnerability to the impacts of climate change (Leichenko & O'Brien, 2002; Spear et al., 2018). Different climate shocks and stressors are considered while aiming at building resilience at multiple levels locally, nationally, and regionally (Jones & Tanner, 2017; World Health Organization, 2022).

The Southern African Development Community (SADC) facilitates sharing knowledge, coordination efforts, and the development of policies and strategies. Since then, the region has adopted climate change resilience initiatives, and recently, the revised climate change resilience framework and the region's experiences and future projections on climate change issues resulted in the implementation of the Southern African Development Community Regional Resilience Framework 2020–2030 (Justo et al., 2022; Coetzee et al., 2023 and Muyambo et al., 2023). This framework aims to provide a broad outline for coordinated and harmonized regional and national actions to address and respond to climate change and its impacts. The adopted strategies to curb climate change-related shocks and stressors fail as a result of factors including poor participation by the countries in the SADC region, which ultimately have negative impacts on climate change resilience (Nemakonde & Van Niekerk, 2023).

Climate change resilience activities can be promoted at the national level. Eswatini experiences droughts. However, the country managed to boost its cash crop production of sugarcane, earning an increased gross domestic product (GDP) per capita through the adoption of irrigation schemes, which has enabled the country to become a middle-income earner despite the changing climate (Mhlanga-Ndlovu, 2022; World Bank, 2019; Maziya, 2023; Kibirige & Singh, 2023). Botswana adopted the 11 Volume 1 Botswana National Development Plan (April 2017–March 2023), which embraces the development of climate-smart agriculture (CSA) programs with a specific focus on increasing climate change resilience in production systems and subsequent outcomes through irrigation schemes (Madebwe, 2019; Kwembeya & Shikangalah, 2023). A study conducted in the Ditladi and Gulushabe fields in Botswana in May 2019 revealed the benefit of irrigation carried out by farmers who perceived an increase in agricultural production (Bosekeng, 2020; Moseki et al., 2019). Together, these actions help build resilience to climate change among rural communities whose livelihoods depend heavily on rain-fed agriculture.

Education and awareness are also critical to building community resilience to climate change and have been widely applied throughout the region. Through climate change awareness programs among rural fishing communities in the Seychelles and Mauritius, shark catches have provided US $1.6 million with the possibility of increasing to US $14 million if the countries fully evolve into climate change resilience (Demirel et al., 2023; Saddington, 2023; Tolotti et al., 2023). The South African government integrated environmental education with the provisions of the National Climate Change Adaptation Plan (NCCAP) (Fambasayi, 2021). The NCCAP aimed at disaster risk management, food and nutrition security, water and the blue economy, forestry, wildlife and tourism, health, sanitation and human settlements, manufacturing, energy, and transport (Gbomagba et al., 2023; Muyambo et al., 2023). Studies conducted in the Gwanda district of Zimbabwe demonstrated the importance of environmental education (EE) by providing the community with the ability to determine the need for sustainable environmental practices to reduce droughts induced by climate change and the severity of related shocks and stressors, including outbreaks of diseases (Dzvimbo et al., 2022; Moyo-Nyoni, 2022; Sibanda & Manik, 2022).

Other climate change resilience initiatives that have helped communities bounce back better from drought include water harvesting, community gardens, and cooperatives. In Zimbabwe, water harvesting and irrigation practices in the Silalatshani irrigation scheme in Insiza district, Matabeleland South Province, significantly help to build resilience to climate change, specifically in agricultural production during prolonged dry spells (Muroyiwa et al., 2023; Nkonya et al., 2023; Wellington et al., 2023). Musasa and Matsa (2022) contend that water harvesting in arid and semiarid regions for communitarian horticultural gardens, apiculture, and climate-smart agriculture contributes to resilience to climate change. A study conducted by Ndebele-Murisa et al. (2020) on the Future Resilience for African Cities and Lands (FRACTAL) project shared experiences that contributed to climate-resilient development extending to regional borders including South Africa, Namibia, Zambia, and Zimbabwe.

Although government-led climate change resilience initiatives are essential, community-led climate resilience initiatives play a crucial role in fostering long-term well-being within communities and sustainability. Community-led initiatives have been promoted for managing natural resources and sustainable land management in rural areas of Zimbabwe to improve local climate change resilience (UNEP, 2019). In South Africa, rural communities play a critical role in improving resilience to climate change. These initiatives have incorporated indigenous knowledge, which has helped preserve seed genotypes and conserve soil, improving climate resilience. In this sense, community-led climate change resilience programs have helped to build the capacity of individuals, communities, and systems to anticipate, respond to, and recover from disruptions. As such, climate change resilience initiatives in the region must emphasize transformative measures to ensure the long-term resilience of local communities to climate change. This will go a long way in helping the region meet the goals of multiple welfare

Sustainable Development Goals (SDG 1, 2, and 3) and Africa Agenda 2063 objectives.

Despite growing regional, national, and community efforts to build climate-resilient rural communities, the region still lags behind in climate resilience. Research by Chilonda et al. and Mutengwa et al. (2023) revealed that more financial resources restrict regional efforts to effect sustainable climate change-resilient rural communities. This hinders the region's capacity to adopt irrigation initiatives, early warning systems, and the latest technologies, resulting in difficulties in coping and surviving in the changing climate. These challenges show the region's vulnerability and expose people to hunger, malnutrition, and diseases. They also reveal weaknesses in building resilience to climate change.

The irony of vulnerability and poverty in rural areas is that rural people in the world and those of Southern Africa, in particular, have, on long-term scales, adapted to climate variability through their respective Indigenous knowledge systems and the diversification of livelihoods through informal institutions for risk sharing and risk management. Given supportive policies and institutions, similar adaptation and coping strategies can form the basis for adaptation to climate change. However, the effectiveness of such approaches will depend on the severity and speed of the unfolding climate change-related impacts. As such, regional leaders, development partners, and all stakeholders must work together to promote long-term resilience to climate change among rural communities within the Southern African region.

1.5 The Rationale and Purpose of the Book

Climate change has exacerbated the socioeconomic and environmental challenges in Southern Africa. It poses significant threats to the biophysical and socioeconomic fabric of societies in the region, with consequent implications for food security, water resources, agriculture, biodiversity, tourism, and poverty. Agro-based societies have suffered the most; rural areas are mainly at the forefront. Climate-induced reduction in crop production significantly correlates with increased food insecurity in the region. Food insecurity has followed an upward trend, increasing from an estimated 23 million in 2011/2012 to 51 million in 2020/2021 (WFP 2021). Science has shown that the highest food-insecure caseloads follow the worst rain seasons. Additionally, extreme weather events associated with climate change have exacerbated existing vulnerabilities in rural Southern Africa.

The need to fortify rural communities against existing and imminent climate catastrophes cannot be overemphasized. Emerging insights on climate change strategies suggest that building resilience in human and environmental systems is ideal for combating dynamic environmental challenges and future uncertainties and addressing socioeconomic and environmental concerns related to climate change. Various stakeholders across sectors have made concerted efforts to build community resilience to climate change. Some interventions have built resilience in at-risk households, communities, and systems, while others have faced significant challenges. However, for rural communities in Southern Africa, the trends and dynamics of climate resilience remain shrouded in obscurity. There needs to be a more scientific understanding of the progress, challenges, prospects, and successes of current and planned interventions to build resilience to climate change in rural areas.

The book, therefore, seeks to unpack the prospects, challenges, and lived experiences related to the construction of resilience to climate change in Southern Africa. It explores the science, economics, and politics of climate resilience-building processes in rural Southern Africa. In the process, the volume is also dedicated to understanding practical and policy-oriented approaches to tackle the challenges related to building resilience to climate change in rural areas. The work brings together researchers, development practitioners, and policymakers from various disciplines to understand the dynamics of resilience to climate change in rural Southern Africa. In addition, it explores the scientific and policy options for enhancing the region's capacity to build resilience.

1.6 Methodology

1.6.1 The Geographical Context

This book covers the southern African region (Fig. 1.2), which lies south of the Equator and includes countries such as Angola, Botswana, the Democratic Republic of the Congo, Lesotho, Malawi, Mozambique, Namibia, South Africa, Eswatini, Tanzania, Zambia, and Zimbabwe. When island nations are not included, the region's population is estimated to be 389.38 million. Only 6.11% of the 9.86 million square kilometers of land these countries include are used for agriculture. More than 60% of the population depends primarily on agriculture for their livelihood and contributes significantly to GDP. Although just over 20 million acres of fertile land, or 6.6% of the total cultivated area, are considered suitable for irrigation, the region's agricultural economy depends highly on rainfall.

1.6.2 Research Approaches

This volume is founded on a comprehensive and multifaceted data collection and analysis approach guided by the mixed methods research design.

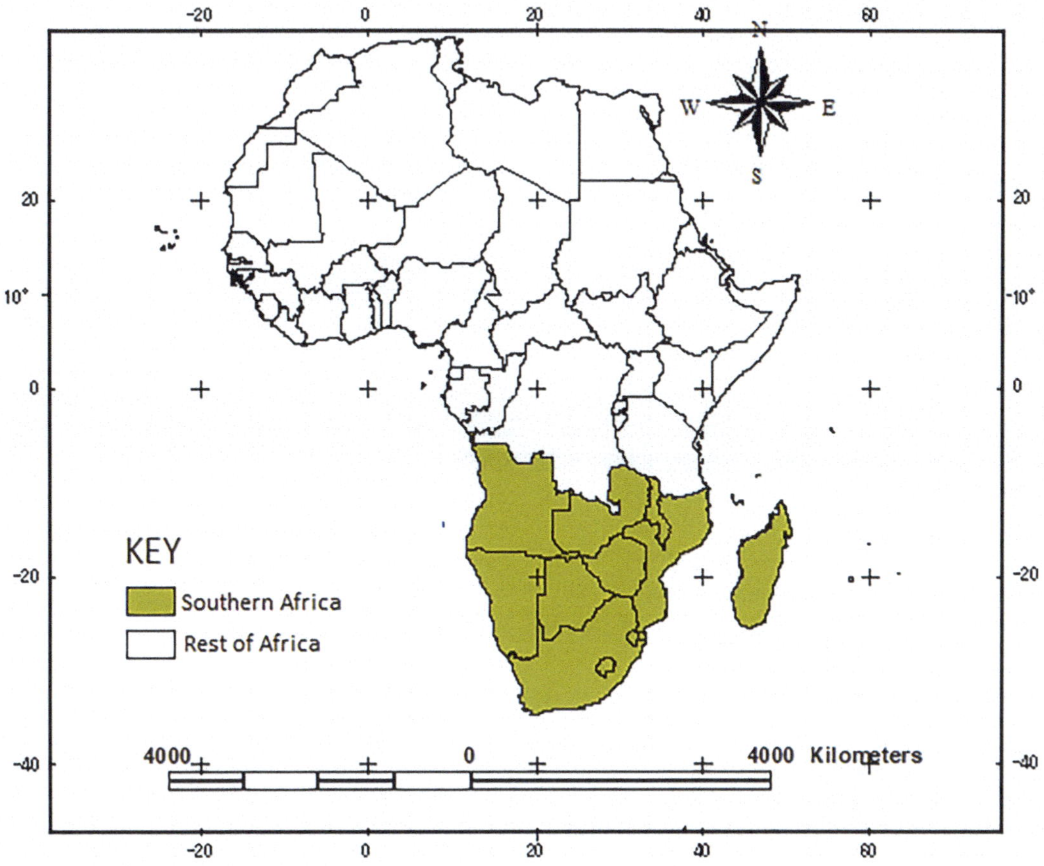

Fig. 1.2 Study area map

Various research techniques and methodologies were used, ranging from qualitative to quantitative methods. The deliberate adoption of case studies was motivated by their ability to provide detailed empirical insight into specific concepts, sectors, or geographical regions. Data collection primarily involved interviews, a systematic review of documents, and the use of existing databases drawn from various geographical areas pertinent to this research. Statistical analyses were performed using a variety of statistical packages to assess trends and variable dynamics. Some chapters used existing circulation models and conducted statistical downscaling to elucidate the landscape of resilience to climate change in Southern Africa. The Preferred Reporting Items for Systematic Reviews and Meta-Analyses (PRISMA) framework was extensively employed in this volume, given its efficacy in providing robust evidence on climate resilience's dynamics, prospects, and challenges. In addition, content and thematic analyses were used in specific chapters to deepen understanding and analysis further.

1.7 Structure of the Book

Given the need to fortify rural communities against the existing and impending ramifications of climate change, this book seeks to shed more light on the status quo, prospects, and challenges of climate resilience in rural Southern Africa. The introductory chapter sets the tone by providing the background and synthesis of climate resilience in rural Southern Africa. Chapter 2 examines the landscape of climate change in rural Southern Africa, especially the nexus between climate characteristics and rural livelihood patterns. The study highlights that with projected changes in temperature and precipitation patterns in different parts of the region, resilience efforts should pay attention to the characteristics of rural parts of the regions. By analyzing established trends and anticipated climate change patterns, research lays the foundations for assessing risk and vulnerability, paving the way for adaptation strategies that can support rural Southern Africa's livelihoods and economic activities.

Chapter 3 contends that understanding the trends and drivers of climate vulnerability is crucial to resilience-building strategies, addressing vulnerabilities, and promoting sustainable development in Southern Africa. The results suggest that the increase in disasters has led to more displaced people and that spatial variations in vulnerability are influenced by economic, sociocultural, political, and geographic factors. Due to the complex relationship between vulnerability and climate drivers, the study suggests creating a collaborative framework involving governments, civil society, corporations, and local communities to improve resilience and protect Southern Africa's populations.

Chapter 4 demonstrates the interconnections between different but interconnected ecological and socioeconomic systems on different spatial scales, providing examples of the impact of unsustainable and sector-based management of resources that include soil, water, and land. Poor resource management negatively affects livelihoods and human and environmental health. The interrelations between resources manifest when developments in one sector affect the efficiency of other related sectors. Initiatives that have shown improved adaptive capacity to climate change and resilience of vulnerable communities are highlighted.

Chapter 5 explores the climate disaster risk reduction (DRR) mechanisms adopted by selected rural communities in the southern African region to identify best practices and gaps in the process. Findings indicate that rural Southern Africa is vulnerable to various climate-related hazards, such as floods, storms, droughts, and heat waves. These climatic shocks have damaged the livelihood assets of rural communities. Formal institutions are seen to take the leading role in DRR initiatives; traditional and local leadership was frequently overlooked in DRR issues despite their significant influence within local communities. On paper, most DRR policies put communities at risk at the center of the DRR process. The DRR process has been devolved throughout the southern African region to the lowest community level. The findings also show that the burden of financing DRR activities falls mainly on the central government and donor agencies in most parts of rural Southern Africa. Understanding disaster risk reduction mechanisms in rural communities in Southern Africa has the potential to significantly reduce disaster losses, accelerate recovery, and build adaptive capacity to climate risks as the region develops policies and strategies tailored to its needs and unique circumstances. This initiative will go a long way toward mitigating poverty in the southern African region and achieving the 2030 Agenda for Sustainable Development and the Africa Agenda 2063.

Chapter 6 considers the progress and challenges that affect governance and resilience-building efforts in rural Zimbabwe, using comparative case studies by Chimanimani and Matopos. Several challenges undermine post-disaster governance and resilience. The chapter argues three primary challenges to stronger climate governance systems. The first is a need for more apparent roles and responsibilities and a better match between the critical role players' scope, mandate, and capacity. The second challenge is a mismatch between the problem's scale and the response mandates' scope. The third challenge is poor incentives for collective individual action or integrated planning to improve governance systems.

Chapter 7 examines the resilience state of communities in rural drylands of Zimbabwe and determines the relationship between climate resilience capacity index components and the total resilience index. The findings show that communities are exposed to frequent and intense drought, compromising their resilience to climate shocks. In rural communities, resilience was significantly affected by the incidence of crop diseases or pests. Farming activities such as crop and livestock production in terms of types and numbers were shown to affect the resilience of rural communities significantly. Owning cattle helps rural communities improve resilience. Communities that relied on indigenous food crops such as sorghum and pearl millet exhibited greater resilience than those that only grew maize. Given the socioeconomic and environmental similarities across the dryland regions of Southern Africa, these findings can be generalized to the region and the main factors influencing resilience, including education level, general training, access to information, land tenure, agro-processing training, distance to market, and income diversification. The study recommends improving climate information systems and mobilizing resources to address the underlying causes of climate vulnerability from a policy and strategic point of view.

Chapter 8 analyzes the potential and challenges of creating climate-friendly rural communities in Southern Africa. It finds that these communities are vulnerable to extreme weather events due to their dependence on the natural environment. Despite existing efforts to transition to climate-smart communities, barriers remain. Stakeholder participation and partnerships are crucial to enhancing adaptive capacity. Education, information dissemination, and research support are essential for promoting sustainable development in rural areas.

Chapter 9 asserts that climate change increases the risk of global pandemics, whose outbreaks pose numerous challenges to communities worldwide and that Southern Africa has not been spared. This situation has led to the need to improve the resilience of communities to these pandemics, the frequency of which is expected to increase due to climate change. The results of the literature analysis show that factors that affect resilience to pandemics in a changing climate can be categorized into four dimensions: institutional, social, economic, and demographic. Based on such climate change resilience mechanisms to pandemics, this study develops an integrated evaluation framework that researchers can implement to determine climate change resilience.

Chapter 10 analyzes climate resilience planning strategies and policies in rural Southern Africa. The literature search findings revealed that resilience strengthens critical livelihood capacities to withstand current and future shocks. The review established some resilience-building initiatives implemented to end extreme poverty among vulnerable communities in Southern Africa. The study concluded that resilience planning is an innovative platform that promotes the development of context-specific practices in Southern Africa and effectively transforms systems to initiate livelihood support systems. The study recommends the adoption of resilience-building interventions that best manage the impacts of climate change in resource-stressed rural communities in developing southern Africa.

Chapter 11 assesses the concept of climate resilience development and the progress rural Southern Africa has recorded while noting the gaps and prospects for catching up with the rest of the world. This chapter brings to light that rural Southern Africa is struggling to mitigate and adapt to the impacts of climate change despite the need for adequate financial, technical, and scientific support from the government and global financial institutions. The quest for climate resilience in rural Southern Africa is evident in agriculture, soil and water conservation sectors, pest and disease control, green energy development, and other climate-proofing efforts. The chapter concludes that climate resilience development is the ultimate solution to the threat of climate change and the harm it brings, hence the need to consolidate the achievements made to date and push for more and better capacity building in the future.

Chapter 12 examines the status quo, the gaps, and the prospects of climate finance to build resilience in selected countries in southern African countries. The study found that Southern Africa faces intricate challenges in balancing the considerable climate financing gap and the urgent climate financing needed for mitigation and adaptation. Strengthening domestic and international resource mobilization thus becomes imperative, creating a favorable environment for private sector investments in climate-smart technologies and renewable energy. It was revealed that very few studies have been conducted to specifically investigate climate finance in rural areas of Southern Africa, which calls for more empirical studies to be conducted in order to show a clear picture of climate finance in rural areas, which are mostly affected by climate change impacts.

Chapter 13 synthesizes the landscape of climate change mitigation policies, focusing on challenges and opportunities in Botswana, Malawi, Mozambique, South Africa, Zambia, and Zimbabwe. It assesses these policies, providing information on their implementation barriers and recommendations within the context of each country. The results demonstrate that climate change mitigation interventions must be more robust, while policy frameworks remain high on a political agenda. The policy implementation gap is marred by fragmented policies, limited resources, a lack of expertise, and a strong dependence on consulting when formulating policies. Therefore, the study proposes a framework to improve the implementation of climate change mitigation policies.

Chapter 14 explores the gendered impacts of climate change in rural Southern Africa and the intricate relationship between gender equality, gender empowerment, and gender resilience building. It also assesses the legal and policy gaps in implementing climate change resilience-building approaches. The findings reveal that the impacts of climate change are gendered and closely related to existing inequalities. Climate resilience cannot exist without gender equality and gender empowerment. Climate change-resilient laws and policy approaches for implementation are not in sync with international, regional, and national gender laws and policies. They do not enter local community-level projects and programs and do not consider the issues of inequality and the gendered nature of resources. The chapter proposes a model for gender resilience by communities in rural Southern Africa.

Chapter 15 explores the available literature based on empirical research to understand the prospects and challenges of digitalization and AI tools to build climate resilience in rural areas in Southern Africa. The findings show that some creditable successes have been made in using digital technologies to fortify communities against climate change, and there are enormous prospects for using existing community structures, learning cultures, and commitment to execute digitalization projects and use artificial intelligence tools to reinforce climate resilience in rural areas in Southern Africa. Rural Southern Africa has the potential to support more effective early warning and emergency response systems, improve food and water security, improve the performance of the power infrastructure, enable citizen participation and participatory adaptation measures, and reduce the effects of climate hazards. The study proposes viable paths for overcoming the barriers to digitalization to maximize the vast potential that exists in support of climate change resilience building.

Chapter 16 assesses the impact of climate change on the livelihoods of small farmers in the Gokwe North district of Zimbabwe, a remote and dry rural area in Region 3 of the Midlands province. It examines the mitigation measures adopted by smallholder farmers in remote rural areas, which are presented as livelihood diversification strategies. The results show that some small farmers in Gokwe North have remained in the mixed agricultural subsistence business. In contrast, others have resorted to seasonal market gardening, paid employment, artisanal mining, and food and vegetable sales, among other activities. The chapter recommends that stakeholders fund more sustainable projects, such as district water harvesting and environmental rehabilitation programs.

Chapter 17 uses the Mount Darwin district in the northern Mashonaland Central province of Zimbabwe to examine the adaptation constraints of climate change in rural Southern Africa. The district is one of the regions in Zimbabwe that have been the worst affected by climate change. The chapter explored several pertinent variables to shed more light on adaptation constraints to climate change in communal areas of Zimbabwe. These include vulnerabilities induced by climate change-induced vulnerabilities; adaptation needs for catalytic economic sectors, livelihoods, and food systems; adaptation responses and associated capital distribution in the context of the United Nations (UN) green development agenda; and domestication of adaptation solutions to meet Southern Africa's needs. The study's findings indicated that the vulnerability of climate change in rural livelihoods, especially agricultural activities, influenced a shift toward artisanal mining as an alternative income activity.

Furthermore, a symbiotic relationship exists between climate change patterns and livelihood activities in communal Zimbabwe. That is, climate change influences changes in rural livelihood activities, and livelihood patterns are also responsible for intensifying the effects of climate change in the region. The study concluded by identifying several measures to adapt to climate change in rural Southern Africa.

References

Adger, W. N. (2000). Social and Ecological Resilience: Are They Related? *Progress in Human Geography, 24*, 347–364. https://doi.org/10.1191/030913200701540465

Ahmed, M. (2020). *Introduction to modern climate change*. Andrew E. Dessler: Cambridge University Press. 2011, 252 pp, ISBN-10: 0521173159.

Bosekeng, G. (2020). *Vegetable production trend and constraints on vegetable farmers: A case of Ditladi and Gulushabe clusters in the northern region of Botswana*. Part.

Busayo, E. T., Kalumba, A. M., & Orimoloye, I. R. (2019). Spatial planning and climate change adaptation assessment: Perspectives from Mdantsane township dwellers in South Africa. *Habitat International, 90*, 101978.

Chapungu, L., & Nhamo, G. (2021). Interfacing vector-borne disease dynamics with climate change: Implications for the attainment of SDGs in Masvingo city, Zimbabwe. *Jamba Journal of Disaster Risk Studies, 13*(1), 1–12. https://doi.org/10.4102/jamba.v13i1.1175

Chivasa, W., Worku, M., Teklewold, A., Setimela, P., Gethi, J., Magorokosho, C., Davis, N. J., & Prasanna, B. M. (2022). Maize varietal replacement in eastern and southern Africa: Bottlenecks, drivers, and strategies for improvement. *Global Food Security, 32*, 100589.

Coetzee, C., Khoza, S., Nemakonde, L. D., Shoroma, L. B., Wentink, G. W., Nyirenda, M., Chikuse, S., Kamanga, T., Maripe, K., Rankopo, M. J., & Mwansa, L. K. (2023). Financing disaster risk reduction: Exploring the opportunities, challenges, and threats within the southern African development community region. *International Journal of Disaster Risk Science*, 1–15.

Defe, R., & Matsa, M. (2021). The contribution of climate-smart interventions to enhance sustainable livelihoods in Chiredzi District. *Climate Risk Management, 33*(7), 100338. https://doi.org/10.1016/j.crm.2021.100338

Demirel, N., Nauen, C. E., & Palomares, M. L. D. (2023). Fishing effort and the evolving nature of its efficiency. *Frontiers in Marine Science, 10*, 1180174.

Dube, K., Nhamo, G., & Chikodzi, D. (2022). Flooding trends and their impacts on coastal communities of Western Cape Province, South Africa. *GeoJournal, 87*(Suppl 4), 453–468.

Dzvimbo, M. A., Mashizha, T. M., Zhanda, K., & Mawonde, A. (2022). Promoting sustainable development goals: Role of higher education institutions in climate and disaster management in Zimbabwe. *Jamba: Journal of Disaster Risk Studies, 14*(1), 1–11.

Edokpayi, J. N., Makungo, R., Mathivha, F., Rivers, N., Volenzo, T., & Odiyo, J. O. (2020). Influence of global climate change on water resources in South Africa: Toward an adaptive management approach. In *Water conservation and wastewater treatment in BRICS nations* (pp. 83–115). Elsevier.

Engelbrecht, F., Adegoke, J., Bopape, M., Naidoo, M., & Garland, R. (2015). Projections of rapidly rising surface temperatures over Africa under low mitigation. *Environmental Research Letters, 10*, 85004.

Fambasayi, R. (2021). Leveraging city-level climate change law and policy for the protection of children. *The International Journal of Children's Rights, 29*(4), 872–896.

Fawzy, S., Osman, A. I., Doran, J., & Rooney, D. W. (2020). Strategies for mitigation of climate change: A review. *Environmental Chemistry Letters, 18*, 2069–2094.

Gbomagba, M., Afinowi, O. A., & Kumi, N. (2023). Climate change adaptation governance in Africa: The legal and institutional frameworks. In *Climate change in Africa: Adaptation, resilience, and policy innovations* (pp. 35–58). Cham.

Jegede, A. O., Yeakula, G. D., Ater, J. M., & Mothohabonoe, M. (2022). Climate displacement and the relevance of climate justice: A trend analysis of South Africa, Zimbabwe, South Sudan and Liberia. *Global Campus Human Rights Journal*, 41.

Jones, L., & Tanner, T. (2017). 'Subjective resilience': Using perceptions to quantify household resilience to climate extremes and disasters. *Regional Environmental Change, 17*, 229–243.

Justo, C. D., Tafula, J. E., & Moura, P. (2022). Planning sustainable energy systems in the Southern African Development Community: A review of power systems planning approaches. *Energies, 15*(21), 7860.

Kapwata, T., Gebreslasie, M. T., Mathee, A., & Wright, C. Y. (2018). Current and potential future seasonal trends of indoor dwelling temperature and likely health risks in rural southern Africa. *International Journal of Environmental Research and Public Health, 15*(5), 952.

Kareem, B., Lwasa, S., Tugume, D., Mukwaya, P., Walubwa, J., Owuor, S., Kasaija, P., Sseviiri, H., Nsangi, G., & Byarugaba, D. (2020). Pathways to resilience to climate change in African cities. *Environmental Research Letters, 15*(7), 073002.

Kibirige, D., & Singh, A. S. (2023). Goals of smallholder sugarcane farmers in Eswatini. *Journal of Agricultural Studies, 9*, 123–152.

Kusangaya, S., Warburton, M. L., Van Garderen, E. A., & Jewitt, G. P. (2014). Impacts of climate change on water resources in southern Africa: A review. *Physics and Chemistry of the Earth, Parts a/b/c, 67*, 47–54.

Kwembeya, E. G., & Shikangalah, R. N. (2023). Towards understanding rainfall variability in Namibia: An analysis of spatial and temporal variations of rainfall from 2010 to 2019. *International Science and Technology Journal of Namibia, 16*, 26–39.

Le Roux, A. (2024). SADC. Published online at futures.issafrica.org. Retrieved from https://futures.issafrica.org/geographic/recs/sadc/ [Online resource] Updated 13 December 2023.

Leal Filho, W., Totin, E., Franke, J. A., Andrew, S. M., Abubakar, I. R., Azadi, H., et al. (2022). Understanding responses to climate-related water scarcity in Africa. *Science of the Total Environment, 806*, 150420.

Leichenko, R. M., & O'Brien, K. L. (2002). The dynamics of rural vulnerability to global change: The case of southern Africa. *Mitigation and Adaptation Strategies for Global Change, 7*(1), 1–18.

Madebwe, T. (2019). *Enhancing Botswana's environmental performance by 2023* (pp. 15–xvii). Law Env't & Dev. J.

Mafongoya, P., Musokwa, M., Mwadzingeni, L., & Phophi, M. M. (2022). Climate change impacts on food and nutrition security on smallholder farmers in southern Africa. *Food Security for African Smallholder Farmers*, 233–249.

Matsa, M., Guti, M., Nyahwa, E., & Dzawanda, B. (2022). Toward sustainable community garden-driven resilience in semiarid areas of Zimbabwe's Shurugwi district. *Local Development and Society., 2*(2), 1–18. https://doi.org/10.1080/26883597.2022.2092767

Maziya, S. A. (2023). *The impact of the European Union Grant on access to credit and production in smallholder sugarcane agriculture in Siphofaneni, Eswatini*. Research thesis, University of Pretoria. https://doi.org/10.22004/ag.econ.334775

Mhlanga-Ndlovu, N. (2022). Socioeconomic effects of neoliberal transformation on irrigated agriculture in Eswatini: A case of sugarcane farmers' groups in the Komati downstream development project. In *Capital penetration and the peasantry in southern and eastern Africa: Neoliberal restructuring* (pp. 203–218). Springer International Publishing.

Moseki, O., Murray-Hudson, M., & Kashe, K. (2019). Crop water and irrigation requirements of Jatropha curcas L. in semiarid conditions of Botswana: Applying the CROPWAT model. *Agricultural Water Management, 225*, 105754.

Moyo-Nyoni, N. (2022). Adopting indigenous knowledge systems to enhance peace education programs for climate change and adaptation in Zimbabwe. In *Indigenous knowledge and climate governance: A sub-Saharan African perspective* (pp. 27–38). Springer International Publishing.

Muroyiwa, G., Mhizha, T., Mashonjowa, E., & Muchuweti, M. (2023). Using the AquaCrop model to derive deficit irrigation schedules for improved irrigation water management for tomato production in Zimbabwe. *African Crop Science Journal, 31*(3), 365–378.

Musasa, T., & Matsa, M. (2022). Impact of resilience building interventions in semiarid rural agro-communities: lessons from Mberengwa and Zvishavane Districts in South Eastern Zimbabwe African Geographical Review. https://doi.org/10.1080/19376812.2021.2023892

Mutengwa, C. S., Mnkeni, P., & Kondwakwenda, A. (2023). Climate-smart agriculture and food security in southern Africa: A review of the vulnerability of smallholder agriculture and food security to climate change. *Sustainability, 15*(4), 2882.

Muyambo, F., Belle, J., Nyam, Y. S., & Orimoloye, I. R. (2023). Climate-change-induced weather events and implications for urban water resource management in the Free State Province of South Africa. *Environmental Management, 71*(1), 40–54.

Ndebele-Murisa, M. R., Mubaya, C. P., Pretorius, L., Mamombe, R., Iipinge, K., Nchito, W., Mfune, J. K., Siame, G., & Mwalukanga, B. (2020). City-to-city learning and knowledge exchange for climate resilience in southern Africa. *PLoS One, 15*(1), e0227915.

Nemakonde, L. D., & Van Niekerk, D. (2023). Enabling conditions for integrating government institutions for disaster risk reduction and climate change adaptation in the SADC region and beyond. *Risk, Hazards & Crisis in Public Policy, 14*(1), 6–26.

Nhemachena, C., Nhamo, L., Matchaya, G., Nhemachena, C. R., Muchara, B., Karuaihe, S. T., & Mpandeli, S. (2020). Climate change impacts on water and agriculture sectors in southern Africa: Threats and opportunities for sustainable development. *Water, 12*(10), 2673.

Nilsson, M., Sie, A., Muindi, K., Bunker, A., Ingole, V., & Ebi, K. L. (2021). Weather, climate, and climate change research to protect human health in sub-Saharan Africa and South Asia. *Global Health Action, 14*(sup1), 1984014.

Nkonya, E., Kato, E., Msimanga, M., & Nyathi, N. (2023). Climate shock response and resilience of smallholder farmers in the drylands of South-Eastern Zimbabwe. *Frontiers in Climate, 5*, 890465.

Ntontis, E., Drury, J., Amlôt, R., Rubin, G. J., & Williams, R. (2020). What lies beyond social capital? The role of social psychology in building community resilience to climate change. *Traumatology, 26*(3), 253.

Saddington, L. (2023). Geopolitical imaginaries in climate and ocean governance: Seychelles and the blue economy. *Geoforum, 139*, 103682.

Scholz, G., & Methner, N. (2020). A social learning and transition perspective on a climate change project in South Africa. *Environmental Innovation and Societal Transitions, 34*, 322–335.

Sharifi, A. (2021). Co-benefits and synergies between urban climate change mitigation and adaptation measures: A literature review. *Science of the Total Environment, 750*, 141642.

Sibanda, A., & Manik, S. (2022). Environmental education efforts in a rural drought-prone district in Zimbabwe: Micro-innovations with growth potential. *The Journal of Geography Education in Africa, 5*, 77–95.

Singh, P. K., & Chudasama, H. (2021). Pathways for climate resilient development: Human well-being within a safe and just space in the 21st century. *Global Environmental Change, 68*, 102277.

Spear, D.D., Haimbili, E., Baudoin, M.A.D., Hegga, S.D., Zaroug, M.D., Okeyo, A., Angola, M. (2018). Vulnerability and adaptation to climate change in semiarid areas in Southern Africa.

Subiyanto, A., Boer, R., Adrian, E., & Kinseng, R. (2020). Climate Resilience: Concepts, Theory and Methods of Measuring, *Environment Asia, 13*(1), 1–13. https://doi.org/10.14456/ea.2020.1

Thomas, T. S., Robertson, R. D., Strzepek, K., & Arndt, C. (2022). Extreme events and production shocks for key crops in southern Africa under climate change. *Frontiers in Climate, 4*, 787582.

Tolotti, M., Guillotreau, P., Forget, F., Capello, M., & Dagorn, L. (2023). Unintended effects of single-species fisheries management. *Environment, Development, and Sustainability, 25*(9), 9227–9250.

Trew, B. T., & Maclean, I. M. (2021). Vulnerability of global biodiversity hotspots to climate change. *Global Ecology and Biogeography, 30*(4), 768–783.

Udeagha, M. C., & Muchapondwa, E. (2022). Investigating the moderating role of economic policy uncertainty in environmental Kuznets curve for South Africa evidence from the novel dynamic ARDL simulations approach. *Environmental Science and Pollution Research, 29*(51), 77199–77237.

Wellington, M., Kuhnert, P., Lawes, R., Renzullo, L., Pittock, J., Ramshaw, P., Moyo, M., Kimaro, E., Tafula, M., & van Rooyen, A. (2023). Decoupling crop production from water consumption at some irrigation schemes in southern Africa. *Agricultural Water Management, 284*, 108358.

World Health Organization. (2022). Health systems resilience toolkit: A WHO global public health toolkit is good for supporting the building and strengthening of sustainable health systems resilience in countries with various contexts.

Wright, C. Y., Kapwata, T., Du Preez, D. J., Wernecke, B., Garland, R. M., Nkosi, V., et al. (2021). Major climate change-induced risks to human health in South Africa. *Environmental Research, 196*, 110973.

Zuñiga, R. A. A., de Lima, G. N., & Quiroga, D. E. (2022). Climate change and population displacement in Africa. In *The Routledge handbook of African demography* (pp. 510–532). Routledge.

The Climate Change Landscape for Rural Southern Africa

2

Mohau J. Mateyisi, Shingirai S. Nangombe, Malebajoa Anicia Maoela, and Vincent Chenzi

2.1 Background

The evaluation of the impacts of climate change is based on studying historical trends and comparing model projections with a baseline while considering the level of adaptation. Research efforts toward the development of National Climate Change Communication and National Adaptation Plans for respective countries in Southern Africa have contributed to this understanding; however, the understanding of climate change impacts for the rural setting still needs to be unpacked further. Southern Africa is identified as a climate change hotspot (e.g. Ayanlade et al., 2018; Mbow, 2020). Its rural areas, characterised by widely dispersed settlements, low population densities and livelihoods relying on agriculture and natural resources, should be addressed in climate change responses (PLAAS, 2022). A sizeable proportion of Southern Africa is classified as rural, mainly in the eastern side of the subcontinent, including the island of Madagascar. Southern Africa is home to approximately 277 million people, 60% of whom live in rural areas, and its economies are highly dependent on agriculture and natural resources (Vincent Itai & Constance, 2014).

The region is highly susceptible to climate change due to its dependence on climate-sensitive economic sectors (Huang et al., 2020; Trisos et al., 2020). During the last decade (2010–2020), the region has faced serious climate-related challenges, including water scarcity and land degradation (Trisos et al., 2020). These challenges have significantly affected food security, energy supply, infrastructure development, health and overall quality of life in the region (Chemura et al., 2022). For example, maise is the primary food source in Southern Africa, accounting for 73% of the total food demand in the region. However, its growth is greatly affected by weather variations, leading to reduced yields when conditions deviate from normal (Lobell et al., 2014; Meng et al., 2016). Specifically, during the growing season, variations in temperature and rainfall significantly impact successful maise production, especially in rain-fed agricultural areas in rural southern Africa (Hossain, 2020). This situation is worsened by high temperatures, which contribute to the water deficit through evapotranspiration.

M. J. Mateyisi (✉)
Smart Places Cluster, Council for Scientific and Industrial Research, Pretoria, South Africa
e-mail: mmateyisi@csir.co.za

S. S. Nangombe · V. Chenzi
Department of Peace, Security and Society, Faculty of Arts and Humanities, University of Zimbabwe, Harare, Zimbabwe

M. A. Maoela
Institute for Corporate Citizenship, University of South Africa, Pretoria, South Africa

M. Matsa et al. (eds.), *Climate Change Resilience in Rural Southern Africa*, Sustainable Development Goals Series, https://doi.org/10.1007/978-3-031-73600-1_2

The impact of global climate change is felt notably by rural communities in Southern Africa in the form of altered weather patterns, rising temperatures and extreme weather events (IPCC, 2014). Southern African communities are particularly vulnerable because a large part of their population lives in impoverished conditions (IPCC, 2023). The region has already witnessed extreme climate events, including severe droughts, floods, wildfires and cyclones. Droughts in Southern Africa are major natural disasters that have affected all economic sectors (Marumbwa et al., 2020). Gore et al. (2020) studied the impact of the El Niño Southern Oscillation (ENSO). They found that El Niño conditions weaken circulation during La Niña events while strengthening it during El Niño events. This variability in circulation patterns results in below-normal rainfall and above-normal temperatures during El Niño, leading to drought conditions in Southern Africa. On the contrary, La Niña conditions bring above-normal precipitation and below-normal temperatures to the region. Since the 1970s, roughly 66% of extreme drought events in the region have been associated with El Niño (Nicholson, 2017).

In Africa, a study found increasing trends in the warmest days and nights, with insignificant trends in Zimbabwe (Aguilar et al., 2009). Mount Mukulu in Zambia also showed insignificant positive trends in high-temperature extremes (Chisanga et al., 2017). Another study found significant increasing trends in extreme temperature indices in Southern Africa (Alexander et al., 2006). Extreme temperature events have substantial impacts on human activities and natural systems, especially in Africa. Changes in these events can significantly affect society and the environment. Agriculture in Africa is particularly vulnerable to these impacts due to its limited adaptive capacity and dependence on rain-fed agriculture.

The evaluation of the impacts of climate change in Southern Africa is based on historical trends and model projections, considering adaptation levels (Aguilar et al., 2009; Peterson & Manton, 2008). Efforts have been made to develop national climate change communication and national adaptation plans, but understanding of the impacts of rural climate change still needs to be improved. Southern Africa is a climate change hotspot, and rural areas are often overlooked in responses (Chemura et al., 2022; Engelbrecht & Monteiro, 2021; Hoegh-Guldberg et al., 2019). Around 60% of the region's population lives in rural areas, which are dependent on agriculture and natural resources (Bansah et al., 2023).

The global climate change crisis has led to varying impacts in different regions, and the climate landscape of Southern Africa (the region south of the equator) is no exception. These changes in global temperatures have caused a range of significant events, including rising sea levels from melting glaciers, increased water scarcity due to intense evapotranspiration, severe droughts, wildfires, melting polar ice, floods, catastrophic storms and declining biodiversity, among other disastrous events (Dellink et al., 2019). As a result, these patterns of climate change have significantly shaped and adversely affected a substantial portion of livelihood activities worldwide. The consequences of these climate changes have been felt in various ways, including changes in fisheries catch, loss of land and assets due to rising sea levels, significant damage caused by hurricanes and wildfires, alterations in tourism patterns, changes in labour productivity and an increase in healthcare costs. Furthermore, some of the health challenges, relevant not just globally but specifically for rural Southern Africa, include issues such as heat stress, malaria, waterborne diseases and potential fluctuations in energy demands for cooling and heating, all likely to be intensified by the effects of climate change (Dellink et al., 2019).

In the pursuit of understanding the intricacies of climate change in rural Southern Africa, it is worth acknowledging that several data and research gaps remain within this field. This is despite the fact that the region has been identified as a climate change hotspot by the latest IPCC AR6 report (Engelbrecht & Monteiro, 2021; IPCC, 2023). First, the need for more reliable and consistent climate data across the region hinders climate research and forecasting. Furthermore, although there is an increasing understanding of the impacts of climate change on agriculture and water resources in the region, a scarcity of accessible socio-economic data on rural households, communities and other sectors hinders resilience-building efforts. Climate information on multiple timescales is critical for long-term and operational adaptation decisions by smallholder and commercial rural farmers.

Understanding projected climate hazards is crucial for effective climate risk management in rural areas. This approach allows quantifying risks, prioritising adaptation efforts and contextualising climate hazards. This chapter discusses past and present climate change, associated extremes and their impacts in the southern African region. The goal is to provide information for decision-making and policy development in rural settings. The focus is on characterising temperature and precipitation patterns and understanding climate change-induced extreme events and their impacts.

2.2 Methods and Materials

2.2.1 Determination of Rural Areas

To explore climate patterns specific to rural areas, we utilise a unique land cover classification termed 'social pixels' or 'village pixels' as introduced by Marumbwa et al. (2020) to represent rural communities. These social pixels represent rural communities and are defined by an index that designates areas as rural when their cattle and goat density exceeds 400 per 10 square kilometres. Remarkably, this index's spatial distribution closely aligns with a general map of rural population density, enhancing its utility in connecting climate model output. This correlation helps to understand the impact of climate change on these communities. Specific data on goat and cattle density were obtained from the Food and Agriculture Organisation (FAO) geoportal accessed on 28 August 2023 (FAO, 2023) (Fig. 2.1).

2.2.2 Determination of the Current Climate and Possible Future Impacts

To evaluate climate changes and associated extremes in rural Southern Africa, an analysis was performed for a baseline period (1961–1990) and over four future levels of global warming (GWL). These GWLs—1.5 °C, 2 °C, 3 °C and 4 °C above preindustrial temperatures—represent distinct scenarios, with the first two aligning with the Paris Agreement thresholds and the latter pair depicting a potential worst-case global warming temperature scenario. To simulate future climate scenarios in this region, ten general circulation models (GCMs) from the Phase 6 Coupled Model Intercomparison Project (CMIP6) were analysed. These models were statistically downscaled to a standard 0.5° horizontal grid.

The GCMs included in the analysis are CAN-ESM 5, CNRM-CM6-1, CNRM-ESM 2-1, EC-EARTH3, MIROC6, GFDL-ESM 4, IPSL-CM6A-LR, MPI-ESM 1, MRI-ESM 2-0 and UKESM1-0-LL. More details about these ten models, including their full names and source of origin, can be found in Lange et al. (2021). The bias adjustment applied to the data was based on the Inter-Sectoral Impact Model Intercomparison Project (ISIMIP) protocol, as described by Stefan Lange (2019). The suite of CMIP6 models represents the latest and widely used set of models for future climate projections. These models have been extensively applied in climate change assessments across Africa due to their demonstrated reliability in the region compared to historical observations. (Aloysius et al., 2016; Ongoma et al., 2019).

2.3 Spatial Patterns of Historical Climate

2.3.1 Temperature

Rural Southern Africa is characterised by a warm climate with an average temperature for a large part of the region above 15 °C throughout the

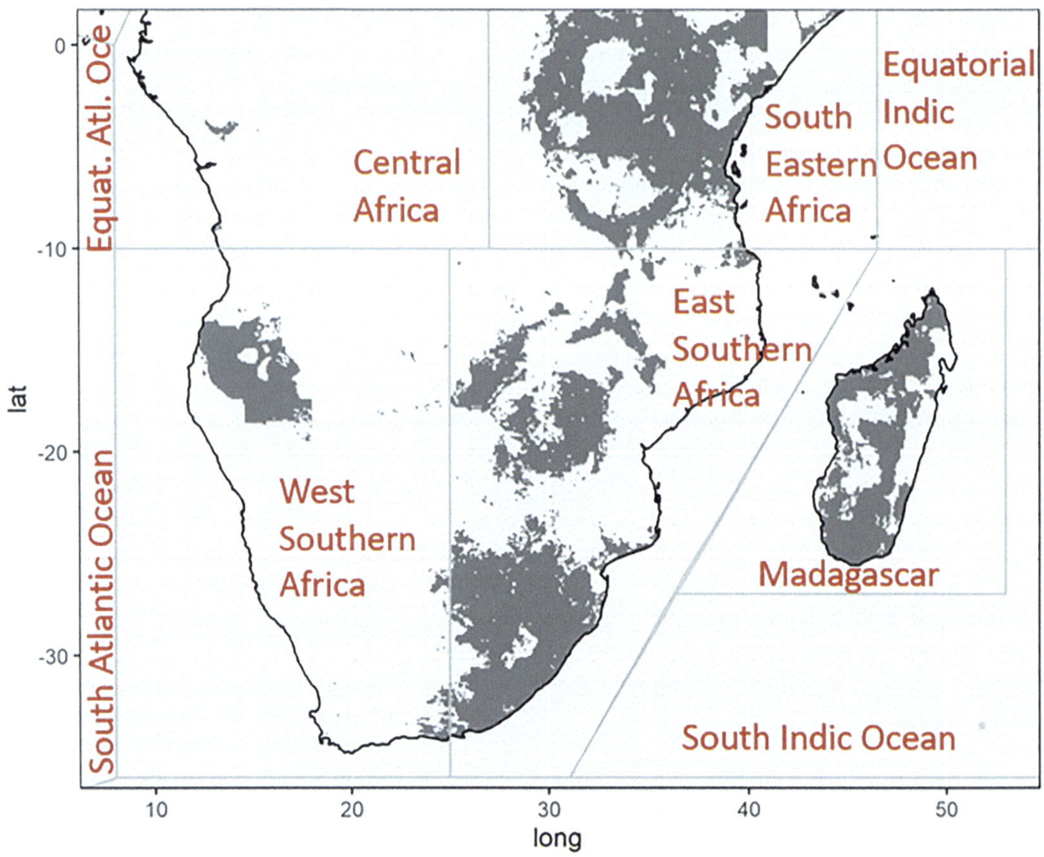

Fig. 2.1 Map showing the rural areas in Southern Africa (grey stipplings) and the respective subregions

year (Fig. 2.2). The average nighttime temperature ranges between 3 and 25 °C, with the coolest temperatures felt in the winter rural interior along the eastern escapement, where the altitude is relatively high. Snowfall is experienced in winter in some parts of the region, especially on the Drakensberg plateau, where subzero temperatures in winter are common and neighbouring rural areas are frequented by frost.

The highest temperatures in summer are recorded near the equator in the desert region of west southern Africa of Botswana, namely, the Kalahari Desert, as well as in eastern, southern Africa within rural areas that constitute the north-eastern interior of South Africa, where the average annual temperature ranges between 15 and 36 °C (Fig. 2.2). In rural highlands, the range of daily temperatures, defined as the difference between the maximum and minimum daily temperature, is relatively pronounced, reaching as high as 19 °C in some areas. Compared to coastal areas, rural areas are characterised by a cooler diurnal temperature range. The warm south-flowing Algalhus current plays a crucial role in regulating summer and winter temperatures in the eastern coastal region of Southern Africa and adjacent rural areas. On the coast of western southern Africa, the north-flowing Benguela current relays a cooler air mass that sweeps the southern interior of the region through cold frontal systems in winter. The two major currents explain the east-to-west latitudinal gradient in temperatures, which becomes more pronounced during the winter season, especially for the southernmost parts of South Africa and Lesotho, as well as Namibia, where the average daily temperature can drop as low as 15 °C during the season.

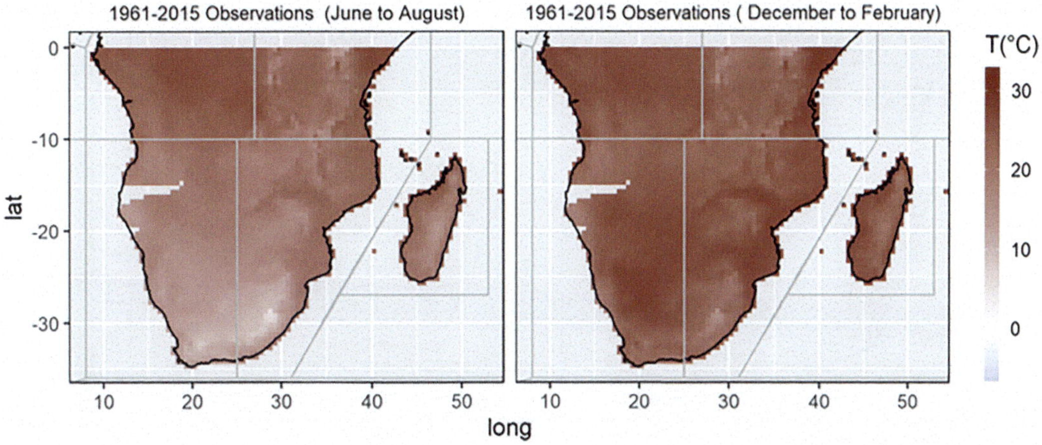

Fig. 2.2 Average temperature (°C) over Southern Africa calculated from the Climate Research Unit (CRU) gridded observational dataset for the period 1980–2015 under different seasons (the data are sourced from the IPCC interactive atlas (Iturbide et al., 2022). (**a**) JJA, (**b**) DJF. The boxes depicted by the grey lines show the subregions in Southern Africa

Extremes in daytime and nighttime temperatures in summer and winter, respectively, make a big distinction between summer and winter seasons despite the gradual transitions into this season. For most rural Southern Africans, as is the case for the whole region, the winter season falls between June and August, while summer falls between December and February.

2.3.2 Rainfall

Rising temperatures and prolonged droughts have resulted in crop failures in rural Southern Africa, putting enormous stress on food security, agriculture livelihoods and economies, such as that experienced during the strong El Niño of 2015/2016 (Engelbrecht & Monteiro, 2021). On the other hand, Cyclone Idai ravaged the region in 2018, leaving more than 1500 people dead, affecting more than three million people (primarily rural dwellers) and causing extensive damage to infrastructure and properties (Nhamo & Chikodzi, 2021). There are also many low-lying coastal cities and towns in Southern Africa that are at risk of flooding and sea level rise (IPCC, 2019). Coastal cities such as Maputo, Durban, Cape Town and those in Madagascar are vulnerable to rising sea levels, which could cause massive economic losses and displacement (Mather & Stretch, 2012). As shown in Fig. 2.1, some rural areas of Southern Africa are located within low-lying coastal regions such as those of KwaZulu-Natal (KZN), Maputo and Madagascar, with settlements and port infrastructure exposed to possible risks associated with sea level rise and precipitation extremes related to the risk.

It is instructive to look at the seasonal rainfall distribution in the rural landscape of Southern Africa and to unpack the seasonality of rainfall as the basis for understanding the baseline climate relative to the measured projected change. Variation in annual rainfall accumulation or temporal distribution has a clear northeast-to-southwest gradient. The precipitation gradient is evident during the March–May (MAM) and during the September–November (SON) seasons. It is worth noting that the rural areas of Southern Africa within Southeast Africa and Eastern Southern Africa are at latitudes that receive comparable amounts of rainfall (Fig. 2.3). The rural landscapes of Southern Africa, on average, receive about 1000 mm of rainfall per year. The high-lying areas of eastern Madagascar receive the highest annual accumulation amount of about 4997 mm per year. During the rainy seasons, it is

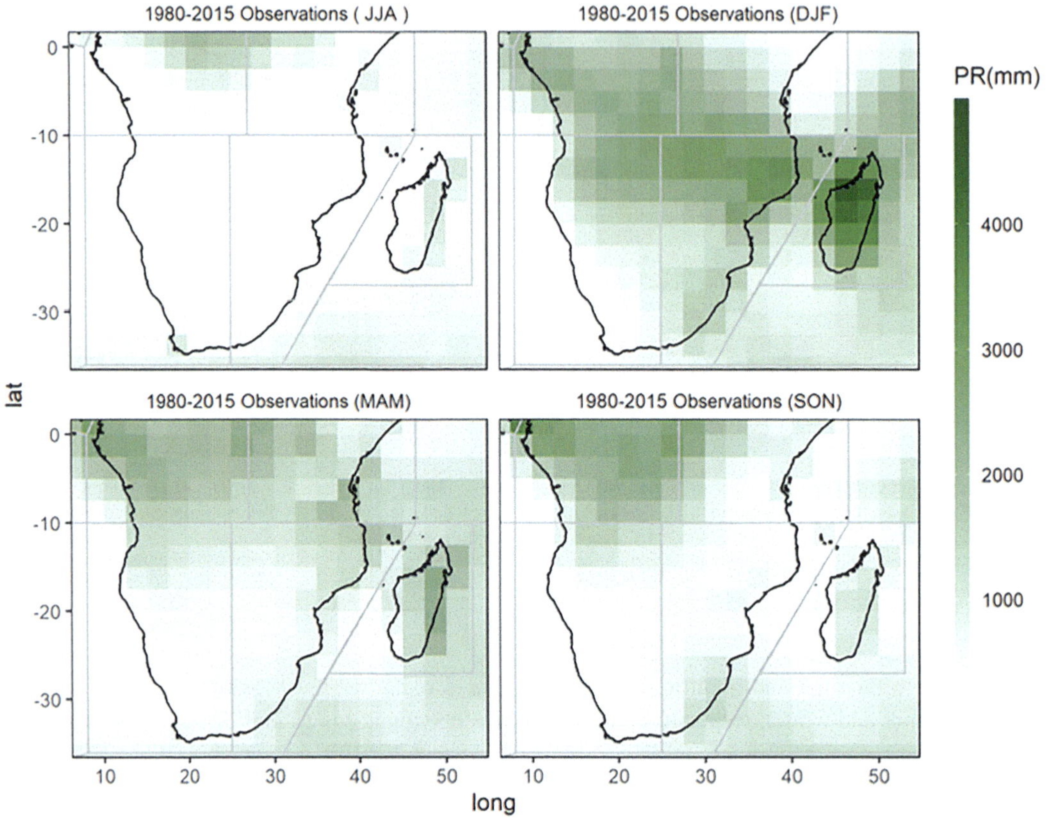

Fig. 2.3 Total precipitation (mm) over Southern Africa was calculated from the Global Precipitation Climatology Project (GPCP) gridded observational dataset for the period 1980–2015 under different seasons (data are sourced from the IPCC interactive atlas) (Iturbide et al., 2022). (**a**) JJA, (**b**) DJF total, and (**c**) MAM, (**d**) SON. The boxes depicted by the grey lines show the subregions of Southern Africa

observed that most rural landscapes receive annual total rainfall in the range of 600–1600 mm per year. The arid rural regions along western South Africa receive as little as 100 mm of rainfall per year.

The South Coast is characterised by a drier climate with a seasonal pattern, which is not the same as the rest of the other regions. Figure 2.3 shows that the highest accumulation of rainfall for most rural pixels occurs at the beginning of the summer, September–November (SON), and late in the summer, December–February (DJF). Rural areas within the tropical region of Central Africa and South Africa receive rain in winter. The effect of the mid-latitude cyclone or disturbance associated with the westerly wind does not affect the region. With global warnings, extreme weather events and extreme precipitation events are expected to become even more frequent and intense (Seneviratne, 2021), causing more damage and affecting more people, especially vulnerable in rural areas on the East Coast, where extremes such as tropical cyclones are part of climate variability. The following section focuses on the projected climate in Southern Africa.

2.4 Projected Climate Changes over Southern Africa

2.4.1 Extremes Based on Precipitation

The frequency and intensity of heavy precipitation in Southern Africa are projected based on the extreme precipitation indices below (i.e. very heavy precipitation with 20 mm or more of precipitation (R20mm), maximum 5-day precipitation (Rx5day) and consecutive dry days (CDD) in the four GWLs (1.5 °C, 2 °C, 3 °C and 4 °C). General increases in R20mm are projected north of 10° S but with increases over the larger southern African region to the south. This pattern increases with higher levels of GWL (Fig. 2.4). An exception is the eastern escarpment areas of South Africa, where increases in R20mm events are projected, consistent with the findings of IPCC AR6 (Ranasinghe et al., 2023). A similar regional trend is seen with Rx5day, albeit with lower magnitudes. CDD, on the other hand, is projected to also intensify and increase in magnitude with increasing GWLs, consistent with the general pattern of drying projected for the region (Pörtner et al., 2022).

Climate change projections of drought conditions suggest that most parts of Africa, including the regions of South East Africa (SEAF) and West Southern Africa (WSAF), are likely to experience drier conditions during the 2021–2050 period. These drier conditions are also expected to extend to the northern regions of Zambia and Zimbabwe during the period. Parts of eastern Southern Africa that intersect with Mozambique are projected to experience an increasing trend in annual wet conditions in the 2021–2050 period. However, the projected change is not statistically significant. The southern part of Mozambique, on the contrary, is projected to experience elevated dry conditions towards the end of the century.

Lately, Southern Africa has been experiencing multiyear droughts, and a recent notable example is the 2015–2017 drought in Western Cape that led to the Cape Town municipality declaring the day zero water crisis in the city of Cape Town (Sousa et al., 2018; Wolski, 2018). This was caused by a poleward shift in westerly moisture fluxes across the South Atlantic and Hadley cell expansion (Sousa et al., 2018). However, heavy rain days in parts of Angola, Botswana, South Africa and Zimbabwe during the summer rainy season have increased in recent decades.

Africa is a semi-arid region characterised by high rainfall variability, which results in frequent droughts and extreme rainfall events (Jury, 2013). The impacts of climate change have already been felt in the region (Nangombe et al., 2019; Seneviratne, 2021). In Fig. 2.5, the general trend in the Standardised Precipitation Index (SPEI) during the 1962–1990 period for most of the interior of Southern Africa suggests a decline in precipitation compounded by the increase in temperature. During the historical period, the model output calculated the trend of the SPEI index in most of Southern Africa, which reflects a shift towards drier conditions; however, the trend is not statistically significant. The dry signal is more pronounced along the north-east part of South Africa, northern Botswana and western Zambia. These trends are consistent with the findings of a study based on observed rainfall and aridity trends in 1960–2007 in Southern Africa, which established that rainfall has decreased by approximately 26% over the region during the period, while aridity has increased by 11% between 1989 and 2007 (Nhamo et al., 2019). A study conducted using CRU observation data-derived SPEI in conjunction with a vegetation condition index (VCI) in the region revealed that the frequency of drought during the most recent period (1998–2018) was highest in the south-western parts of the region (Marumbwa et al., 2020). The same study revealed that during the years 2015–2016, there was extreme drought in Southern Africa. Specifically, the values of the VCI index were the lowest among the land cover classes occupied by rural communities, reflecting that the exposure of rural communities to drought impacts was relatively more extreme.

The modelled trends calculated suggest a statistically significant decrease over the Western Coastal region of South Africa and parts of South Africa's central interior, namely, the Mpumalanga province. The Western Cape in South Africa is a region that, between 2015 and 2018, experienced a hydrological drought that led to a historic 'DAY ZERO', which is a level 7 water restriction under which the municipal water supplied had to be predominantly switched off. In particular, in the north and northwest parts of Botswana to the western regions of Zimbabwe, as well as central Mozambique to southern Malawi and eastern Zambia, the trend of the SPI suggests a decrease

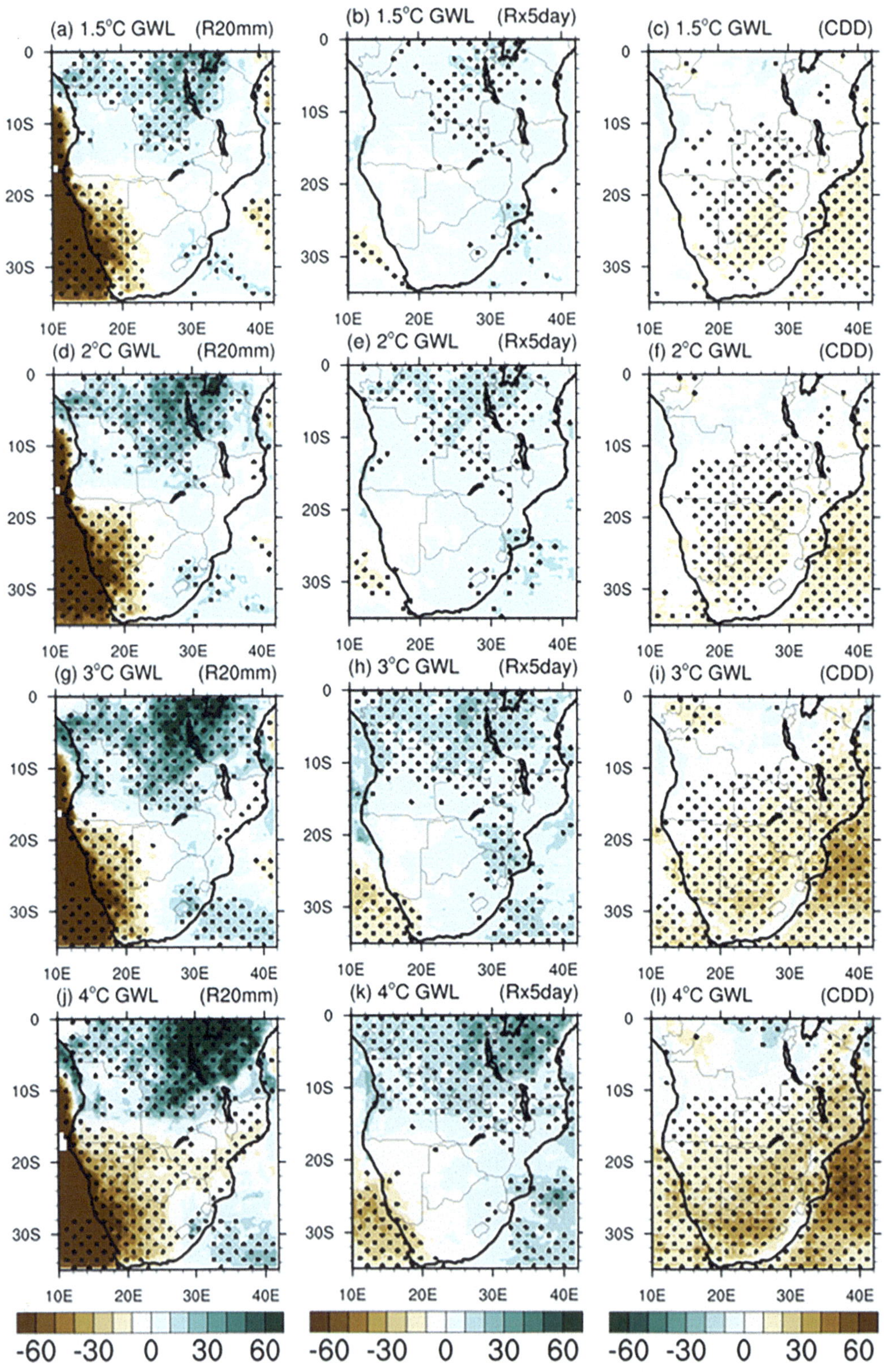

Fig. 2.4 Changes in extreme precipitation indices (%) in Southern Africa projected by the average of the CMIP6 SSP5-8.5 ISIMIP3b ensemble at various levels of global warming reached concerning the 1961–1990 baseline period. (**a**–**c**) Extreme precipitation indices R20mm, Rx5day and CDD below 1.5 °C GWL. (**d**–**f**), (**g**–**i**) and (**j**–**k**) same as (**a**–**c**) but under 2 °C, 3 °C and 4 °C GWLs, respectively. Units are in %. The two upper panels directly represent the members of the global mitigation goal expressed in the Paris Agreement. Stippling shows areas that are significant at a 95% level

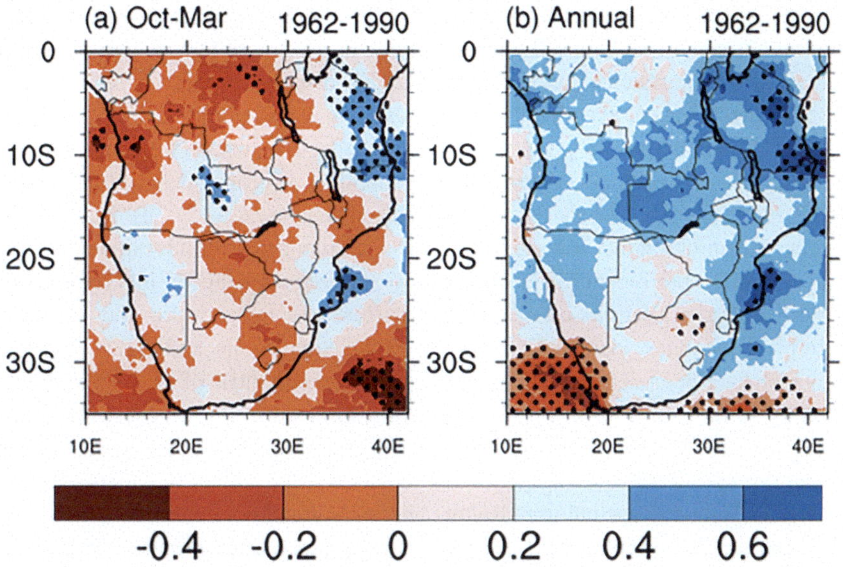

Fig. 2.5 Distribution of Kendall's trend analysis for 6 months calculated for the (**a**) October–March season and (**b**) annual time series in the Southern Africa region for the period 1962–1990. Stippling shows the areas where the trend is significant at a 95% level

in rainfall. These are regions where the novel livestock density-based village pixel index developed by Marumbwa et al. (2020) reflects a predominance of habitation by rural communities.

The north-east parts of South Africa, specifically the north-east regions of Tanzania and the southern part of the Mozambique SPEI index (Fig. 2.5), reflect a statically significant increasing trend in the SPEI, which is indicative of an increase in rainfall, especially during the summer season. According to the village pixel map, the north-east part of Tanzania is also inhabited mainly by rural communities. The landfall of tropical cyclones emerging from the Atlantic Ocean has historically been rare in Tanzania, apart from the recent Kenneth tropical cyclone of 2019 that became felt south of the country, claiming approximately 38 and destroying approximately 35,000 homes. However, these storms produce flooding that affects most parts of the two countries, including rural areas. Mozambique, on the other hand, is frequented by some of the most intense tropical cyclones to the detriment of livelihoods for both coastal and rural Mozambique, exacerbating social vulnerability that is also a burden for most communities.

The experienced changes in climate conditions are expected to persist for a long time, including increases in temperature, heatwave frequency, intensity and duration (Schlenker & Lobell, 2010). Rising temperatures and prolonged droughts have resulted in crop failures, putting huge stress on food security, rain-fed agriculture livelihoods and economies, as demonstrated during the strong El Niño of 2015/2016 (Engelbrecht & Monteiro, 2021). The region is widely recognised as vulnerable to these changes due to low adaptive capacity and heavy reliance on rain-fed agriculture (Bansah et al., 2023). These extreme weather events are expected to become more frequent and intense (Seneviratne et al., 2021), causing more damage and affecting more people (Fig. 2.5).

The standardised evapotranspiration index calculated from model projections suggests a spatial increase in drought in the south-western parts of southern Africa that extends into parts of the interior of the subcontinent during the summer season and throughout the year. The projected trend is statistically significant at the 95% level. Future drought conditions in the Botswana region are likely to be detrimental to livestock farming. East Southern Africa, specifically central and northern Mozambique, is projected to experience an increase in precipitation during the future periods (Fig. 2.6). This is also a region prone to extreme precipitation that is contributed by tropical cyclones. In the recent past (March 2019, March 2022), the region has experienced category 4 tropical cyclones Idai and Gombe, respectively, which have been detrimental to lives and properties.

The region is widely recognised as vulnerable to these changes due to low adaptive capacity and a heavy reliance on rain-fed agriculture (Rahut et al., 2021). During the period (1962–2020), rural Southern Africa experienced some rates of climate-driven food insecurity. Mbow et al. (2020) discovered that an increase in temperatures of 2 °C would significantly reduce crop yields such as wheat while altering the growth suitability of crops such as maise and sorghum (Nhemachena et al., 2020). Maise is recognised as a particularly prolific staple crop in most parts of the region and is inherently sensitive to increases in temperature and moisture (IFPRI, 2013). Small-scale and rural farmers represent approximately 60% of the population and are the most affected by the impacts of climate change and associated extreme impacts (Clapp et al., 2018). Rural communities are likely to experience not only the direct impacts of climate but also the resulting increases in food prices and extreme cli-

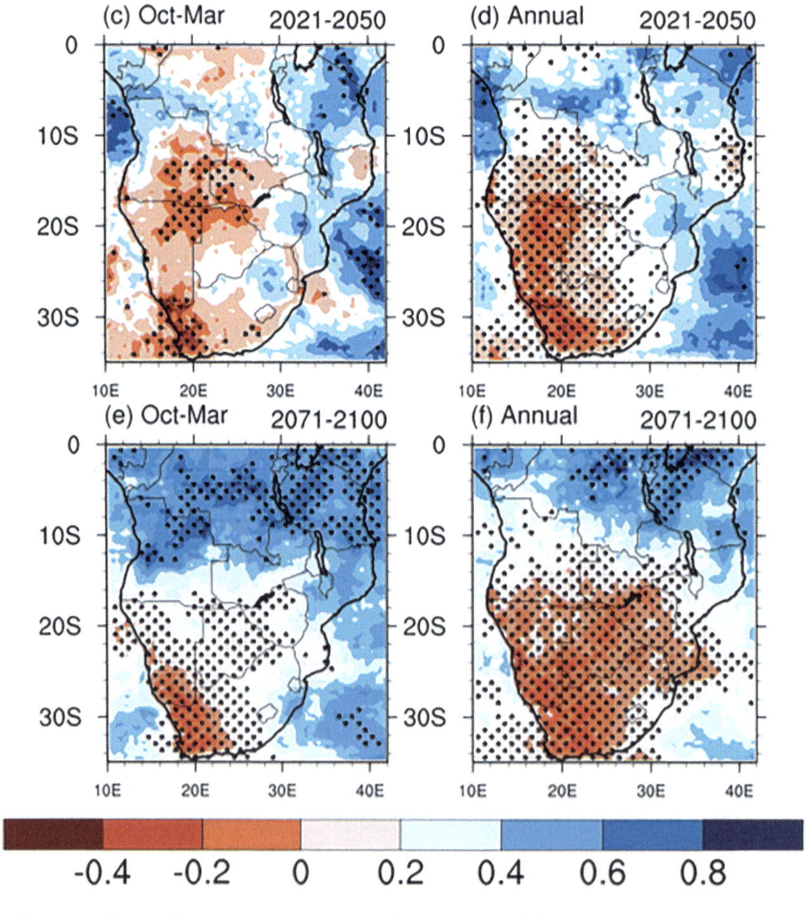

Fig. 2.6 Distribution of Kendall's trend analysis for the 6-month (c–f) SPEI series in the southern African region for the projection periods. Stippling shows the areas where the trend is significant at a 95% level

mate events induced by post-harvest losses, which are at risk of undermining food security in the region.

2.4.2 Extremes Based on Temperature

To explore changes in extreme heat conditions per year, Fig. 2.7 shows the projected changes in the annual average count of the heatwave days index under 1.5, 2, 3 and 4 GWL. The number of days of heatwaves has been increasing significantly in places that include rural Southern Africa since the 1960s' historical periods (Alexander et al., 2006; Intergovernmental Panel on Climate Change, 2023; Kusangaya et al., 2021). Furthermore, Hoffmann et al. (2002) pointed out that the tropical savanna climate, where most of rural Southern Africa lies, is projected to become more impacted by extreme heatwaves in the future. The days of heatwaves, as well as the hottest day of the year index (TXx), are projected to increase in magnitude as the GWLs increase, as highlighted in Fig. 2.7. Changes are more amplified in the western and northern parts of Southern Africa in countries such as Botswana, Namibia, Angola, the Democratic Republic of the Congo and Tanzania. Specifically, in Angola and Tanzania, the highest changes in the GWL are projected to have about 30 more days of heatwave relative to the 1961–1990 baseline at 1.5 °C. This change increases to approximately 100 days and more than 150 days under 3 °C and 4 °C GWL, respectively. The lowest chance is found in the coastal areas of South Africa, Lesotho, Swaziland and the southern parts of Zimbabwe and Mozambique. In most of South Africa, TXx is projected to be around 0.5 °C below 1.5 °C GWL and will increase to about 4.5 °C below 4 °C GWL. The extreme temperature changes under different GWLs are also corroborated by previous studies (e.g. Allen et al., 2018; Dosio et al., 2018; Maúre et al.,

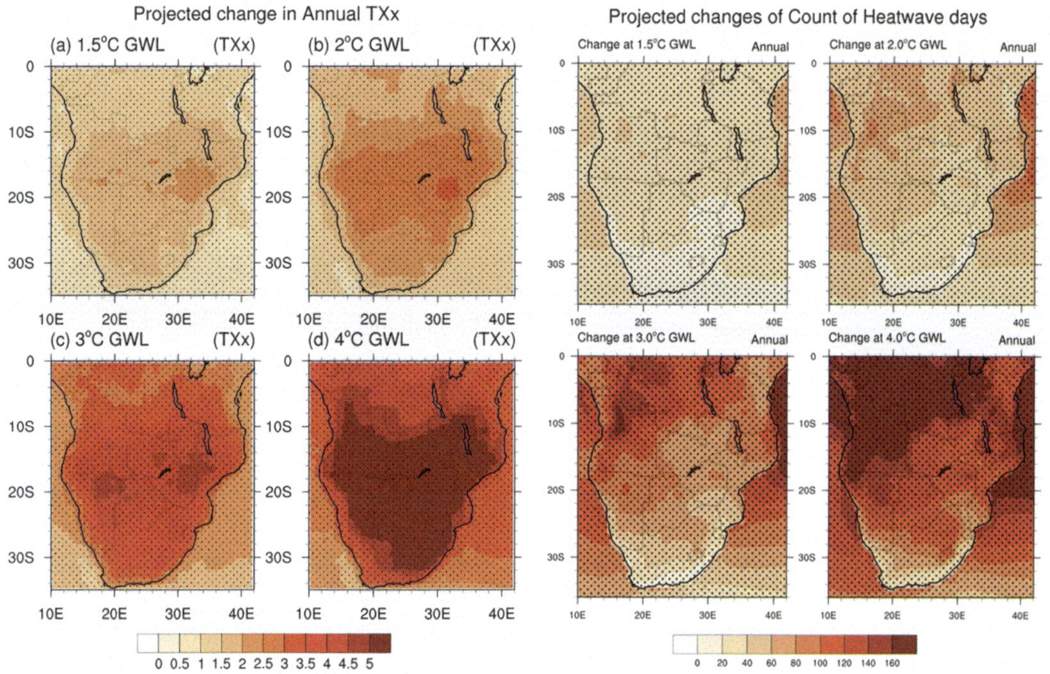

Fig. 2.7 Changes in the average temperature of the hottest day (TXx) and the count of heatwave days over Southern Africa projected by the average of the CMIP6 SSP5-8.5 ISIMIP3b ensemble at various levels of global warming. Anomalies are calculated under (**a**) 1.5 °C, (**b**) 2 °C, 3 °C and (**d**) 4 °C GWLs. The anomalies are related to the 1961–1990 baseline. Units are in °C. Stippling shows areas that are significant at a 95% level

2018; Nangombe et al., 2018, 2019; Schleussner et al., 2016).

The extreme temperature results shown in this study are comparable to other previous studies that examined future extreme temperatures in areas including Southern Africa (Dosio et al., 2018; Engelbrecht et al., 2015; Hoffmann et al., 2002; Mbokodo et al., 2020; Nangombe et al., 2019; Schleussner et al., 2016). All these studies drew a similar conclusion of a significant increase in extreme heat events such as heatwaves in African places, especially in Southern Africa. These increases in heat extremes are linked to rapid increases in maximum temperatures in the future in Southern Africa, as noted by Engelbrecht et al. (2015) and (Mbokodo et al., 2020). Engelbrecht et al. (2015) projected a temperature increase of 3–5 °C by the end of the century in Southern Africa under a low mitigation scenario is corroborated by Singo et al.'s (2023) study, which projected a rapid maximum temperature increase of about 4.5 °C–6 °C in the Limpopo Savana area relative to the 1961–1980 baseline period. Dynamically, these increases in the rate of occurrence of extreme temperatures could imply strengthening as well as the persistence of the middle-level high-pressure systems that generally prevail over the Southern Africa region under the influence of the poleward shift in the positioning of the Hadley cell in connection with climate change (Engelbrecht et al., 2011; Mahlobo et al., 2019).

2.5 Conclusion

The urgent climate crisis in Southern Africa presents a formidable challenge for the most vulnerable, particularly the rural population. The stark variability in temperatures, ranging from scorching heat near the equator in the Kalahari Desert of Botswana to cooler rural highlands in the north-east of South Africa, highlights the diverse climate conditions. This, coupled with significant variability in rainfall patterns, has led to distressing droughts, floods and crop failures in the region, especially impacting agricultural activities.

Without proactive efforts to mitigate global warming, temperature-related extremes in rural Southern Africa are projected to intensify significantly, exacerbating the challenges faced by these communities. The region, as part of Africa in general, faces lower adaptation support and limited institutional responses compared to other continents. This situation, combined with high community exposure, particularly in rural Southern Africa, risks preventing sustainable development progress, particularly in the treatment of poverty and malnutrition. Such climate changes could potentially prompt rural community migration to other countries or overcrowded urban settlements, magnifying vulnerabilities and social strains.

Increasing occurrences of drier conditions and temperature extremes, such as heatwaves, are expected to disproportionately impact poorer communities through reduced incomes, agricultural production failures and subsequent food shortages. These vulnerabilities are not only influenced by settlement and agricultural sensitivities but also by the limited adaptive capacity of communities. The shift in climate patterns, the increase in extreme weather events and the increased variability in precipitation, including the projected increase in the frequency of heavy precipitation, underscore the urgent need for comprehensive, community-inclusive strategies to address the vulnerabilities faced by rural populations in the region and mitigate the adverse impacts of climate change in Southern Africa.

Although the 'social pixels' or 'village pixels' index is instrumental in analysing the exposure of the rural population to physical climate hazards for most rural communities under climate change, it may not be representative of rural communities relying on ecotourism and other activities other than stock farming. An expansion of this study could consider additional spatial data covering rural areas dominated by other forms of rural livelihood, such as ecotourism and subsistence crop or stock farming, to define an alternative 'village index'. The study presents the climate change signal through temperature and precipitation indices, which presents a natural starting point to understand the implications of the impacts of climate change. Further analysis of impacts should consider investigating the combined impacts of temperature, humidity, fire weather, wind and rainfall. Such effects could amplify the discussed impacts of climate extremes in a way that an analysis of climate change extremes in isolation could easily miss.

Acknowledgments The analysis presented here was supported by the Department of Science and Innovation parliamentary grant (P1FCM00) awarded to the CSIR, and the FOCUS-AFRICA project, which has recieved funding from the European H2020 Research and Innovation program (Grant Agreement 869575). The content of this documents reflects only the authors' views. The European Commission and South Africa Department of Science and Innovation are not responsible for for any use that may be mode of the information that the chapter contains.

References

Aguilar, E., Aziz Barry, A., Brunet, M., Ekang, L., Fernandes, A., Massoukina, M., et al. (2009). Changes in temperature and precipitation extremes in western Central Africa, Guinea Conakry, and Zimbabwe, 1955–2006. *Journal of Geophysical Research: Atmospheres, 114*(D2). https://doi.org/10.1029/2008JD011010

Alexander, L. V., Zhang, X., Peterson, T. C., Caesar, J., Gleason, B., Klein Tank, A. M. G., et al. (2006). Global observed changes in daily climate extremes of temperature and precipitation. *Journal of Geophysical Research: Atmospheres, 111*(D5). https://doi.org/10.1029/2005JD006290

Allen, M., Dube, O. P., Solecki, W., Aragón-Durand, F., Cramer, W., Humphreys, S., & Kainuma, M. (2018). Special report: Global warming of 1.5 C.

Aloysius, N. R., Sheffield, J., Saiers, J. E., Li, H., & Wood, E. F. (2016). Evaluation of historical and future simulations of precipitation and temperature in Central Africa from CMIP5 climate models. *Journal of Geophysical Research: Atmospheres, 121*(1), 130–152. https://doi.org/10.1002/2015JD023656

Ayanlade, A., Radeny, M., & Akin-Onigbinde, A. I. (2018). Climate variability/change and attitude to adaptation technologies: A pilot study among selected rural farmers' communities in Nigeria. *GeoJournal, 83*(2), 319–331. https://doi.org/10.1007/s10708-017-9771-1

Bansah, K. J., Arthur-Holmes, F., & Assan, E. (2023). Climate-induced transformation of agriculture to artisanal mining economy in dry regions. *Journal of Rural Studies, 99*, 11–19. https://doi.org/10.1016/j.jrurstud.2023.02.005

Chemura, A., Nangombe, S. S., Gleixner, S., Chinyoka, S., & Gornott, C. (2022). Changes in climate extremes and their effect on maise (Zea mays L.) suitability over southern Africa. *Frontiers in Climate, 4*. https://doi.org/10.3389/fclim.2022.890210

Chisanga, C., Phiri, E., & Chinene, V. (2017). Trends of Extreme Events in Precipitation and Temperature during the 1963–2012 Period at Mt Makulu, Zambia. *Journal of Scientific Research and Reports, 15*(4), 1–19. https://doi.org/10.9734/JSRR/2017/34815

Clapp, J., Newell, P., & Brent, Z. W. (2018). The global political economy of climate change, agriculture and food systems. *The Journal of Peasant Studies, 45*(1), 80–88. https://doi.org/10.1080/03066150.2017.1381602

Dellink, R., Lanzi, E., & Chateau, J. (2019). The sectoral and regional economic consequences of climate change to 2060. *Environmental and Resource Economics, 72*(2), 309–363. https://doi.org/10.1007/s10640-017-0197-5

Dosio, A., Mentaschi, L., Fischer, E. M., & Wyser, K. (2018). Extreme heat waves under 1.5 °C and 2 °C global warming. *Environmental Research Letters, 13*(5), 054006. https://doi.org/10.1088/1748-9326/aab827

Engelbrecht, F. A., & Monteiro, P. M. S. (2021). The IPCC Assessment Report Six Working Group 1 report and Southern Africa: Reasons to take action. *South African Journal of Science, 117*(11/12). https://doi.org/10.17159/sajs.2021/12679

Engelbrecht, F. A., Landman, W., Engelbrecht, C., Landman, S., Bopape, M., Roux, B., et al. (2011). Multi-scale climate modelling over southern Africa using a variable-resolution global model. *Water SA, 37*(5). https://doi.org/10.4314/wsa.v37i5.2

Engelbrecht, F., Adegoke, J., Bopape, M.-J., Naidoo, M., Garland, R., Thatcher, M., et al. (2015). Projections of rapidly rising surface temperatures over Africa under low mitigation. *Environmental Research Letters, 10*(8), 085004. https://doi.org/10.1088/1748-9326/10/8/085004

FAO. (2023). *FAO Geoportal*. http://www.fao.org/geonetwork/srv/en/. Accessed 28 Aug 2023.

Gore, M., Abiodun, B. J., & Kucharski, F. (2020). Understanding the influence of ENSO patterns on drought over southern Africa using SPEEDY. *Climate Dynamics, 54*(1–2), 307–327. https://doi.org/10.1007/s00382-019-05002-w

Hoegh-Guldberg, O., Jacob, D., Taylor, M., Guillén Bolaños, T., Bindi, M., Brown, S., et al. (2019). The human imperative of stabilising global climate change at 1.5 °C. *Science, 365*(6459). https://doi.org/10.1126/science.aaw6974

Hoffmann, W. A., Schroeder, W., & Jackson, R. B. (2002). Positive feedbacks of fire, climate, and vegetation and the conversion of tropical savanna. *Geophysical Research Letters, 29*(22). https://doi.org/10.1029/2002GL015424

Hossain, A. (2020). In A. Hossain (Ed.), *Maise—Production and use*. IntechOpen. https://doi.org/10.5772/intechopen.82912

Huang, J., Zhang, G., Zhang, Y., Guan, X., Wei, Y., & Guo, R. (2020). Global desertification vulnerability to climate change and human activities. *Land Degradation & Development, 31*(11), 1380–1391. https://doi.org/10.1002/ldr.3556

IFPRI. (2013). *Southern African agriculture and climate change: A comprehensive analysis*. https://doi.org/10.2499/9780896292086

Intergovernmental Panel on Climate Change. (2023). *Climate change 2021—The physical science basis*. Cambridge University Press. https://doi.org/10.1017/9781009157896

IPCC. (2014). Regional context. In V. R. Barros, C. B. Field, D. J. Dokken, M. D. Mastrandrea, & K. J. Mach (Eds.), *Climate change 2014: Impacts, adaptation and vulnerability* (pp. 1133–1198). Cambridge University Press. https://doi.org/10.1017/CBO9781107415386.001

IPCC. (2019). *History*. https://www.ipcc.ch/about/history/. Accessed 2 Mar 2023.

IPCC. (2023). *Climate change 2021—The physical science basis*. Cambridge University Press. https://doi.org/10.1017/9781009157896

Iturbide, M., Fernández, J., Gutiérrez, J. M., Pirani, A., Huard, D., Al Khourdajie, A., et al. (2022). Implementation of FAIR principles in the IPCC: The WGI AR6 Atlas repository. *Scientific Data, 9*(1), 629. https://doi.org/10.1038/s41597-022-01739-y

Jury, M. R. (2013). Climate trends in southern Africa (with erratum). *South African Journal of Science, 109*(1/2), 111. https://doi.org/10.1590/sajs.2013/980

Kusangaya, S., Mazvimavi, D., Shekede, M. D., Masunga, B., Kunedzimwe, F., & Manatsa, D. (2021). Climate change impact on hydrological regimes and extreme events in southern Africa. In *Climate change and water resources in Africa* (pp. 87–129). Springer International Publishing. https://doi.org/10.1007/978-3-030-61225-2_5

Lange, S. (2019). *WFDE5 over land merged with ERA5 over the ocean (W5E5). V. 1.0*. GFZ Data Services.

Lange, S., Menz, C., Gleixner, S., Cucchi, M., Weedon, G. P., Nicolas, A. A., et al. (2021). *WFDE5 over land merged with ERA5 over the ocean (W5E5 v2.0)*. ISIMIP Repository.

Lobell, D. B., Roberts, M. J., Schlenker, W., Braun, N., Little, B. B., Rejesus, R. M., & Hammer, G. L. (2014). Greater sensitivity to drought accompanies maise yield increase in the U.S. Midwest. *Science, 344*(6183), 516–519. https://doi.org/10.1126/science.1251423

Mahlobo, D. D., Ndarana, T., Grab, S., & Engelbrecht, F. (2019). Integrated climatology and trends in the subtropical Hadley cell, sunshine duration and cloud cover over South Africa. *International Journal of Climatology, 39*(4), 1805–1821. https://doi.org/10.1002/joc.5917

Marumbwa, F. M., Cho, M. A., & Chirwa, P. W. (2020). An assessment of remote sensing-based drought index over different land cover types in southern Africa. *International Journal of Remote Sensing, 41*(19), 7368–7382. https://doi.org/10.1080/01431161.2020.1757783

Mather, A. A., & Stretch, D. D. (2012). A perspective on sea level rise and coastal storm surge from southern and eastern Africa: A case study near Durban, South Africa. *Water, 4*(1), 237–259. https://doi.org/10.3390/w4010237

Maúre, G., Pinto, I., Ndebele-Murisa, M., Muthige, M., Lennard, C., Nikulin, G., et al. (2018). The southern African climate under 1.5 °C and 2 °C of global warming as simulated by CORDEX regional climate models. *Environmental Research Letters, 13*(6), 065002. https://doi.org/10.1088/1748-9326/aab190

Mbokodo, I., Bopape, M.-J., Chikoore, H., Engelbrecht, F., & Nethengwe, N. (2020). Heatwaves in the future warmer climate of South Africa. *Atmosphere, 11*(7), 712. https://doi.org/10.3390/atmos11070712

Mbow, C., Rosenzweig, C. E., Barioni, L. G., Benton, T. G., Herrero, M., Krishnapillai, M., et al. (2020). Food security. In *Climate change and land* (pp. 437–550). Cambridge University Press. https://doi.org/10.1017/9781009157988.007

Meng, Q., Chen, X., Lobell, D. B., Cui, Z., Zhang, Y., Yang, H., & Zhang, F. (2016). Growing sensitivity of maise to water scarcity under climate change. *Scientific Reports, 6*(1), 19605. https://doi.org/10.1038/srep19605

Nangombe, S., Zhou, T., Zhang, W., Wu, B., Hu, S., Zou, L., & Li, D. (2018). Record-breaking climate extremes in Africa under stabilised 1.5 °C and 2 °C global warming scenarios. *Nature Climate Change, 8*(5), 375–380. https://doi.org/10.1038/s41558-018-0145-6

Nangombe, S. S., Zhou, T., Zhang, W., Zou, L., & Li, D. (2019). High-temperature extreme events over Africa under 1.5 and 2 °C of global warming. *Journal of Geophysical Research: Atmospheres, 124*(8), 4413–4428. https://doi.org/10.1029/2018JD029747

Nhamo, G., & Chikodzi, D. (2021). The catastrophic impact of tropical cyclone Idai in southern Africa (pp. 3–29). https://doi.org/10.1007/978-3-030-72393-4_1

Nhamo, L., Mabhaudhi, T., & Modi, A. (2019). Preparedness or repeated short-term relief aid? Building drought resilience through early warning in southern Africa. *Water SA, 45*(1 January). https://doi.org/10.4314/wsa.v45i1.09

Nhemachena, C., Nhamo, L., Matchaya, G., Nhemachena, C. R., Muchara, B., Karuaihe, S. T., & Mpandeli, S. (2020). Climate change impacts on water and agriculture sectors in southern Africa: Threats and opportunities for sustainable development. *Water, 12*(10), 2673. https://doi.org/10.3390/w12102673

Nicholson, S. E. (2017). Climate and climatic variability of rainfall over eastern Africa. *Reviews of Geophysics, 55*(3), 590–635. https://doi.org/10.1002/2016RG000544

Ongoma, V., Chen, H., & Gao, C. (2019). Evaluation of CMIP5 twentieth century rainfall simulation over the equatorial East Africa. *Theoretical and*

Applied Climatology, 135(3–4), 893–910. https://doi.org/10.1007/s00704-018-2392-x

Peterson, T. C., & Manton, M. J. (2008). Monitoring changes in climate extremes: A tale of international collaboration. *Bulletin of the American Meteorological Society, 89*(9), 1266–1271. https://doi.org/10.1175/2008BAMS2501.1

PLAAS. (2022). Climate change and rural livelihoods in southern Africa: A research and policy agenda.

Pörtner, H. O., Roberts, D. C., Adams, H., Adler, C., Aldunce, P., Ali, E., et al. (2022). *Climate change 2022: Impacts, adaptation and vulnerability*. IPCC Sixth Assessment Rep.

Rahut, D. B., Aryal, J. P., & Marenya, P. (2021). Understanding climate-risk coping strategies among farm households: Evidence from five countries in eastern and southern Africa. *Science of the Total Environment, 769*, 145236. https://doi.org/10.1016/j.scitotenv.2021.145236

Ranasinghe, R., Ruane, A. C., Vautard, R., Arnell, N., Coppola, E., Cruz, F. A., et al. (2023). Climate change information for regional impact and risk assessment. In *Climate change 2021—The physical science basis* (pp. 1767–1926). Cambridge University Press. https://doi.org/10.1017/9781009157896.014

Schlenker, W., & Lobell, D. B. (2010). Robust negative impacts of climate change on African agriculture. *Environmental Research Letters, 5*(1), 014010. https://doi.org/10.1088/1748-9326/5/1/014010

Schleussner, C.-F., Lissner, T. K., Fischer, E. M., Wohland, J., Perrette, M., Golly, A., et al. (2016). Differential climate impacts for policy-relevant limits to global warming: The case of 1.5 °C and 2 °C. *Earth System Dynamics (Pembroke, Ont.), 7*(2), 327–351. https://doi.org/10.5194/esd-7-327-2016

Seneviratne, S., Xuebin, Z., Muhammad, A., Wafae, B., Claudine, D., Alejandro, D. L., et al. (2021). *Weather and climate extreme events in a changing climate (Chapter 11)* (pp. 1513–1766). Cambridge University Press.

Singo, M. V., Chikoore, H., Engelbrecht, F. A., Ndarana, T., Muofhe, T. P., Mbokodo, I. L., et al. (2023). Projections of future fire risk under climate change over the South African savanna. *Stochastic Environmental Research and Risk Assessment, 37*(7), 2677–2691. https://doi.org/10.1007/s00477-023-02412-5

Sousa, P. M., Blamey, R. C., Reason, C. J. C., Ramos, A. M., & Trigo, R. M. (2018). The 'day zero' Cape Town drought and the poleward migration of moisture corridors. *Environmental Research Letters, 13*(12), 124025. https://doi.org/10.1088/1748-9326/aaebc7

Trisos, C. H., Merow, C., & Pigot, A. L. (2020). The projected timing of abrupt ecological disruption from climate change. *Nature, 580*(7804), 496–501. https://doi.org/10.1038/s41586-020-2189-9

Vincent Itai, T., & Constance, H. (2014). Challenges and opportunities for regional integration in Africa: The case of Sadc. *IOSR Journal of Humanities and Social Science, 19*(12), 103–115. https://doi.org/10.9790/0837-19124103115

Wolski, P. (2018). How severe is Cape Town's "day zero" drought? *Significance, 15*(2), 24–27. https://doi.org/10.1111/j.1740-9713.2018.01127.x

3 The Trends and Drivers of Climate Change Vulnerability in Rural Southern Africa

Beauty Dzawanda, Lazarus Chapungu, Malebajoa Anicia Maoela, Mark Matsa, and Caitlin Mapitsa

3.1 Introduction

Across the world, evidence of climate change and its hazards has been documented. Climate change is impacting the lives of millions and exacerbating existing challenges in all spheres of life. The Intergovernmental Panel on Climate Change (IPCC) report shows solid evidence of human interference with the climate system (IPCC, 2022). The changes that have been observed since 1750 include warming of the atmosphere and a rise in sea level. The IPCC report also reported the continued increase in global emissions of greenhouse gases and other climate agents that result in dangerous climate change. The impacts of climate change are region-specific, with unequal distribution both within and between countries, regions and even communities (IPCC, 2022; Shrikhande et al., 2023). The consequences of climate change hazards are varied depending on the differential vulnerability of regions and their capacity to cope, adapt and respond to climatic hazards (Shrikhande et al., 2023). However, many are projected to be negative, particularly in regions such as Southern Africa.

Many countries have committed to developing national climate action and adaptation plans after the Paris Agreement in 2015 (IPCC, 2021, 2022). However, regional differences in vulnerabilities, capacities and consequent impacts remain a significant challenge in developing these plans. Regional and national indicators such as infrastructure development, access to healthcare care, socio-economic factors and even awareness determine individual vulnerability to climate change, especially in resource-limited low and middle-income regions such as Southern Africa (Matsa, 2021; Shrikhande et al., 2023). High-income countries govern the global dialogue on climate change. At the same time, their continued use of extractive policies intensifies climate hazards and impacts economies in regions that are not responsible for damage. For example, Africa is responsible for only 2–3% of greenhouse gas emissions worldwide, yet a report by the World Meteorological Organization shows that it suffers a disproportionate share of the impact, including drought, rising temperatures and water stress due to flooding (United Nations, 2022). Worse still,

B. Dzawanda (✉) · M. Matsa
Department of Geography, Environmental Sustainability and Resilience Building, Midlands State University, Gweru, Zimbabwe
e-mail: dzawandab@staff.msu.ac.zw; matsam@staff.msu.ac.zw

L. Chapungu · M. A. Maoela
Institute for Corporate Citizenship, University of South Africa, Pretoria, South Africa
e-mail: tchapul@unisa.ac.za; maoelma@unisa.ac.za

C. Mapitsa
Wits School of Governance, University of the Witwatersrand, Johannesburg, South Africa
e-mail: Caitlin.Mapitsa@wits.ac.za

M. Matsa et al. (eds.), *Climate Change Resilience in Rural Southern Africa*, Sustainable Development Goals Series, https://doi.org/10.1007/978-3-031-73600-1_3

population expansion in African nations places more people in increasingly vulnerable locations (Harrington & Otto, 2018). These recent findings agree with findings discovered more than a decade earlier, such as the 2011 African Development Bank report, which indicated that climate damage in Africa is expected to be greater than in any other region worldwide, more than 10% higher than the next exposed Indian region, and more than twice as high as in Eurasia, Russia, the USA and Latin America (Fonds Africain de Developpement, 2011).

The term 'vulnerability' is widely used in both social and physical sciences. Therefore, more than a dozen scholarly definitions of vulnerability have emerged in disciplines and practices (Mavhura, 2018). To promote a common understanding of vulnerability, the Office of Disaster Risk Reduction of the United Nations defined vulnerability as the circumstances and characteristics of a system, community or asset that make it susceptible to the adverse effects of a hazard (United Nations, 2022). The degree to which communities are susceptible to hazards is not exclusively dependent upon their exposure to a hazard but also on their socioeconomic and demographic characteristics (Marin et al., 2021). Such factors are independent of the hazards triggering their vulnerability but significantly impact their capability to prepare for, respond to and recover from hazards or disasters. Because many communities live in different economic, social, cultural, political and institutional settings, they have different levels of vulnerability.

Rural areas in Africa are particularly vulnerable to climate change due to their dependence on rainfed agriculture, which is highly sensitive to climate variables and extreme weather events (Tawodzera & Ncube, 2019; Matsa & Marambanyika, 2020; Chirisa et al., 2021; Mucherera & Mavhura, 2020). Floods and droughts, the most common hazards in Southern Africa, exacerbate the vulnerability of rural households dependent on rainfed agriculture and livestock, compounded by limited infrastructure and institutional support (Mucherera & Mavhura, 2020). The projections indicate an increase in flood frequency and intensity, further increasing the vulnerability of small farmers (Nhemachena et al., 2020). This is worsened by the high sensitivity of social-ecological systems and the limited capacity of institutional actors to respond to emerging threats (Jiri & Mafongoya, 2018). High levels of poverty and limited employment opportunities further intensify the vulnerability of these farmers (Mucherera & Mavhura, 2020). Understanding the extent of exposure and vulnerability among different rural groups is crucial, as socio-economic status and geographical location create systematic differences in climatic risk exposure. This chapter examines spatial and temporal trends and drivers of climate change vulnerability in rural Southern Africa.

3.2 Materials and Methods

3.2.1 Description of the Study Area

A total of ten countries located in the southern part of Africa were included in the study, namely, Zimbabwe, South Africa, Malawi, Lesotho, Eswatini, Angola, Botswana, Namibia, Zambia and Mozambique. The region comprises seasonal, semi-arid, temperate and tropical climates. Droughts occur frequently and are becoming more severe (Mucherera & Mavhura, 2020). IPCC (2021) predicts that the region will become drier due to the changing climate system. Climate change is also exacerbating the proliferation of weather extremes, such as tropical cyclones, heatwaves, heavy rains, cold snaps, floods and dust storms.

3.2.2 Research Design

The parallel convergent design of the mixed-methods approach was adopted as the inquiry strategy (Fig. 3.1). It involved the collection and analysis of qualitative data from the Scopus database and quantitative data from the Global Internal Displacement Database (GIDD), which is maintained by the Internal Displacement Monitoring Centre (IDMC) (https://www.

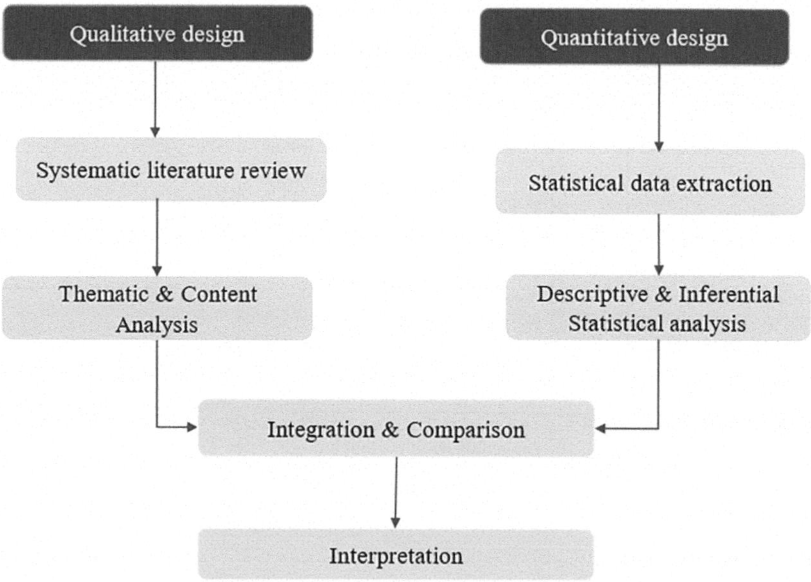

Fig. 3.1 The parallel convergent design of the mixed methods adopted in this study. (Source: Authors, adapted from Chapungu et al. 2024)

Table 3.1 The terms searched in the title, abstract or keywords within Scopus and Science Direct databases

TITLE-ABS-KEY ("climate change" AND vulnerability AND drivers) AND (LIMIT-TO (OA, "all")) AND (LIMIT-TO (LANGUAGE, "English")) AND (LIMIT-TO (DOCTYPE, "ar")) AND (LIMIT-TO (AFFILCOUNTRY, "South Africa") OR LIMIT-TO (AFFILCOUNTRY, "Malawi") OR LIMIT-TO (AFFILCOUNTRY, "Zimbabwe") OR LIMIT-TO (AFFILCOUNTRY, "Zambia") OR LIMIT-TO (AFFILCOUNTRY, "Madagascar") OR LIMIT-TO (AFFILCOUNTRY, "Botswana") OR LIMIT-TO (AFFILCOUNTRY, "Namibia"))

internal-displacement.org/database). The results of the parallel processes were compared to draw conclusions.

3.2.2.1 Qualitative Design (PRISMA)

For qualitative analysis, the chapter adopted a systematic literature review approach, focusing on primary peer-reviewed articles from the Scopus and Science Direct databases. The selection of the PRISMA methodology was based on its ability to improve the transparency and scientific rigour of the systematic review process. The primary rationale for selecting articles mainly from the Scopus and Science Direct databases was their standing as the largest repositories of citations and curated abstracts from peer-reviewed research literature and reputable sources on the Web. With its extensive coverage of the scientific and social science literature, it emerged as an ideal resource for study. Keywords that were used to initiate the literature search strategy are shown in Table 3.1.

The search was applied to the abstract, title and keywords of published papers spanning from 2011 to 2023, resulting in 1331 papers from Scopus and 13 from Science Direct. Following a review based on accessibility criteria, 1111 articles were excluded due to factors such as limited availability, non-English language and relevance to the study area. A final set of 33 articles was selected for in-depth analyses (Fig. 3.2). The coefficient of agreement of 0.724 among the five authors who performed the screening was obtained, which falls into the substantial agreement class of 0.61–0.80. Data analysis was carried out in an interpretative manner in which all sources that were gathered were individually analysed and then synthesised according to the

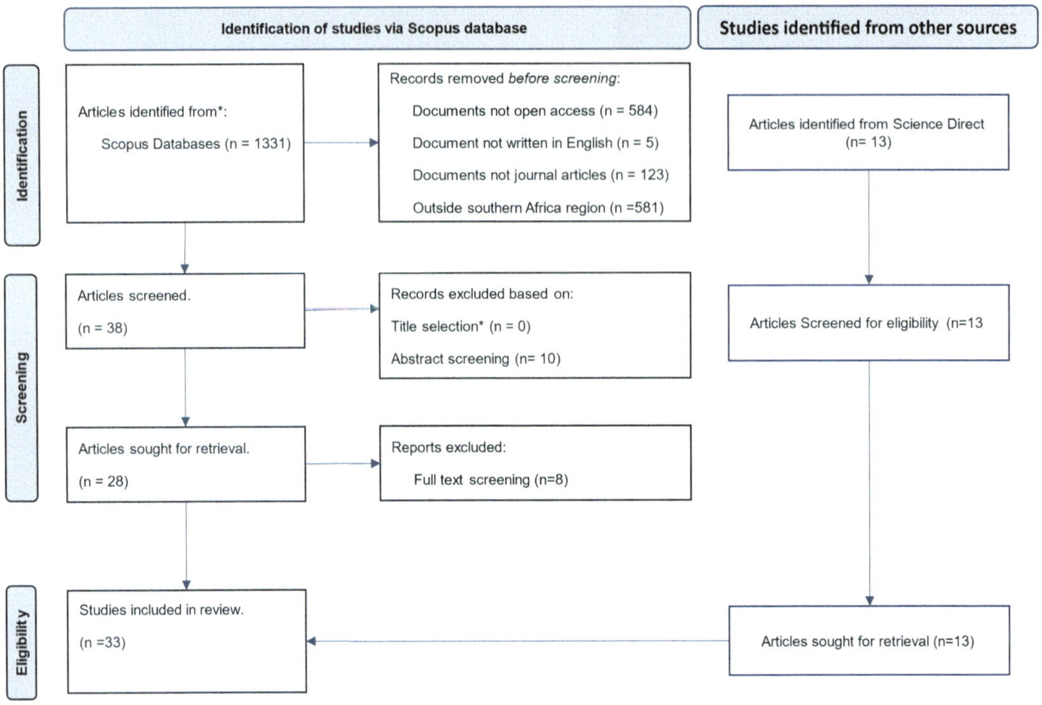

Fig. 3.2 Selection and screening of literature from databases. (Source: Authors)

objectives of the study. After collecting the data, the researchers read the literature to get a sense of what it contains. Then, the data was connected by creating and identifying common ideas/codes, which were then put into themes. The themes were then presented cohesively to answer the research objectives.

The limitation of this methodology is its reliance on specific search terms, which may have overlooked a considerable number of articles that did not explicitly mention these terms in their titles, abstracts or keywords. As a result, there is a possibility that relevant studies meeting the research criteria were unintentionally excluded from the search results.

3.2.2.2 Quantitative Design

Data for quantitative analysis were extracted from the existing Global Internal Displacement Database (GIDD) developed and managed by the Internal Displacement Monitoring Center (IDMC) (https://www.internal-displacement.org/database). The data, which comprised the frequencies of climate-related hazards and the number of people displaced (vulnerable) by hazards between 2008 and 2022, were initially subjected to the Kolmogorov-Smirnov test to determine whether they deviate from normal distribution or not. Following the resultant normal distribution, the trend of displaced individuals, as well as the frequency of disasters, was examined using the non-parametric Mann-Kendall (M-K) test. A non-parametric technique called the M-K test is frequently employed to find monotonic trends in collections of hydrometeorological data (Chapungu & Nhamo, 2021). The test is straightforward, reliable and tolerant of missing values, seasonality and values below the detection threshold. This test was conducted using the Microsoft Excel XLSTAT 2020 add-in because it can account for and eliminate the impact of autocorrelations. The Mann-Kendall test statistic is computed using the following formula:

$$S = \sum_{i=1}^{n-1} \sum_{j=i+1} \operatorname{sgn}(xj - xi)$$

where:

S is the Kendall score
n is the number of observations
xj and xi are the data values at times j and i

sgn($xj - xi$) is the sign function which returns the following:

$$Sgn\left(xj - xi\right) = \left\{1\,if\ x > 0{,}0\,if\ x = 0, -1\,if\ x < 0\right\}$$

3.2.3 Determination of Vulnerability

This study defined climate vulnerability as the degree to which a country is exposed, sensitive to and capable of mitigating the adverse effects of climate change. The global vulnerability ranking platform, https://gain-new.crc.nd.edu/ranking/vulnerability, was used to derive country-based spatial vulnerability ratings. The ND-GAIN index score, which is made up of preparedness and vulnerability scores, is used to rank overall vulnerability. Six life-supporting areas are considered when measuring the overall vulnerability by the Notre Dame Global Climate Initiative (ND-GAIN): food, water, health, ecological services, human habitat and infrastructure. Temporal trends of vulnerability were computed using the Mann-Kendall trend test based on temporal exposure to climatic hazards, sensitivity and adaptive capacity. The sum of climate disasters and disaster frequency per country between 2008 and 2022 was used as proxy indicators for exposure, while the number of displaced people per year was used as the proxy indicator for sensitivity based on the assumption that highly climate-sensitive countries have the highest number of displaced people.

3.3 Presentation of the Findings

3.3.1 Trends in Climate Vulnerability in Southern Africa

The results indicate an increasing trend of climate-induced vulnerability in Southern Africa between 2008 and 2022. The ten countries under study have been increasingly exposed to different types of climate disasters, including storms, floods, mass movement, wildfires and droughts, among others. Exposure to climate-induced hazards has been compounded by an increasing trend in sensitivity to climate change and a reduction in adaptive capacity, leading to an exacerbation of vulnerability over time. Figure 3.3 shows trends in the number of people displaced by climate-induced disasters, the frequency of climate disasters and the sensitivity to climate and vulnerability in the region.

As shown in Fig. 3.3a, although the trend is not statistically significant, there is an increase in the number of people displaced by climate-induced disasters between 2008 and 2022. Furthermore, the region is increasingly being exposed to climate-induced disasters, as shown in Fig. 3.3b. The trend of climate-induced disaster frequency is statistically significant ($p = 0.007$, $\alpha = 0.05$). The results also show that there is a statistically significant increase ($p = 0.023$, $\alpha = 0.05$) in vulnerability to climate-induced disasters. Although the trend of sensitivity is not statistically significant ($p = 0.52$, $\alpha = 0.05$), there is an increase in sensitivity to climate-induced disasters over time, which contributes to increasing the vulnerability of Southern Africa to climate change.

The existing literature confirms the trends observed in this study. It reveals an extensive spatial and temporal extent of climate vulnerability in Southern Africa, particularly in countries and regions exposed to intense and highly frequent climate hazards. For example, rural areas in South Africa have been exposed to prevalent extreme weather events such as droughts and floods (Flatø et al., 2017; Samuels et al., 2022; Spear et al., 2018). In Zimbabwe, the proliferation of cyclical and random droughts has been found to become more severe and destructive over time, exposing smallholder farmers and all farming communities to harsh conditions that increase their vulnerability to climate change (Jiri et al., 2017; Mugandani et al., 2022; Mwadzingeni et al., 2021). A significant number of studies indicated that the entire region of Southern Africa has experienced an increased occurrence of drought episodes, making rural

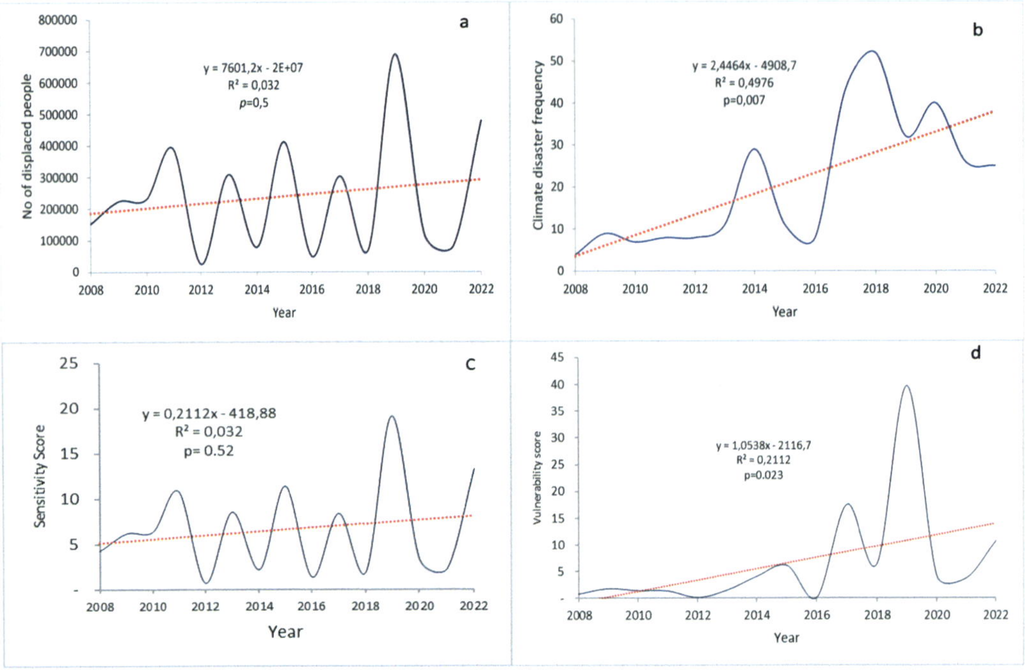

Fig. 3.3 Temporal trends of climate vulnerability in Southern Africa. (**a**) Climate-induced displacements between 2008 and 2022. (**b**) Frequency of climate disasters in Southern Africa between 2008 and 2022. (**c**) Southern Africa's sensitivity to climate disasters between 2008 and 2022. (**d**) Southern Africa's trend of climate vulnerability between 2008 and 2022. (Source: Authors)

areas prone to challenges associated with drought hazards (Conradie et al., 2022; Hawkins et al., 2022; Mpandeli et al., 2019).

In tandem with the increasing frequency and severity of droughts, the Southern African region has been subjected to flood hazards (Mafofo et al., 2021; Mucherera & Mavhura, 2020; Nhamo & Chapungu, 2021). Floods amplified the region's susceptibility to the impacts of climate change. Although urban locations have experienced a considerable proportion of flood hazards attributed to deficiencies in urban planning and infrastructure inadequacies (Kita, 2017), rural domains have borne a disproportionate burden, with numerous floods inflicting substantial damage to essential elements of rural infrastructure, including roads, educational facilities, bridges, healthcare centres and various other amenities relied upon by rural inhabitants (Mafofo et al., 2021; Mucherera & Mavhura, 2020). The genesis of most flood events lies in tropical cyclones, which precipitate heavy rainfall, leading to extensive surface run-off. In particular, cyclones Idai and Kenneth, which struck Mozambique, Malawi and Zimbabwe in 2019, triggered catastrophic flooding, precipitating a humanitarian crisis of unprecedented magnitude within the region, resulting in over 1500 deaths and the displacement of several thousand individuals from their residences (Chapungu, 2020; Mafofo et al., 2021).

Although the temporal trajectory of vulnerability to climate change demonstrates an overall increase, there is spatial diversity in vulnerability levels throughout the Southern African region. Using country index data sourced from the Notre Dame Global Climate Initiative (ND-GAIN) (https://gain-new.crc.nd.edu/ranking/delta/vulnerability), it becomes evident that there exists significant spatial heterogeneity in climate vulnerability within the region. The data set reveals that vulnerability is not only dependent on exposure to climate-related hazards; instead, it reflects a complex interplay of economic, social, political and environmental dynamics, many of which are shaped at the national level. Figure 3.4 illustrates the spatial patterns of vulnerability to climate change in Southern Africa.

As depicted in Fig. 3.4a, vulnerability, determined by a country's exposure, sensitivity and adaptive capacity, is most pronounced in Angola, Malawi and Zimbabwe, with vulnerability indices ranging between 0.5 and 0.55. Subsequently, Mozambique and Zambia exhibit a slightly lower vulnerability index, ranging between 0.48 and 0.49. Namibia occupies a middling position within the vulnerability spectrum, with an index within the range of 0.45–0.47. South Africa and Botswana emerge as the least vulnerable countries in the region, with vulnerability indices ranging from 0.39 to 0.44. Compared to other global regions, this region is one of the most vulnerable due to the rapid and ad hoc proliferation of diverse climate-induced hazards, which have become increasingly intense, resulting in severe loss and damage.

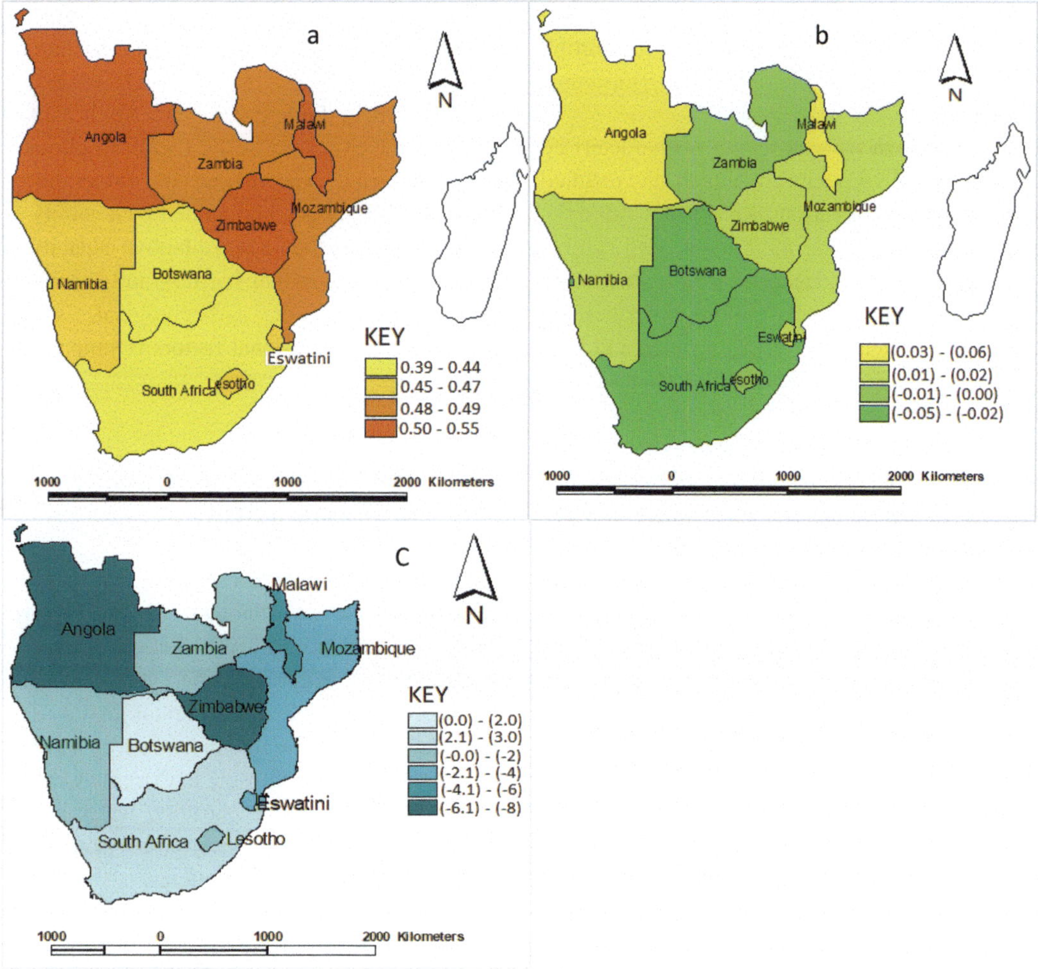

Fig. 3.4 Spatial patterns of climate vulnerability in Southern Africa: (**a**) Vulnerability determined by a country's exposure, sensitivity and ability to adapt to the negative impact of climate change. (**b**) GDP-adjusted vulnerability (**c**) Overall vulnerability by considering vulnerability in six life-supporting sectors—food, water, health, ecosystem service, human habitat and infrastructure. (Source: Authors' elaboration based on data from ND-GAIN 2024)

When considering the vulnerability adjusted for GDP to climate change, as illustrated in Fig. 3.4b, South Africa and Botswana continue to maintain their status as the least vulnerable countries in the region. However, Zambia occupies the second position, while Zimbabwe, Mozambique and Namibia fall into the category above Angola, which is designated as the most vulnerable nation in the region. However, the comprehensive vulnerability index, which encompasses a wide variety of variables such as infrastructure, healthcare, ecosystem services, food security, water resources and habitat condition, delineates Angola and Zimbabwe as the most vulnerable nations, followed by Malawi, Mozambique, Eswatini, Namibia, Zambia, South Africa and Botswana, respectively.

The Southern African region is characterised by widespread vulnerability, which has resulted in some impacts in various sectors. For example, extreme weather events in Zimbabwe and Zambia have affected the education sector, disrupting learning in rural areas by making schools inaccessible, particularly for young women and girls (Tanner et al., 2022). Additionally, when schools serve as temporary shelter from severe weather, there have been allegations of increased exposure to physical and sexual abuse, as well as an increase in the gender roles that differ between boys and girls. Furthermore, Tanner et al. (2022) reiterate that the effects of climate change have increased child, early and forced marriages, as well as absenteeism and dropout rates from school, especially in lower-class homes in Zambia and Zimbabwe.

The region has experienced an increase in the incidence of wildfires, and South Africa has the highest number of cases. Wildfires have increased in intensity and frequency in South Africa, the Western and Eastern Cape provinces being the most affected due to the Mediterranean climate and dense vegetation (Chapungu et al., 2024; Kraaij et al., 2013, 2018). The Western Cape has faced significant fires, such as those in Table Mountain National Park in April 2021 (Forsyth & van Wilgen, 2008). Namibia experiences wildfires in its arid regions, often ignited by lightning or human activities (Eastment et al., 2022; Kapuka & Hlásny, 2020). Botswana's savanna and woodland ecosystems are prone to fires, with notable incidents in protected areas such as the Okavango Delta (Dube, 2013). Zimbabwe also faces recurrent wildfires, driven by factors such as land use changes and climate variability (Mpakairi et al., 2020), impacting national parks such as Matobo and Hwange. Although the incidence of wildfires in Mozambique is lower, grassland and savanna ecosystems are still vulnerable, with reports of fires in areas such as the Niassa Reserve (Nhongo et al., 2019; Shirvani et al., 2023). There is compelling evidence in the literature on the vulnerability of rural areas in Southern Africa to climate change (Table 3.2).

As shown in Table 3.2, several studies have indicated the exposure of the region to climate-related hazards, exacerbating its vulnerability. This vulnerability is compounded by a variety of factors, including the limited adaptive capacity of households, institutional barriers and the interconnected nature of agroecological, socio-economic and institutional factors (Quinn et al., 2011).

3.3.2 The Drivers of Vulnerability in Southern Africa

Several factors interact in a complex way to affect the state of the environment, which, in turn, affects the susceptibility of Southern Africa to climate change. Factors can be categorised into social, economic and environmental drivers. The study findings indicate that these factors include, among other things, exposure to extreme weather events, lack of adaptive capacity, water stress, reliance on agriculture, poverty, environmental degradation and economic policies.

Extreme Weather Events

The results indicate that Southern Africa is experiencing an increase in extreme weather events that have exacerbated the degree of vulnerability of the region. Events include droughts, floods, storms, wildfires and, in some cases, mass movement. The data show that there is spatial

Table 3.2 Trends in vulnerability to climate change, as depicted in the literature

Vulnerability trends	Countries	Evidence (Studies)
Increased climate-induced displacement of people and communities	Mozambique, South Africa, Zimbabwe, Angola, Zambia, Namibia, Botswana	Balsari et al. (2020), Chapungu (2020), Ghosh and Orchiston (2022), Milán-garcía et al. (2021)
Increased exposure and sensitivity to drought hazards	Zimbabwe, Malawi, South Africa, Zambia	Dias et al. (2023), Hawkins et al. (2022), Makwinja et al. (2021), Nhemachena et al. (2020), Quinn et al. (2011), Tanner et al. (2022)
Increased exposure and sensitivity to tropical storms/cyclones	Mozambique, Zimbabwe, Malawi, South Africa (Durban)	Chapungu (2020), Lokotola (2023), Mafofo et al. (2021), Reliefweb (2021), Tanner et al. (2022)
Increased exposure and sensitivity to flooding	Zimbabwe, Mozambique, South Africa	Nhamo and Chapungu (2021), Dias et al. (2023), Tanner et al. (2022), Mucherera and Mavhura (2020), Makwinja et al. (2021)
Increased exposure and sensitivity to wildfires	South Africa, Botswana, Zimbabwe, Mozambique, Namibia	Cassidy et al. (2022), Chapungu et al. (2024), Curto et al. (2024), Dube (2013), Eastment et al. (2022), Kavhu and Ndaimani (2022), Mpakairi et al. (2019, 2020), Nieman et al. (2021)
Increased exposure to climate-induced vector-borne diseases	Zimbabwe, Botswana, South Africa	Amegah et al. (2016), Chapungu (2017), Giesen et al. (2020), Rogers and Randolph (2006), Wilcox et al. (2019)
Increased climate-induced environmental degradation, e.g. soil erosion, pollution and deforestation	South Africa, Zimbabwe, Mozambique	Akadiri et al. (2019), Borrelli et al. (2020), Chapungu et al. (2023), Gulistan et al. (2020), Makuvara et al. (2022)

variability in vulnerability and that some countries experience more frequent events than others over time. Figure 3.5 shows the trends of extreme events in selected countries in the region and the trend of displacement as an indication of high vulnerability.

The displacement of communities from their places of origin by climate-induced disasters is a proxy indicator of vulnerability to climate change. After exposure to the hazard, if communities do not cope with the disaster due to high sensitivity and low adaptive capacity, they are forced to migrate to safe areas. The results show that in Southern Africa, between 2008 and 2022, various countries have exhibited high levels of vulnerability to climate change-induced hazards through internal displacement. Figure 3.6 shows the cumulative total number of people displaced due to climate-induced hazards in Southern Africa between 2008 and 2022.

As shown in Fig. 3.6, exposure to catastrophic weather events has led to increased vulnerability in the region. Mozambique has the highest number of people displaced by extreme weather events. Malawi is second, followed by Angola, South Africa, Zimbabwe and Zambia. Lesotho has the least number of people displaced by extreme weather events. Floods associated with cyclonic events are the leading weather catastrophes that are displacing people.

Water Stress

The situation of water availability in Southern Africa exacerbates its vulnerability to climate change. The region is already prone to water stress due to its semi-arid and arid climates. Climate change exacerbates this stress through changes in precipitation patterns, leading to an increase in the frequency and intensity of droughts. The existing literature presents countries such as Namibia and Botswana as highly vulnerable to water scarcity, which threatens agriculture, food security and livelihoods (Akinlabi et al., 2023; Banda et al., 2022; Gandidzanwa & Togo, 2022; Lee et al., 2023; Luyt & Sealy, 2023; Mapani et al., 2023; Nelson et al., 2022; Scholes et al., 2023.

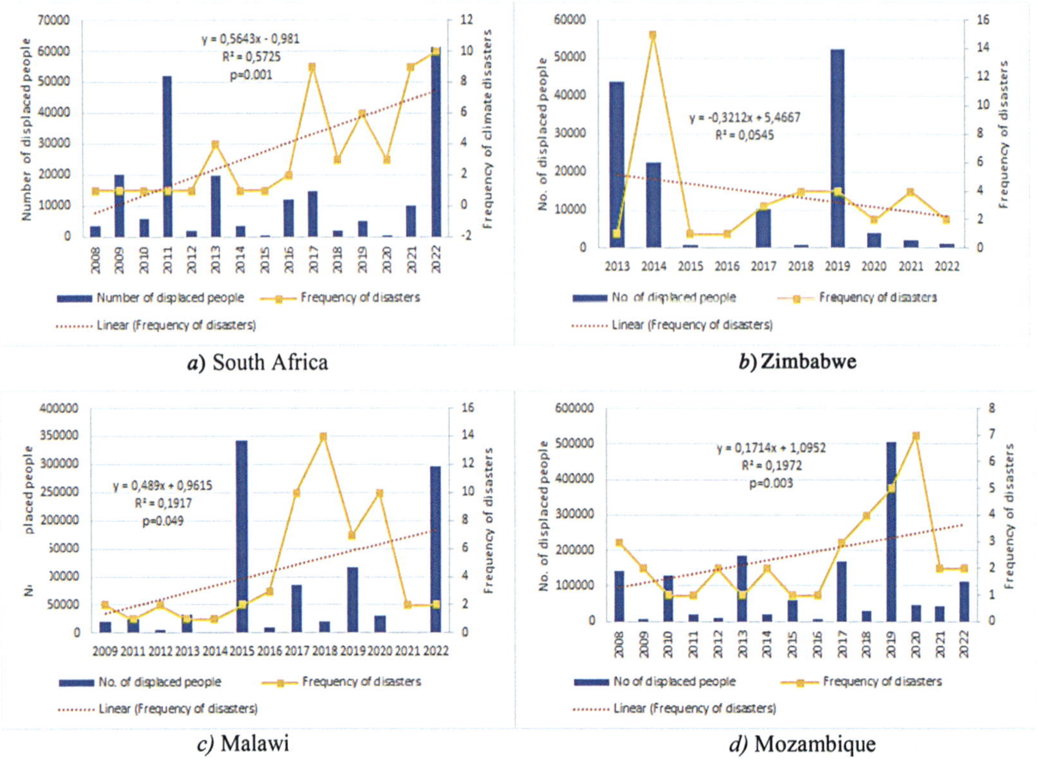

Fig. 3.5 Trends in climate-induced displacements in selected countries in Southern Africa

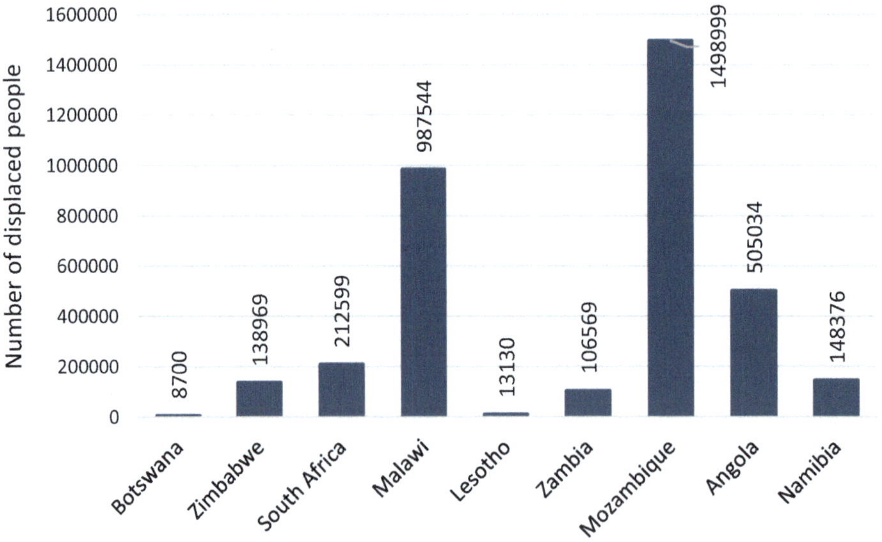

Fig. 3.6 Country-specific displacements between 2008 and 2022 in Southern Africa. (Source: Authors, IDMC data 2023)

Agricultural Dependence

Many Southern African countries rely heavily on rainfed agriculture for their livelihoods and food security. The impacts of climate change, such as erratic rainfall patterns, prolonged droughts and extreme weather events, such as floods and heatwaves, disrupt agricultural production. Zimbabwe, for example, has experienced significant crop failures and food shortages due to climate-related factors. Crop failure exposes communities in the region to the impacts of climate change because agriculture is the primary source of fo.od and livelihood. Thus, they do not have a diversity of adaptation and resilience options and mechanisms.

Economic Vulnerability

Southern African countries with limited economic diversification and high dependence on natural resources are particularly vulnerable to the impacts of climate change. For example, reliance on tourism in countries such as South Africa and Namibia exposes them to risks such as habitat loss, loss of biodiversity and reduced attractiveness due to climate-related factors such as water scarcity and extreme weather events. Zimbabwe has been under economic sanctions for more than two decades, resulting in economic challenges that exacerbate its ability to overcome the deleterious effects of persistent droughts, heatwaves, floods and cyclones. The same applies to Mozambique and Malawi, which do not implement comprehensive climate resilience frameworks due to economic vulnerability. Climate change exacerbates the economic challenges of these countries, creating a vicious climate-underdevelopment nexus.

Other Drivers

The existing body of literature suggests that the determinants of susceptibility to climate change effects are multifaceted, encompassing structural elements such as gender disparity, patriarchal frameworks and deeply ingrained cultural ideologies (Tanner et al., 2022). Gender and other forms of inequality manifest themselves in differential susceptibilities and exposures to hazards arising from climate change. Furthermore, patriarchal systems, social norms and established cultural ideologies contribute to gender-based violence. At the same time, the convergence of climate vulnerabilities with socio-economic deprivation and gendered decision-making within households impacts the experiences of girls regarding these effects.

However, coastal regions in countries such as Mozambique and South Africa face increased risks due to the upward trend in sea levels (Dube et al., 2021; Dube & Chikodzi, 2023), resulting in erosion along coastlines, intrusion of saline water into freshwater sources and increased susceptibility to storm surges. Table 3.3 shows some of the drivers of climate change, the countries affected and the source documents.

3.4 Discussion of Findings

The findings indicate a consistent increase in climate-related displacements in Southern Africa between 2008 and 2022, signalling a region characterised by high sensitivity, low adaptive capacity and consequently elevated vulnerability to climate change. The displacement scale serves as a barometer for gauging the level of vulnerability to climate-induced phenomena. Previous research (Nhemachena et al., 2020; Saarinen et al., 2023) has underscored the increasing frequency and severity of climate-related events in Southern Africa, including floods, droughts, storms, heatwaves and sea level rise, with increased intensity (Dube & Chikodzi, 2023). These factors have progressively exacerbated the vulnerability of rural Southern Africa and contributed significantly to the increase in displacements within the region. Some of the more significant displacement has come in the form of transborder migrations, and the biggest casualty in terms of receiving these migrants is South Africa, which is the economic core of the subregion. Significant numbers of Zimbabweans, Mozambicans and Namibians have been forced to emigrate to South Africa because of the vicissitudes of climate (Matsa, 2021; Matsa & Marambanyika, 2020). Rural communities in the southern border regions of these countries have contributed most of the

Table 3.3 Key drivers of climate vulnerability in Southern Africa

Key drivers of climate vulnerability	Countries	Evidence
Ineffective disaster risk reduction systems	Malawi, Zimbabwe, South Africa, Mozambique	Coetzee et al. (2023), Leal Filho et al. (2020), Mucherera and Mavhura (2020), Ngxongo (2021)
Environmental degradation, e.g. siltation of water bodies and deforestation	Malawi, Mozambique, South Africa, Zimbabwe	Dias et al. (2023), Quinn et al. (2011), Makwinja et al. (2021)
Weak institutions with respect to resources, skills and knowledge	Malawi, Zimbabwe, Mozambique	Coetzee et al. (2023), Mucherera and Mavhura (2020)
Environmentally unsafe practices, e.g. illegal gold mining and abandoned mines	Malawi, Mozambique, Zimbabwe, South Africa	Dutton (2022), Mhlongo (2023),
Conflicts over natural resources	Malawi, Zimbabwe, Mozambique, Zambia, South Africa	Dias et al. (2023), Tanner et al. (2022), Mucherera and Mavhura (2020)
Extreme weather events	Mozambique, South Africa, Zimbabwe, Namibia, Zambia	Chapungu (2020), Mafofo et al. (2021), Mucherera and Mavhura (2020), Reliefweb (2021)
Agricultural dependence	Eswatini, Zimbabwe, South Africa,	Apraku et al. (2021), Mutimukuru-Maravanyika et al. (2022), Nhemachena et al. (2020), Shongwe et al. (2013)
Water stress	Botswana, Namibia, Zimbabwe, South Africa	Akinlabi et al. (2023), Banda et al. (2022), Gandidzanwa and Togo (2022), Lee et al. (2023), Luyt and Sealy (2023), Mapani et al. (2023), Nelson et al. (2022), Scholes et al. (2023)
Rise in sea level	South Africa, Mozambique	Cooper and Green (2023), Dube and Chikodzi (2023), Hancke et al. (2023), Struwig et al. (2024), Wells et al. (2023)
Economic vulnerability	Zimbabwe, South Africa, Namibia, Zambia	Coetzee et al. (2023), Kollipara et al. (2023), Mlambo (2023), Wekeza et al. (2022)
Health impacts	Zimbabwe, South Africa, Malawi, Mozambique	Chapungu and Nhamo (2021), Giesen et al. (2020), Mordecai et al. (2020), Wilcox et al. (2019)

brawn power due to proximity (Matsa, 2021; Matsa & Marambanyika, 2020).

Climate change exerts pressure on livelihoods beyond those dependent on agriculture and, as a result, imposes challenges to the achievement of the Sustainable Development Goals (SDGs) in the Southern African region. Hermans and McLeman (2021) emphasise the impact of drought on resource-dependent rural populations, which can result in loss of livelihood and subsequent migration out of affected areas. Southern Africa hosts populations relying on ecosystem goods such as food, feed, fibre, medicines and wood, with additional ecosystem services encompassing water purification, erosion regulation, pollination, water and nutrient cycling, habitats and recreation. However, adverse effects of climate change lead to loss of biodiversity, alterations in vegetation structure and composition and threats to specific animal species, thus diminishing the ecosystem's contribution to provisioning, regulatory, supporting and cultural services.

A further risk posed by the increase in climate vulnerability is the adaptive capacity of indigenous knowledge to integrate the current pace of change, as well as local preferences around staple crops. Farming techniques for local staple foods have often been refined over generations, and highly specialised and locally appropriate methods are widely known. As rainfall patterns change and other crops become more suitable, knowledge about their growth in local conditions is limited (Nyadzi et al., 2021). Furthermore, while new crops may be established relatively quickly in response to climate shocks, diets are slower to evolve, and the loss of a preferred staple crop can increase household-level vulnerability, with the need to purchase this staple rather than adapt to a different diet.

This trend has the potential to fuel climate-induced displacements (SDG 13) and affect the efforts made to reduce poverty and end hunger (SDG 1 and 2). The impact becomes more significant given that many rural communities in Southern Africa, for example, significant parts of Zimbabwe, Zambia, Botswana and Namibia and parts of South Africa, are semi-arid to arid and traditionally depend on rainfed crop farming and livestock such as cattle, goats, sheep and donkeys as their safety nets (Matsa & Mini, 2014; Matsa & Mupepi, 2021). Many crops and livestock perish from such shocks and stressors, significantly reducing the resilience of rural communities in the subregion and affecting the achievement of the SDGs.

Furthermore, environmentally displaced populations often settle in pristine areas, where they rely entirely on the environment for food. Such resettlements frequently result in further environmental degradation due to various anthropogenic activities, some of which may exacerbate climate change. This is evident in recent research that illustrates the expansion of cropland into protected areas in the region (Meng et al., 2023). An example is Zimbabwe's 2000 fast-track land reform programme, which had historical, political and other reasons. However, it also had a climate change lining, as most of the peasant beneficiaries who left their traditional rural homes were essentially moving away from environments that had been ravaged by the changing climate, manifesting itself through increased atmospheric temperatures, reduced precipitation and in the southern and northwest provinces, increasing desertification (Matsa & Marambanyika, 2020). Without an intelligent regional climate framework, climate change will perpetuate a cycle that triggers additional displacements.

Southern Africa has experienced a discernible temporal increase in vulnerability to climate change, with varying degrees influenced by geographical factors that determine exposure to climate-induced disasters. Nations located near oceans are subjected to increased exposure to cyclones and floods, leading to a significant number of climate-induced displacements. However, socio-economic and political factors also shape vulnerability levels. For example, GDP emerges as a significant influencer of climate vulnerability, with economic fragility exacerbating susceptibility in some Southern African countries.

Given the multifaceted nature of climate vulnerability factors in the region, there is a pressing need to establish a comprehensive, multisectoral framework to enhance climate resilience across the region. A practical approach would be for each country to develop its nationally determined climate action plan in line with the Paris Agreement and develop actionable measures at the national level, taking into account its national peculiarities and vulnerabilities. These country-specific action plans would then be discussed at a Southern Africa subregional climate vulnerability workshop for integration into a regional framework to improve resilience in the subregion. Without such measures, climate change-induced disasters are likely to continue to impede the region's development trajectory, significantly affecting the achievement of the SDGs.

3.5 Conclusions and Recommendations

This study aimed to analyse the trends and factors driving climate change in the Southern African region. Findings reveal a significant increase in temporal climate vulnerability, with climate change exacerbating existing inequalities and intersectional vulnerabilities. One notable indicator of increasing vulnerability is the increase in displacement due to climate-induced events. The impacts of climate change across various sectors are intricately linked to social, economic, political and geographical factors. Although the region remains highly vulnerable, there are disparities both between and within countries due to varying adaptation strategies, economic conditions and sociopolitical dynamics.

The study identifies numerous drivers of vulnerability to climate change, including the effectiveness of disaster management systems and other services, the reliance on agriculture, poverty rates and social protection mechanisms, exposure to disasters, economic fragility and water scarcity, among others. Additionally, vulnerabilities are deeply rooted in social structures and intersectional marginalisation, such as gender disparities, highlighting the need for structural changes to support gender-inclusive and climate-resilient development. However, Southern Africa as a whole is increasingly vulnerable to climate change; due to its increasing scope, more localised research is needed to understand the localised dynamics of this vulnerability better.

To address these vulnerabilities, comprehensive approaches are essential, integrating poverty reduction, sustainable development and social equity initiatives with climate adaptation and mitigation efforts. Areas that are already near or past a tipping point in water availability and are heavily dependent on rainfed agriculture have already been faced with greater uncertainty in their dependence on crops. It is becoming increasingly evident that these households must depend on multilocal coping strategies, such as receiving remittances from family members who migrated or migrating themselves in order to survive. It will become an increasingly important question in climate vulnerability programming how best to reduce vulnerability in communities that have past tipping points on the sustainability of rainfed agriculture.

Collaboration between governments, civil society, corporations and local communities is crucial to building resilience and protecting the well-being of southern Africa's populations in the face of climate change. Addressing climate vulnerability in rural Southern Africa requires a holistic strategy that combines climate adaptation and mitigation measures with sustainable development objectives. This involves improving adaptive capacity, strengthening ecosystem resilience, promoting livelihood diversification, strengthening governance mechanisms and fostering regional cooperation to mitigate the impacts of climate change on rural communities.

More studies are recommended to focus on determining communities past tipping points that may need specific resilience interventions, as well as characterising loss and damage due to climate change vulnerability in the region. This information will be key in determining insurance mechanisms to reduce the overall impact of climate change on Southern African society. Additionally, a similar methodology could be applied to more localised vulnerability studies for countries in the region.

References

Akadiri, S. S., Bekun, F. V., & Sarkodie, S. A. (2019). Contemporaneous interaction between energy consumption, economic growth and environmental sustainability in South Africa: What drives what? *Science of the Total Environment, 686*, 468–475. https://doi.org/10.1016/j.scitotenv.2019.05.421

Akinlabi, F. M., Pirie, M. D., & Oskolski, A. A. (2023). Fire, frost, and drought constrain the structural diversity of wood within southern African Erica (Ericaceae). *Botanical Journal of the Linnean Society, 203*(4), 370–389. https://doi.org/10.1093/botlinnean/boad033

Amegah, A. K., Rezza, G., & Jaakkola, J. J. K. (2016). Temperature-related morbidity and mortality in Sub-Saharan Africa: A systematic review of the empirical evidence. *Environment International, 91*, 133–149. https://doi.org/10.1016/j.envint.2016.02.027

Apraku, A., Morton, J. F., & Apraku Gyampoh, B. (2021). Climate change and small-scale agriculture in Africa: Does indigenous knowledge matter? Insights from Kenya and South Africa. *Scientific African, 12*, e00821. https://doi.org/10.1016/j.sciaf.2021.e00821

Balsari, S., Dresser, C., & Leaning, J. (2020). Climate Change, Migration, and Civil Strife. *Current Environmental Health Reports, 7*(4), 404–414. https://doi.org/10.1007/s40572-020-00291-4

Banda, V. D., Dzwairo, R. B., Singh, S. K., & Kanyerere, T. (2022). Hydrological modelling and climate adaptation under changing climate: A review with a focus in Sub-Saharan Africa. *Water (Switzerland), 14*(24). https://doi.org/10.3390/w14244031

Borrelli, P., Robinson, D. A., Panagos, P., Lugato, E., Yang, J. E., Alewell, C., Wuepper, D., Montanarella, L., & Ballabio, C. (2020). Land use and climate change impacts on global soil erosion by water (2015–2070). *Proceedings of the National Academy*

of Sciences, 117(36), 21994 LP–22001 LP. https://doi.org/10.1073/pnas.2001403117
Cassidy, L., Perkins, J. S., & Bradley, J. (2022). Too much, too late: fires and reactive wildfire management in northern Botswana's forests and woodland savannas. *African Journal of Range and Forage Science, 39*(1), 160–174. https://doi.org/10.2989/10220119.2022.2033833
Chapungu, L. (2017). *Impact of climate change on vegetative species diversity in Masvingo Province, Zimbabwe* (Issue August) [UNISA]. https://uir.unisa.ac.za/handle/10500/23781?
Chapungu, L. (2020, May). *Mitigating the impact of cyclone disasters: Lessons from Cyclone Idai*. South African Institute of International Affairs. https://www.africaportal.org/series/climate-change-and-migration-africa/
Chapungu, L., & Nhamo, G. (2021). Interfacing vector-borne disease dynamics with climate change: Implications for the attainment of SDGs in Masvingo city, Zimbabwe. *Jamba: Journal of Disaster Risk Studies, 13*(1), 1–12. https://doi.org/10.4102/jamba.v13i1.1175
Chapungu, L., Nhamo, G., Dube, K., & Chikodzi, D. (2023). Soil erosion in the savanna biome national parks of South Africa. *Physics and Chemistry of the Earth, 130*(March 2022), 103376. https://doi.org/10.1016/j.pce.2023.103376
Chapungu, L., Nhamo, G., Chikodzi, D., & Dube, K. (2024). Trends and impacts of temperature and fire regimes in South Africa's coastal national parks: implications for tourism. *Natural Hazards*. https://doi.org/10.1007/s11069-023-06384-1
Chirisa, I., Gumbo, T., Gundu-Jakarasi, V. N., Zhakata, W., Karakadzai, T., Dipura, R., & Moyo, T. (2021). Interrogating climate adaptation financing in Zimbabwe: proposed direction. *Sustainability, 13*(12), 6517.
Coetzee, C., Khoza, S., Nemakonde, L. D., Shoroma, L. B., Wentink, G. W., Nyirenda, M., Chikuse, S., Kamanga, T., Maripe, K., Rankopo, M. J., Mwansa, L.-K., & Van Niekerk, D. (2023). Financing disaster risk reduction: Exploring the opportunities, challenges, and threats within the southern african development Community Region. *International Journal of Disaster Risk Science, 14*(3), 398–412. https://doi.org/10.1007/s13753-023-00499-6
Conradie, W. S., Wolski, P., & Hewitson, B. C. (2022). Spatial heterogeneity of 2015–2017 drought intensity in South Africa's winter rainfall zone. *Advances in Statistical Climatology, Meteorology and Oceanography, 8*(1), 63–81. https://doi.org/10.5194/ascmo-8-63-2022
Cooper, J. A. G., & Green, A. N. (2023). Southern African sandy coasts in the context of near-future sea-level rise. *Transactions of the Royal Society of South Africa, 78*(3), 149–166. https://doi.org/10.1080/0035919X.2023.2272829
Curto, A., Nunes, J., Milà, C., Nhacolo, A., Hänninen, R., Sofiev, M., Valentín, A., Saúte, F., Kogevinas, M., Sacoor, C., Bassat, Q., & Tonne, C. (2024). Associations between landscape fires and child morbidity in southern Mozambique: A time-series study. *The Lancet Planetary Health, 8*(1), e41–e50. https://doi.org/10.1016/S2542-5196(23)00251-6
Dias, A. C. E., Armitage, D., Nayak, P. K., Akintola, S. L., Arizi, E. K., Chuenpagdee, R., Kumar Das, B., Diba, S. A., Ghosh, R., Isaacs, M., Islam, G. M. N., Kane, A., Li, Y., Manase, M. M., Mbaye, A. A., Onyango, P., Pattanaik, S., Sall, A., Susilowati, I., et al. (2023). From vulnerability to viability: A situational analysis of small-scale fisheries in Asia and Africa. *Marine Policy, 155*(July), 105731. https://doi.org/10.1016/j.marpol.2023.105731
Dube, O. P. (2013). Challenges of wildland fire management in Botswana: Towards a community inclusive fire management approach. *Weather and Climate Extremes, 1*, 26–41. https://doi.org/10.1016/j.wace.2013.08.001
Dube, K., & Chikodzi, D. (2023). Examining current and future challenges of sea level rise on coastal national parks. *Geojournal of Tourism and Geosites, 50*(4), 1573–1580. https://doi.org/10.30892/GTG.50436-1154
Dube, K., Nhamo, G., & Chikodzi, D. (2021). Rising sea level and its implications on coastal tourism development in Cape Town, South Africa. *Journal of Outdoor Recreation and Tourism, 33*, 100346. https://doi.org/10.1016/j.jort.2020.100346
Dutton, J. (2022, July). South Africa's abandoned mines endanger local communities. *African Business*. https://african.business/2022/07/energy-resources/south-africas-poisonous-mines-spark-acid-crisis#
Eastment, C., Humphrey, G., Hoffman, M. T., & Gillson, L. (2022). The influence of contrasting fire management practice on bush encroachment: Lessons from Bwabwata National Park, Namibia. *Journal of Vegetation Science, 33*(2). https://doi.org/10.1111/jvs.13123
Flatø, M., Muttarak, R., & Pelser, A. (2017). Women, weather, and woes: The triangular dynamics of female-headed households, economic vulnerability, and climate variability in South Africa. *World Development, 90*, 41–62. https://doi.org/10.1016/j.worlddev.2016.08.015
Forsyth, G. G., & van Wilgen, B. W. (2008). The recent fire history of the table mountain national park and implications for fire management. *Koedoe, 50*(1), 3–9. https://doi.org/10.4102/koedoe.v50i1.134
Fonds Africain de Developpement. (2011). Rapport sur le developpement en Afrique 2011, Banque Africaine de Developpement, http://www.afdb.org
Gandidzanwa, C. P., & Togo, M. (2022). Adaptive responses to water, energy, and food challenges and implications on the environment: An exploratory study of harare. *Sustainability (Switzerland), 14*(16). https://doi.org/10.3390/su141610260
Ghosh, R. C., & Orchiston, C. (2022). A systematic review of climate migration research: Gaps in existing

literature. *SN Social Sciences, 2*(5), 1–22. https://doi.org/10.1007/s43545-022-00341-8

Giesen, C., Roche, J., Redondo-Bravo, L., Ruiz-Huerta, C., Gomez-Barroso, D., Benito, A., & Herrador, Z. (2020). The impact of climate change on mosquito-borne diseases in Africa. *Pathogens and Global Health, 114*(6), 1–15. https://doi.org/10.1080/20477724.2020.1783865

Gulistan, A., Tariq, Y. B., & Bashir, M. F. (2020). Dynamic relationship among economic growth, energy, trade openness, tourism, and environmental degradation: Fresh global evidence. *Environmental Science and Pollution Research., 27*, 13477. https://doi.org/10.1007/s11356-020-07875-5

Hancke, L., Smeed, D., Roberts, M., Russo, C., Rayner, D., & Jebri, F. (2023). Atmospheric and advective forcing of upwelling on South Africa's central Agulhas Bank. *Deep-Sea Research Part II: Topical Studies in Oceanography, 209*, 105293. https://doi.org/10.1016/j.dsr2.2023.105293

Harrington, L. J., & Otto, F. E. (2018). Adapting attribution science to the climate extremes of tomorrow. *Environmental Research Letters, 13*(12), 123006.

Hawkins, P., Geza, W., Mabhaudhi, T., Sutherland, C., Queenan, K., Dangour, A., & Scheelbeek, P. (2022). Dietary and agricultural adaptations to drought among smallholder farmers in South Africa: A qualitative study. *Weather and Climate Extremes, 35*, 100413. https://doi.org/10.1016/j.wace.2022.100413

Hermans, K., & McLeman, R. (2021). Climate change, drought, land degradation and migration: exploring the linkages. *Current opinion in environmental sustainability, 50*, 236–244.

IPCC. (2021). *Climate change 2021, the physical science basis, working group I contribution of to the sixth assessment report of the IPCC*. IPCC.

IPCC. (2022). Impacts, adaptation and vulnerability. Working group II to the IPCC sixth assessment report. IPCC.

Jiri, O., & Mafongoya, P. L. (2018). Velvet Bean and cowpea residual effects on maize crop in smallholder farming areas of Zimbabwe. *Sustainable Agriculture Research, 7*(1), 54–63.

Jiri, O., Mafongoya, P. L., & Chivenge, P. (2017). Contextual vulnerability of rainfed crop-based farming communities in semi-arid Zimbabwe: A case of Chiredzi District. *International Journal of Climate Change Strategies and Management, 9*(6), 777–789. https://doi.org/10.1108/IJCCSM-03-2017-0070

Kapuka, A., & Hlásny, T. (2020). Social vulnerability to natural hazards in Namibia: A district-based analysis. *Sustainability (Switzerland), 12*(12). https://doi.org/10.3390/SU12124910

Kavhu, B., & Ndaimani, H. (2022). Analysing factors influencing fire frequency in Hwange National Park. *South African Geographical Journal, 104*(2), 177–192. https://doi.org/10.1080/03736245.2021.1941219

Kita, S. M. (2017). Urban vulnerability, disaster risk reduction and resettlement in Mzuzu city, Malawi. *International Journal of Disaster Risk Reduction, 22*, 158–166. https://doi.org/10.1016/j.ijdrr.2017.03.010

Kollipara, A., Moonasar, D., Balawanth, R., Silal, S. P., Yuen, A., Fox, K., Njau, J., Pillay, Y. G., & Blecher, M. (2023). Mobilising resources with an investment case to mitigate cross-border malaria transmission and achieve malaria elimination in South Africa. *Global Health Action, 16*(1). https://doi.org/10.1080/16549716.2023.2205700

Kraaij, T., Baard, J. A., Cowling, R. M., Van Wilgen, B. W., & Das, S. (2013). Historical fire regimes in a poorly understood, fire-prone ecosystem: Eastern coastal fynbos. *International Journal of Wildland Fire, 22*(3), 277–287. https://doi.org/10.1071/WF11163

Kraaij, T., Baard, J. A., Arndt, J., Vhengani, L., & van Wilgen, B. W. (2018). An assessment of climate, weather, and fuel factors influencing a large, destructive wildfire in the Knysna region, South Africa. *Fire Ecology, 14*(2), 1–12. https://doi.org/10.1186/s42408-018-0001-0

Leal Filho, W., Taddese, H., Balehegn, M., Nzengya, D., Debela, N., Abayineh, A., Mworozi, E., Osei, S., Ayal, D. Y., Nagy, G. J., Yannick, N., Kimu, S., Balogun, A.-L., Alemu, E. A., Li, C., Sidsaph, H., & Wolf, F. (2020). Introducing experiences from African pastoralist communities to cope with climate change risks, hazards and extremes: Fostering poverty reduction. *International Journal of Disaster Risk Reduction, 50*, 101738. https://doi.org/10.1016/j.ijdrr.2020.101738

Lee, T. T., Dalvie, M. A., Röösli, M., Merten, S., Kwiatkowski, M., Mahomed, H., Sweijd, N., & Cissé, G. (2023). Understanding diarrhoeal diseases in response to climate variability and drought in Cape Town, South Africa: a mixed methods approach. *Infectious Diseases of Poverty, 12*(1), 76. https://doi.org/10.1186/s40249-023-01127-7

Lokotola, C. L. (2023). Towards a climate-resilient primary health care service. *South African Family Practice, 65*(1), 1–6. https://doi.org/10.4102/safp.v65i1.5749

Luyt, J., & Sealy, J. (2023). Carnivore stable isotopes as environmental integrators in southern African winter rainfall ecosystems. *Quaternary International, 650*, 4–11. https://doi.org/10.1016/j.quaint.2022.03.008

Mafofo, L., Kanosvamhira, T. P., & Olanyo, J. (2021). Media discourses on natural disasters and management: A case of cyclones Idai and Kenneth and Floods in Four Southern African Countries. Cyclones in Southern Africa. In G. Nhamo & K. Dube (Eds.), *Cyclones in Southern Africa. (Sustainabl)*. Springer.

Makuvara, Z., Marumure, J., Chapungu, L., Machingura, J., & Siwela, A. H. (2022). Comparison of oxidative stress biomarkers in Oreochromis mossambicus in minimally and highly disturbed aquatic environments in the Matabeleland region, Zimbabwe. *African Journal of Aquatic Science, 47*(4), 499–506. https://doi.org/10.2989/16085914.2022.2086846

Makwinja, R., Kaunda, E., Mengistou, S., Alemiew, T., Njaya, F., Kosamu, I. B. M., & Kaonga, C. C. (2021). Lake Malombe fishing communities' livelihood, vulnerability, and adaptation strategies. *Current Research in Environmental Sustainability, 3*(December 2020), 100055. https://doi.org/10.1016/j.crsust.2021.100055

Mapani, B. S., Shikangalah, R. N., & Mwetulundila, A. L. (2023). A review on water security and management under climate change conditions, Windhoek, Namibia. *Journal of African Earth Sciences, 197*, 104749. https://doi.org/10.1016/j.jafrearsci.2022.104749

Marin, M., Feng, M., Phillips, H. E., & Bindoff, N. L. (2021). A global, multiproduct analysis of coastal marine heatwaves: Distribution, characteristics, and long-term trends. *Journal of Geophysical Research: Oceans, 126*(2), p.e2020JC016708.

Matsa, M. (2021). Climate change and agriculture in Zimbabwe. In *Sustainability in minority farming communities* (96 pages). Springer Nature, Switzerland AG. ISBN 978-3-030-51346-7.

Matsa, M., & Marambanyika, T. (2020). *Climate change, environmental change and community development in Zimbabwe* (276 pages). MSU Press. ISBN 978 1-77925 338-5.

Matsa, M., & Mini, S. (2014). Grappling climate change in Southern Zimbabwe: The experiences of the BaKalanga minority farmers. *Journal of Environmental Studies, 4*(1), 34–52.

Matsa, M., & Mupepi, O. (2021). Flood risk and damage analysis in urban areas of Zimbabwe. A case of 2020/21 rain season floods in the city of Gweru. *International Journal of Disaster Risk Reduction, 67*, 102638. https://doi.org/10.1016/j.ijdrr.2021.102638

Mavhura, E., (2018). Disaster risk reduction policy and management in Zimbabwe. In Handbook of disaster risk reduction & management (pp. 589–612).

Meng, Z., Dong, J., Ellis, E. C., Metternicht, G., Qin, Y., Song, X. P., & Xiao, X. (2023). Post-2020 biodiversity framework challenged by cropland expansion in protected areas. *Nature Sustainability, 6*(7), 758–768.

Mhlongo, S. E. (2023). Physical hazards of abandoned mines: A review of cases from South Africa. *Extractive Industries and Society, 15*(June), 101285. https://doi.org/10.1016/j.exis.2023.101285

Milán-garcía, J., Caparrós-martínez, J. L., Rueda-lópez, N., & Valenciano, J. D. P. (2021). Climate change-induced migration: A bibliometric review. *Globalization and Health, 3*, 1–10.

Mlambo, D. N. (2023). Regional integration, COVID-19, and the fundamental development challenges of the 21st century: Reference from the Southern African Development Community (SADC). *Journal of African Union Studies, 12*(3), 105–128. https://doi.org/10.31920/2050-4306/2023/12n3a5

Mordecai, E. A., Ryan, S. J., Caldwell, J. M., Shah, M. M., & LaBeaud, A. D. (2020). Climate change could shift disease burden from malaria to arboviruses in Africa. *The Lancet Planetary Health, 4*(9), e416–e423. https://doi.org/10.1016/S2542-5196(20)30178-9

Mpakairi, K. S., Tagwireyi, P., Ndaimani, H., & Madiri, H. T. (2019). Distribution of wildland fires and possible hotspots for the Zimbabwean component of Kavango-Zambezi Transfrontier Conservation Area. *South African Geographical Journal, 101*(1), 110–120. https://doi.org/10.1080/03736245.2018.1541023

Mpakairi, K. S., Kadzunge, S. L., & Ndaimani, H. (2020). Testing the utility of the blue spectral region in burned area mapping: Insights from savanna wildfires. *Remote Sensing Applications: Society and Environment, 20*, 100365. https://doi.org/10.1016/j.rsase.2020.100365

Mpandeli, S., Nhamo, L., Moeletsi, M., Masupha, T., Magidi, J., Tshikolomo, K., Liphadzi, S., Naidoo, D., & Mabhaudhi, T. (2019). Assessing climate change and adaptive capacity at local scale using observed and remotely sensed data. *Weather and Climate Extremes, 26*, 100240. https://doi.org/10.1016/j.wace.2019.100240

Mucherera, B., & Mavhura, E. (2020). Flood survivors' perspectives on vulnerability reduction to floods in Mbire district, Zimbabwe. *Jamba: Journal of Disaster Risk Studies, 12*(1), 1–12. https://doi.org/10.4102/JAMBA.V12I1.663

Mugandani, R., Muziri, T., Murewi, C. T. F., Mugadza, A., Chitata, T., Sungirai, M., Zirebwa, F. S., Manhondo, P., Mupfiga, E. T., Nyamutowa, C., Mudereri, B. T., Mugari, Z. E., Mwadzingeni, L., & Mafongoya, P. (2022). Mapping and managing livelihoods vulnerability to drought: A case study of Chivi District in Zimbabwe. *Climate, 10*(12). https://doi.org/10.3390/cli10120189

Mutimukuru-Maravanyika, T., Chapungu, L., & Majeke, F. (2022). Food systems in Dryland communities: Challenges and opportunities in Gutu District, Masvingo Province, Zimbabwe. In X. Poshiwa & G. R. Chary (Eds.), *Climate change adaptations in dryland agriculture in Semi-Arid Areas* (pp. 157–172). Springer Singapore. https://doi.org/10.1007/978-981-16-7861-5_11

Mwadzingeni, L., Mugandani, R., & Mafongoya, P. L. (2021). Assessing vulnerability to climate change in smallholder irrigation schemes of Zimbabwe. *Sustainability (Switzerland), 13*(18). https://doi.org/10.3390/su131810023

Nelson, W. C. D., Hoffmann, M. P., May, C., Mashao, F., Ayisi, K., Odhiambo, J., Bringhenti, T., Feil, J. H., Yazdan Bakhsh, S., Abdulai, I., & Rötter, R. P. (2022). Tackling climate risk to sustainably intensify smallholder maise farming systems in southern Africa. *Environmental Research Letters, 17*(7). https://doi.org/10.1088/1748-9326/ac77a3

Ngxongo, N. A. (2021). The impact of climate change on visitor destination selection: A case study of the Central Drakensberg Region in KwaZulu-Natal. *Jamba: Journal of Disaster Risk Studies, 13*(1), 1–10. https://doi.org/10.4102/jamba.v13i1.1161

Nhamo, G., & Chapungu, L. (2021). *The increasing risk of floods and tornadoes in Southern Africa*. https://link.springer.com/10.1007/978-3-030-74192-1

Nhemachena, C., Nhamo, L., Matchaya, G., Nhemachena, C. R., Muchara, B., Karuaihe, S. T., & Mpandeli, S. (2020). Climate change impacts on water and agriculture sectors in southern africa: Threats and opportunities for sustainable development. *Water (Switzerland), 12*(10), 1–17. https://doi.org/10.3390/w12102673

Nhongo, E. J. S., Fontana, D. C., Guasselli, L. A., & Bremm, C. (2019). Probabilistic modelling of wildfire occurrence based on logistic regression, Niassa Reserve, Mozambique. *Geomatics, Natural Hazards and Risk, 10*(1), 1772–1792. https://doi.org/10.1080/19475705.2019.1615559

Nieman, W. A., Van Wilgen, B. W., & Leslie, A. J. (2021). A review of fire management practices in African savanna-protected areas. *Koedoe, 63*(1), 1–13. https://doi.org/10.4102/koedoe.v63i1.1655

Nyadzi, E., Ajayi, O. C., & Ludwig, F. (2021). Indigenous knowledge and climate change adaptation in Africa: A systematic review. In *CAB reviews: Perspectives in agriculture, veterinary science, nutrition and natural resources* (p. 2021).

Quinn, C. H., Ziervogel, G., Taylor, A., Takama, T., & Thomalla, F. (2011). Coping with multiple stresses in rural South Africa. *Ecology and Society, 16*(3), 10. https://doi.org/10.5751/ES-04216-160302

Reliefweb. (2021). *Southern Africa – Tropical cyclone Eloise flash update no. 9.* United National Office for the Coordination of Humanitary Affairs, 7, np. https://reliefweb.int/report/mozambique/southern-africa-tropical-cyclone-eloise-flash-update-no9-26-january-2021

Rogers, D. J., & Randolph, S. E. (2006). Climate change and vector-borne diseases. *Advances in Parasitology, 62*(February), 345–381. https://doi.org/10.1016/S0065-308X(05)62010-6

Saarinen, M., Heikkinen, J., Ketoja, E., Kyttä, V., Hartikainen, H., Silvennoinen, K., Valsta, L., & Lång, K. (2023). Soil carbon plays a role in the climate impact of diet and its mitigation: the Finnish case. *Frontiers in Sustainable Food Systems, 7*, 904570.

Samuels, M. I., Masubelele, M. L., Cupido, C. F., Swarts, M. B. V., Foster, J., De Wet, G., Links, A., Van Orsdol, K., & Lynes, L. S. (2022). Climate vulnerability and risks to an indigenous community in the arid zone of South Africa. *Journal of Arid Environments, 199*, 104718. https://doi.org/10.1016/j.jaridenv.2022.104718

Scholes, R. J., Coetzer, K. L., Matsika, R., Coetzee, B. W. T., Ernst, Y., Etale, A., Kubanza, N. S., Moyo, K., Nkrumah, B., Engelbrecht, F. A., Simatele, M. D., & Vogel, C. H. (2023). A Delphi assessment of climate change risks in southern Africa in the 21st century. *Climate Risk Management, 42*, 100566. https://doi.org/10.1016/j.crm.2023.100566

Shirvani, Z., Abdi, O., & Goodman, R. C. (2023). High-resolution semantic segmentation of woodland fires using residual attention UNet and time series of Sentinel-2. *Remote Sensing, 15*(5). https://doi.org/10.3390/rs15051342

Shrikhande, S. S., Merten, S., Cambaco, O., Lee, T., Lakshmanasamy, R., Röösli, M., Dalvie, M.A., Utzinger, J., & Cissé, G. (2023). "Climate change and health?": Knowledge and perceptions among key stakeholders in Puducherry, India. International *Journal of Environmental Research and Public Health, 20*(6), 4703.

Shongwe, P., Masuku, M. B., & Manyatsi, A. M. (2013). Cost benefit analysis of climate change adaptation strategies on crop production systems: A case of Mpolonjeni Area Development Programme (ADP) in Swaziland. *Sustainable Agriculture Research, 3*(1), 37. https://doi.org/10.5539/sar.v3n1p37

Spear, D., Haimbili, E., Angula, M., Baudoin, M.-A., Hegga, S., Zaroug, M., & Okeyo, A. (2018). *Vulnerability and adaptation to climate change in the Semi-Arid Regions of Southern Africa.* 1–117. http://www.uct.ac.za/

Struwig, M., Van den Berg, A., & Hadi, N. (2024). Challenges in the ocean economy of South Africa. *Development Southern Africa, 41*(1), 1–15. https://doi.org/10.1080/0376835X.2023.2232396

Tanner, T., Mazingi, L., & Muyambwa, D. F. (2022). Youth, gender and climate resilience: Voices of adolescent and young Women in Southern Africa. *Sustainability (Switzerland), 14*(14), 1–19. https://doi.org/10.3390/su14148797

Tawodzera, M., & Ncube, A. (2019). Communities' perceptions of health hazards induced by climate change in Mount Darwin district, Zimbabwe. *Jàmbá: Journal of Disaster Risk Studies, 11*(1), 1–11.

United Nations. (2022). United Nations Climate Change Annual Report 2022. UNFCCC.

Wekeza, S. V., Sibanda, M., & Nhundu, K. (2022). Prospects for organic farming in coping with climate change and enhancing food security in Southern Africa: A systematic literature review. *Sustainability (Switzerland), 14*(20). https://doi.org/10.3390/su142013489

Wells, C., Pringle, J., & Stretch, D. D. (2023). Cold water pathways and regional flow patterns in the southwest Indian ocean associated with temperature anomalies on the Sodwana reef system, South Africa. *Estuarine, Coastal and Shelf Science, 287*, 108337. https://doi.org/10.1016/j.ecss.2023.108337

Wilcox, B. A., Echaubard, P., De Garine-Wichatitsky, M., & Ramirez, B. (2019). Vector-borne disease and climate change adaptation in African dryland social-ecological systems. *Infectious Diseases of Poverty, 8*(1), 1–12. https://doi.org/10.1186/s40249-019-0539-3

4 Pathways to Enhance the Adaptive Capacity and Resilience to Climate Change in Southern Africa

Stanley Liphadzi, Sylvester Mpandeli, Walter Svinurai, Jerikias Marumure, and Luxon Nhamo

4.1 Introduction

More than 3.4 billion people, accounting for almost 45% of the global population, reside in rural areas of developing countries, dependent on natural systems for their livelihoods, making them highly vulnerable to climate change and other natural disasters (Mpandeli et al., 2018; Wilmoth et al., 2023). In Southern Africa alone, agriculture employs more than 60% of the population but is mostly rainfed (Nhamo et al., 2018). Most rural areas are remote, marginalised and arid and generally lack the resources required to sustain livelihoods (Nhamo et al., 2022; Wilmoth et al., 2023). They are areas with poor quality natural resources, limited communication and transportation networks and weak institutions, which pose a significant risk of vulnerability to climate change (Wilmoth et al., 2023). The increasing frequency and intensity of extreme weather events, such as droughts, heatwaves and floods, already impact the livelihoods of rural communities with evidence of low crop and livestock yields, exposure to new infectious diseases and poor nutrition (Alexandratos & Bruinsma, 2012; Nhamo et al., 2019b). Recent studies have shown that adaptation strategies and resilience-building initiatives are most effective when local communities are involved from the beginning (IPCC, 2014; Mpandeli et al., 2019).

Over the centuries, rural communities have developed the capacity to adapt to climate change, including practising shifting cultivation, adopting new crop varieties, adopting climate-smart technologies and modifying grazing patterns (Nelson et al., 2007). However, these strategies are often stretched to extremes, and rural communities are forced to change and adopt new adaptation pathways with support from external sources (Nelson et al., 2007). At the same time, the intensity and high frequency of extreme weather events in recent years have outpaced the speed of autonomous adaptations,

S. Liphadzi · S. Mpandeli
Water Research Commission of South Africa, Pretoria, South Africa

Faculty of Science, Engineering and Agriculture, University of Venda, Thohoyandou, South Africa

W. Svinurai
Department of Livestock Sciences, University of Zimbabwe, Harare, Zimbabwe

J. Marumure
Department of Physics, Geography and Environmental Science, School of Natural Sciences, Great Zimbabwe University, Masvingo, Zimbabwe

L. Nhamo (✉)
Water Research Commission of South Africa, Pretoria, South Africa

Centre for Transformative Agricultural and Food Systems (CTAFS), School of Agricultural, Earth and Environmental Sciences, University of KwaZulu-Natal, Scottsville, Pietermaritzburg, South Africa
e-mail: luxonn@wrc.org.za

M. Matsa et al. (eds.), *Climate Change Resilience in Rural Southern Africa*, Sustainable Development Goals Series, https://doi.org/10.1007/978-3-031-73600-1_4

risking the ability of poor rural and marginalised groups to cope with the vagaries of climate change (IPCC, 2022). This highlights the need for research and development to provide novel and coherent context-based adaptation strategies and pathways that lead to timely and informed change from the norm, enhancing resilience-building initiatives (Biagini et al., 2014; Mpandeli et al., 2019). Context-based interventions in climate change adaptation and resilience-building initiatives are critical as climate change impacts differ from place to place (Mpandeli et al., 2019). Furthermore, adaptation strategies that are effective in one area are not always effective in other areas (Mushore et al., 2021). Therefore, context-based adaptation should be preferred in adaptation strategies rather than national or global strategies.

However, adaptation occurs in various types and forms: (a) in terms of time, adaptation can be anticipatory or reactive, and (b) in terms of preparation and outside intervention, it can be planned or autonomous (Biagini et al., 2014). Adaptation is described as reactive when it happens within natural and ecological systems; however, when it occurs at the individual and societal levels, it can be both anticipatory and reactive on the basis of it being observed and expected climate change (Darjee et al., 2023).

Given the inevitable and evident changes in the global climate, it is urgently necessary to develop coherent adaptation strategies that enhance the resilience of communities while exploiting new opportunities. In broad terms, adaptation is defined as an adjustment in socio-ecological and economic systems in response to actual, present or anticipated climatic stimuli or their effects, which moderates harm or exploits beneficial opportunities (IPCC, 2022). Importantly, improved adaptive capacity is achieved through positive and innovative changes in wealth, human capital, information and technology, material resources, infrastructure, institutions and entitlements (IPCC, 2022; Mpandeli et al., 2018). The resource-poor rural communities in Southern Africa require a whole range of adaptation strategies to be implemented on time (Darjee et al., 2023). However, as already alluded to, the implementation of these strategies faces multiple challenges, including limited financial resources, inadequate infrastructure and insufficient access to information and technology. Given these limitations, the impacts of climate change are felt more acutely in poor and marginalised communities.

Current observed and projected global environmental changes point to substantial challenges that require significant interventions to reduce risk, including total societal transformations and the adoption of resource-use efficiency smart technologies (Dwivedi et al., 2022). A combination of artificial intelligence (AI), the Internet of Things (IoT) and Big Data platforms have started to catalyse transformative change by providing adaptive pathways and integrated solutions to the multiple challenges brought about by climate change (Padmapriya & Sujatha, 2023; Rauter, 2021). Artificial intelligence could immensely contribute to resource security and sustainability, as technology is expected to increase productivity by up to 40% by 2035 (Deranty & Corbin, 2022). As automation and productivity continue to develop, supply chains and economic growth are projected to multiply through resource use efficiency, a process that promotes the circular economy (Ghobakhloo et al., 2022). There is an anticipation of a greater growth rate due to technological advancements. AI is projected to drive better socioecological and economic outcomes that are efficient and more predictable in a shorter period, promoting resilience-building initiatives.

Based on these challenges and the prospects for improved productivity and technological advancements, this chapter explores the opportunities to enhance climate change resilience building in Southern Africa through case studies. The chapter further assesses the risk levels and the vulnerability of the region due to the slow uptake of novel technologies, limited human capacity and poor resources. The challenges faced by communities in adopting novel and transformative norms are explored, highlighting the importance of integrated and cross-sectoral approaches in achieving comprehensive adaptation to climate change and sustainability. The chapter recommends context-based climate change adaptation practices to respond to local challenges. The aim is to explore proactive responses of local communities in Southern Africa against locally experienced impacts and existential threats of climate change.

4.2 The Geography and Vulnerability of Southern Africa

Southern Africa refers to inland countries south of the Equator, namely, Angola, Botswana, Democratic Republic of Congo, Lesotho, Malawi, Mozambique, Namibia, South Africa, Eswatini, Tanzania, Zambia and Zimbabwe, with a total population of about 389.38 million people, and excludes island countries (Fig. 4.1).

The countries of Southern Africa have a total land area of 9.86 million km^2, of which only 6.11% is under cultivation (Nhamo et al., 2019b). Agriculture contributes the most to the gross domestic product (GDP) and sustains the livelihoods of more than 60% of the population, but it is predominantly rainfed, yet arable land with irrigation potential is approximately 20 million ha, accounting for approximately 6.6% of the total cultivated area (Mabhaudhi et al., 2019b). The region has vast irrigation potential, which needs to be more utilised.

Climate, particularly rainfall, is highly variable in space and time, varying widely from desert, temperate, savannah and equatorial. About 75% of Southern Africa is arid to subhumid and faces acute water scarcity challenges (Nhamo et al., 2019a). The region is prone to extreme climate events that include droughts, flooding, tropical cyclones and heatwaves, among other natural disasters (Davis & Vincent, 2017). In recent years, the region has experienced a downward trend in rainfall, evidenced by below-normal rainfall and increasing intensity and frequency of droughts and cyclonic floods (Ayanlade et al., 2022). Extreme weather conditions are mainly due to the effects of the El Niño Southern Oscillation (ENSO).

Climate, particularly rainfall, is highly variable in space and time, varying widely from desert, temperate, savannah, and equatorial. A combination of declining rainfall totals and increased temperatures accompanied by high variability due to climate change contribute to the region becoming a climate hotspot region (Nhamo et al., 2019a). Rainfall patterns are generally and primarily shaped by the intertropical convergence zone (ITCZ) that occurs around the Equator, the high-pressure belt found in the south of the 20° latitude and the predominance of cold fronts that originate from the southern tip of the African continent (Chiang et al., 2002). Therefore, rainfall is unevenly distributed throughout the region, resulting in highly variable rainfall averages that oscillate between 100 and 2000 mm/year (Nhamo et al., 2019a).

The changing climate increases vulnerability in Southern Africa, with evidence of worsening livelihood conditions, increased food insecurity and increased migration and displacement, compounded by resource mismanagement (Connolly-Boutin & Smit, 2016). Therefore, it is paramount to comprehend the specific socioecological and economic vulnerabilities of the region and how these changes are unfolding in order to provide context-based adaptation interventions.

4.2.1 Extreme Climate Events in the Past Century in Southern Africa

Southern Africa suffered the most severe droughts that occurred between 1900 and 2022 (Nhamo et al., 2019a). Of a total of 702 drought events recorded during this period worldwide, 312 of them occurred on the continent (Table 4.1) (Guha-Sapir et al., 2023). This has resulted in about 12 million drought-related deaths since 1900 to date and the displacement of almost three billion people (Guha-Sapir et al., 2023). The economic expenditure from drought-related interventions in Southern Africa since 1900 is estimated at US$146 billion (Table 4.1). The region experienced six cyclones that affected more than 2.5 million people in Madagascar, Malawi, Mozambique and Zimbabwe between 2021 and 2022.

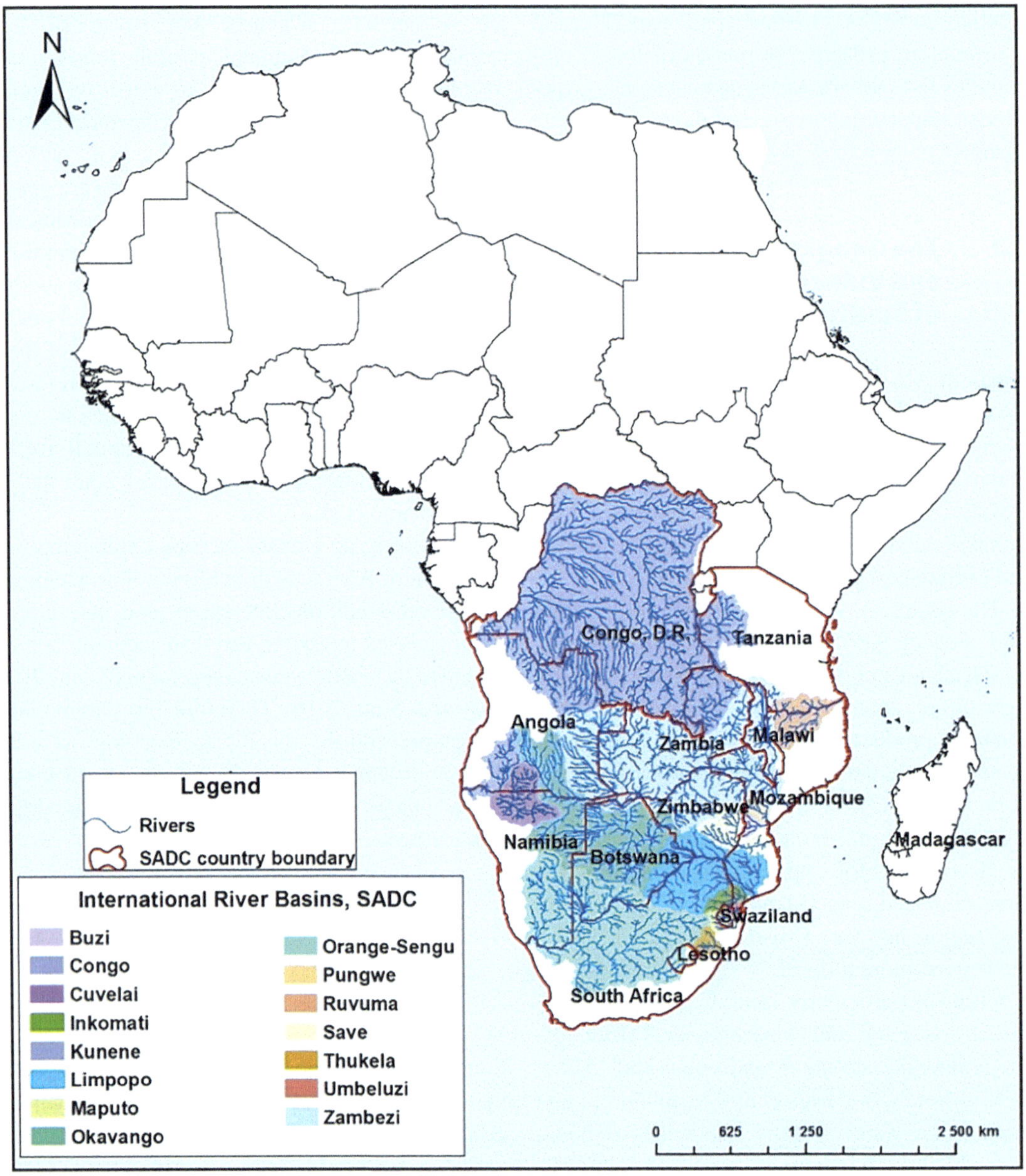

Fig. 4.1 Southern African countries and their transboundary river basins. (Source: Produced by authors)

Table 4.1 Number of droughts and their impact throughout the world during 1900–2022

Continent	Drought events	People killed	People displaced	Economic loss ('000 US$)
Africa	312	867,131	414,235,329	6,428,593
Americas	152	77	109,375,181	59,671,139
Asia	167	9,663,400	2,107,113,716	42,718,264
Europe	42	1,200,002	15,488,769	25,481,309
Oceania	29	684	1,074,2019	11,586,000
Total	702	11,731,294	2,656,955,014	145,885,305

Source: EM-DAT: The International Disaster Database Guha-Sapir et al. (2023)

4.2.2 Spatio-Temporal Changes in Rainfall Patterns in Southern Africa (1960–2007)

Since 1960, rainfall patterns in Southern Africa have significantly changed, with rainfall totals declining from 1200 to 890 mm (Fig. 4.2). Nearly half of the land area has become arid, receiving less than 650 mm/year, presenting high spatio-temporal variability in water resources. Southern Africa has become a drying region, increasing the vulnerabilities of a region grappling with a host of other socio-economic and ecological disasters.

4.2.3 Increasing Aridity and Water Scarcity in Southern Africa

Aridity is also increasing in Southern Africa due to decreasing rainfall totals and increasing temperatures (Fig. 4.3). Land area with severe arid conditions has almost trebled between 1980 and 2007. The spatio-temporal increase in aridness is affecting the predominantly rainfed agriculture sector, which is primarily rainfed. The energy sector is also negatively impacted as water also plays a critical role as an input to the production system. In general, the level of aridity in Southern Africa increased by 10.8% between 1980 and 2007, further exposing the region to adverse impacts of climate change.

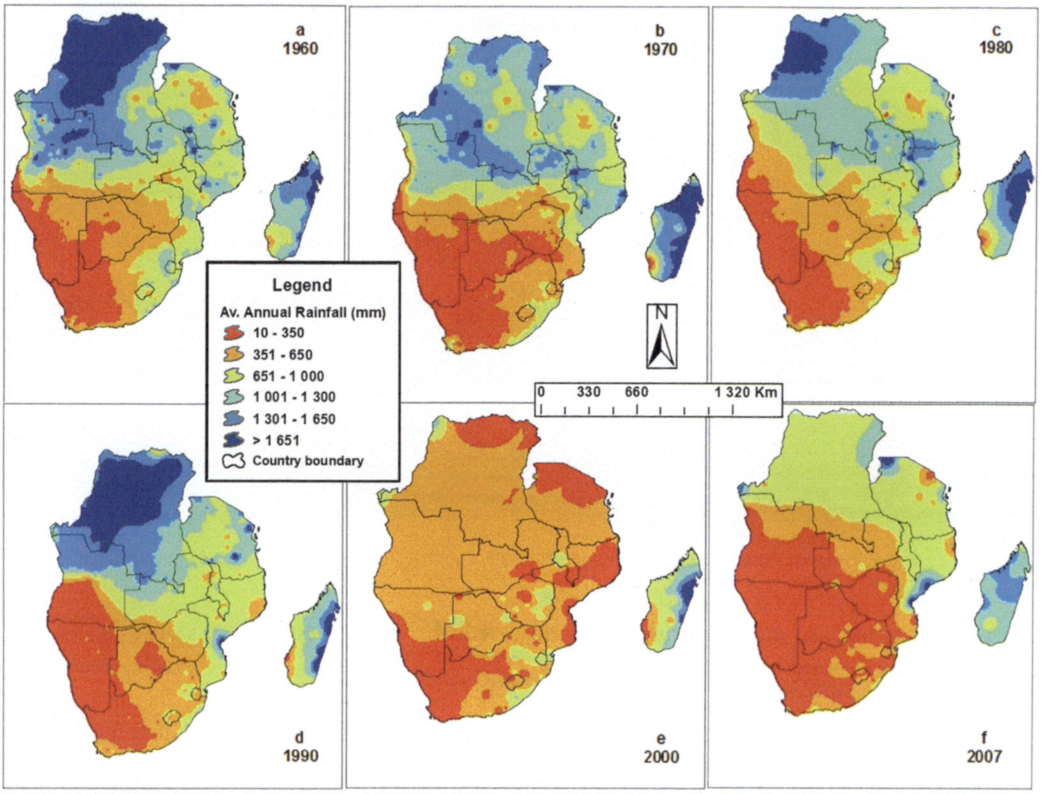

Fig. 4.2 Changes in rainfall patterns in Southern Africa between 1900 and 2007 in space-time. (Source: Produced by authors)

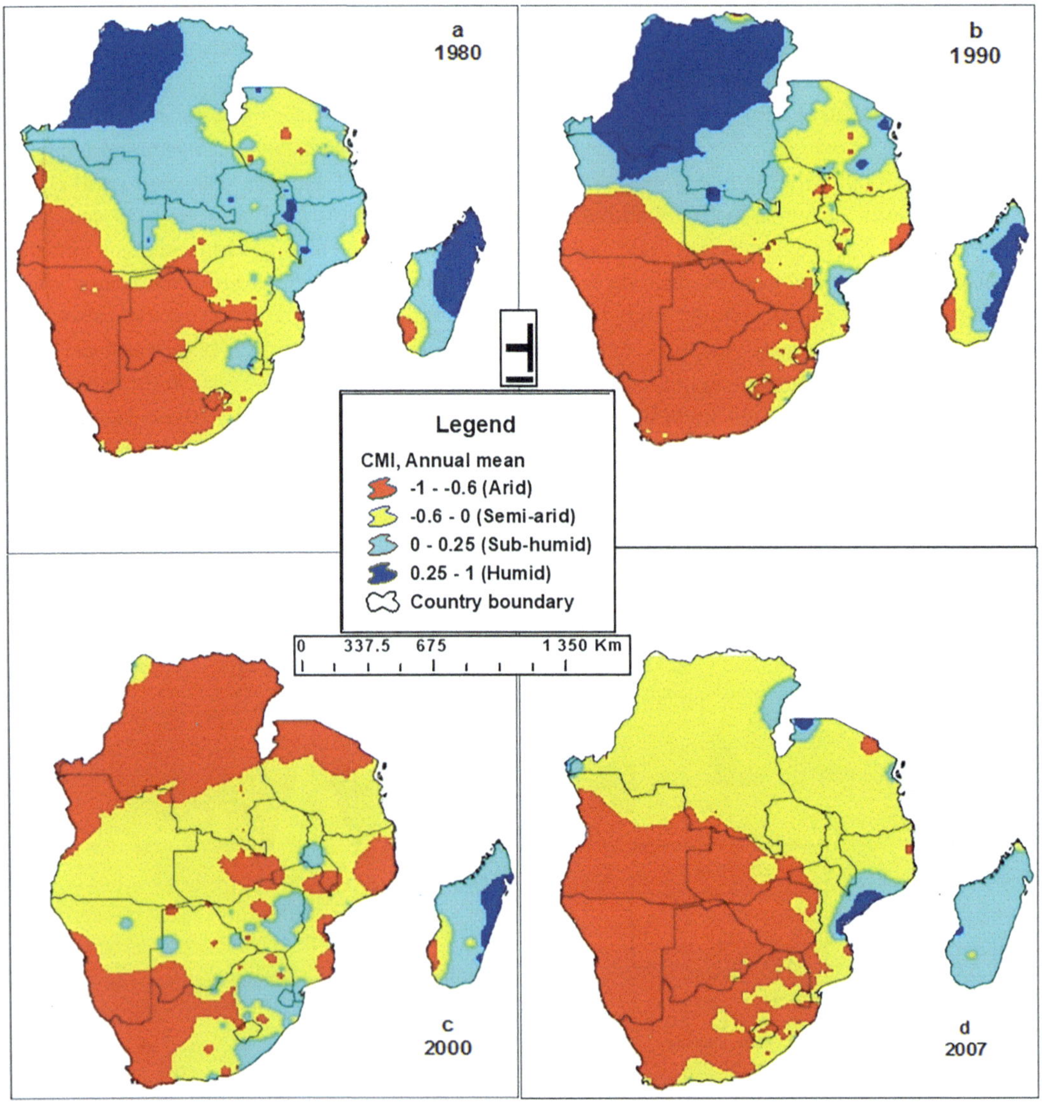

Fig. 4.3 Changes in aridity levels and water scarcity in Southern Africa over time. (Source: Produced by authors)

The trends indicate an increasingly warming region with dire consequences for the agriculture sector, which is the source of livelihood for most of the population, as it is rainfed. Therefore, warming increases the vulnerability of socio-economic and ecological systems with the risk of increased frequency and intensity of extreme climate events and the emergence of novel pests and diseases.

4.2.4 Socioecological Vulnerability

Socioecological vulnerability is the extent to which environmental degradation and climate change cause negative changes in exposure, susceptibility and the capacity of the socioecological system to anticipate, cope with and recover from the hazard (Depietri, 2020). In the case of Southern Africa, and as already mentioned above, the region is highly vulnerable to climate change due to its exposure to extreme climate events, low adaptive capacity and high dependence on climate-sensitive livelihoods and natural resources (Connolly-Boutin & Smit, 2016). Rising temperatures are already contributing to prolonged dry spells and droughts. Climate change is accentuating existing challenges in already vulnerable, poor rural communities (Nhamo et al., 2019a). These challenges include overreliance on natural resources compounded by factors such as widespread poverty and various socio-economic, demographic and policy trends such as limited production capacity and income, poor land tenure arrangements and unstable prices for commodities (Williams et al., 2018). Gender inequality in the region also increases the vulnerability of women and girls who are unduly neglected in the distribution of productive assets such as land and water and are excluded from decision-making processes (Ferrant et al., 2014). The challenges of resource depletion and degradation occur at a time when the agriculture sector is under pressure to meet the food requirements of a growing population. There is a demand for more land for agriculture, which can require more water for irrigation (Mabhaudhi et al., 2022).

Therefore, the challenge of resource insecurity and degradation is exacerbated by socio-economic and environmental challenges. Warning signs of resource insecurity manifest themselves through severe environmental degradation, including environmental pollution, threatening human health (Nhamo & Ndlela, 2021). The demand for resources is exceeding supply, creating a socio-economic imbalance (Boretti & Rosa, 2019). The expansion of both settlements and economic sectors has a negative impact on the landscape and aquatic environment. Therefore, it is essential to understand and manage change and protect ecosystems (Boretti & Rosa, 2019).

Achieving socioecological sustainability requires integrating and cross-systems assessment of ecological processes with socio-economic and political dynamics in various spatio-temporal scales (Peng et al., 2023). Therefore, an interdisciplinary dialogue between human and social sciences, together with the natural sciences, is necessary to achieve the desired environmental outcomes that catalyse the desired transformational change (Deutsch et al., 2023). This integrated approach facilitates the formulation and development of an inclusive, adaptive governance framework that drives the resilience-building agenda. As part of the transformational agenda, achieving this objective requires the involvement of interdisciplinary actors and actions that allow the transfer of knowledge on nature-society interactions and create compatibilities between distinct but related concepts, sources and methods (Nesshöver et al., 2017). This informs policy and decision-making on formulating coherent strategies that promote climate change adaptation practices, including the adoption of climate-smart agriculture, reduction of greenhouse gas emissions, improved forest cover and least change in land use and water management (Fig. 4.4).

4.2.5 Adaptation at the Local Level: The Maloti-Drakensberg Region Case Study

This section illustrates the importance of the participatory contribution of rural communities at the local level, in collaboration with development partners and other stakeholders, to sustainable management of natural resources. This case study was conducted in the Maloti-Drakensberg mountain bioregion that covers almost the whole of Lesotho and parts of the Free State, Eastern Cape and KwaZulu-Natal provinces in South Africa. The Maloti-Drakensberg mountain bioregion is one of the important mountain ranges in Southern Africa, where local communities rely directly on natural resources for their livelihoods (Adelabu et al., 2020). However, natural resource management is currently unsustainable, causing severe land degradation and water depletion, compounded by rapid population growth and establishing poverty cycles (Adelabu et al., 2020).

The land tenure systems in the Maloti-Drakensberg bioregion are diverse, and the climate is heterogeneous, attributes that shape and determine agricultural practices in the region (Nhamo et al., 2022). Almost 79% of the land is

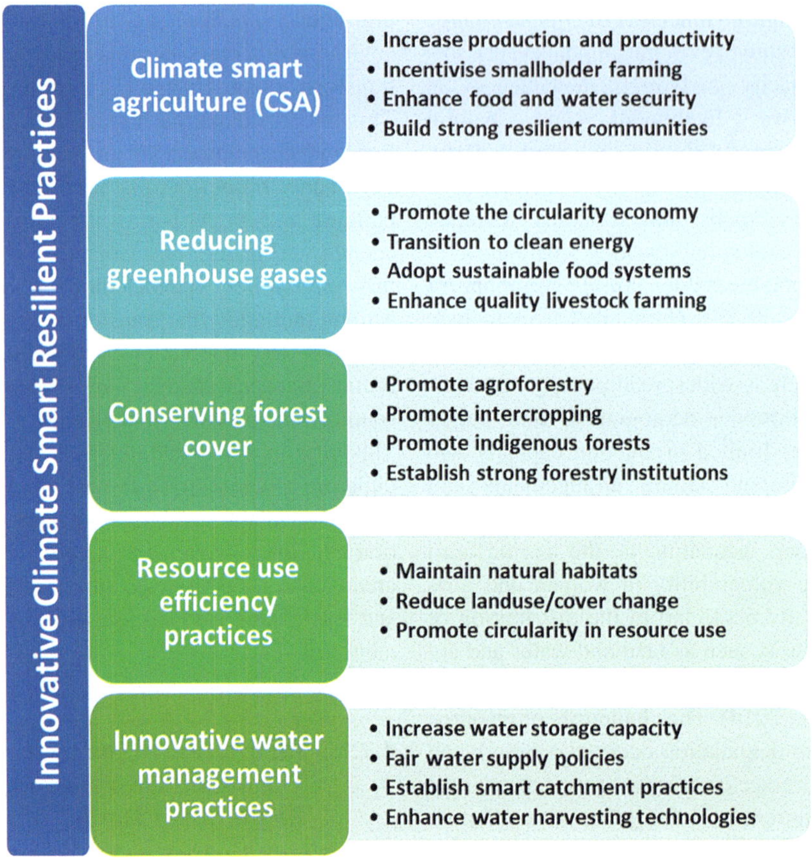

Fig. 4.4 Climate-resilient strategies for sustainable water and agriculture resource management. (Source: Produced by authors)

used for agropastoralism, supported by the abundant natural grasses in this mountain region. However, unsustainable management of rangelands remains a concern, as manifested by severe land degradation and siltation of water bodies (Adelabu et al., 2020). Crop cultivation is mainly subsistence and is practised on approximately 12% of the land, while the remaining land is used for game farming (Adelabu et al., 2020). There is a need for a multidisciplinary study to understand and address the complexity of interlinked socio-ecological challenges related to soil, water and environmental degradation and how these are impacting the base of agricultural natural resources. Innovative and adaptive strategies implemented through participatory research and the inclusion of local communities that provide indigenous knowledge play a pivotal role in sustainable resource management, improving water and rangeland conservation and promoting sustainable food production (Adelabu et al., 2020).

The bioregion has global significance as it is within a prime mountain range of Southern Africa, rich in biological endemism and a diversity of plant species with significant cultural and social value. It is also a primary water source for both Lesotho and South Africa (Zunckel, 2003). The population of the riparian areas of the Maloti-Drakensberg bioregion is approximately two million. It is highly vulnerable to climate change and generally food insecurity due to its high dependence on climate-sensitive natural resources for its livelihoods (Adelabu et al., 2020). Local communities participate widely in sustainable conservation and management practices of natural resources, supported by institutional interventions and interventions that promote sustainable mountain farming, food security and sustainability. In recent years, mountain farming systems have been supported by governments, financial and development partners and technological innovations (Zunckel, 2003).

Major Challenges in the Maloti-Drakensberg Bioregion

Consultations with communities and the various fieldworks being conducted during studies identified multiple challenges in the Maloti-Drakensberg region that include the following (Adelabu et al., 2020; Zunckel, 2003):

- Overstocking: This leads to loss of biodiversity, soil erosion and run-off as some communities are resistant to changing animal husbandry practices.
- Transboundary nature of resources: Resources in the region are generally transboundary, resulting in complex levels of governance, including shared international structures between Lesotho and South Africa. The governance of the Maloti-Drakensberg is controlled by the South African system of Traditional Authorities and elected government, and in Lesotho, through a chieftainship system.
- Water transfer schemes: Water is exported from Lesotho to industrial areas of South Africa, resulting in limited water available in the Maloti-Drakensberg in both countries, with an increasing reliance on catchment and river water.
- Poor waste management: Rapid population growth results in poor waste management, leading to land pollution.

The challenges call for interdisciplinary research that informs transformational change to achieve desired outcomes.

Strategies for Improving Resource Management in the Maloti-Drakensberg Region

Resource management in the Maloti-Drakensberg bioregion is complex as it has a multi-tier governance system linked to both traditional and national governance structures in both Lesotho and South Africa (Adelabu et al., 2020). Sustainable resource management in the region requires the participation and cooperation of smallholder farmers, traditional leaders and communities. Some strategies that enhanced resource management include the following:

- Revitalised terraced agricultural land: As the region is generally mountainous, there should be governance structures that promote the rejuvenation of terrace farming to reduce soil erosion and siltation of water bodies.
- Minimised soil degradation: The bioregion's ecosystems are dynamic and highly vulnerable to climatic change, accelerated soil erosion and landslides that lead to rapid loss of habitats, agricultural land and biodiversity, soil degradation and reduced soil water availability (Adelabu et al., 2020). Extensive tillage and overgrazing compound existing vulnerabilities. Interventions to address these challenges require context-based management strategies that facilitate the mapping of current erosion-prone areas and implement innovative measures that reduce loss of vegetation cover. These will go a long way to achieving the desired outcomes and sustainability.
- Refined agronomic management practices through sustainable agriculture systems: Smallholder farmers should be empowered to adopt crop intensification systems that improve crop productivity and resilience by optimising the benefits of available resources for improved plant growth and development. This is critical to improving crop water productivity.
- They promoted and mainstreamed the cultivation of indigenous, under-used crops. The Maloti-Drakensberg region is vulnerable to climate change, and small farmers focus on adapted crops such as maize (Adelabu et al., 2020). This shift to adapted crops is exacerbating the vulnerabilities of rural communities. Farmers should be encouraged to cultivate indigenous crops that are underused, adapt to harsh local conditions and have high nutritional value (Mabhaudhi et al., 2019a). Underused indigenous crops are critical for food and

nutrition security and improve favourable health, income generation and agro-ecological and environmental services through increased productivity, better marketing and home consumption (Nhamo et al., 2019b).

- Adopting climate-smart agriculture (CSA): Adopting climate-smart agriculture facilitates the development of agricultural strategies that are essential for identifying and securing sustainable food and water security in the face of climate change (Mpandeli et al., 2018). The practice attracts farming insurance and reduces risk through appropriate land use planning, management and weather forecasting (FAO-IFAD-UNICEF-WFP-WHO, 2018), which are critical for climate change adaptation and resilience building. An example of a CSA is the Machobane system in Lesotho, which integrates organic fertilisers, improved indigenous breeds, crop rotation, intercropping and relay crops to protect against climate change and unpredictability (Adelabu et al., 2020). These are being enhanced by a combination of soil, crop and weather sensors that facilitate the provision of precision data to enhance adaptive decision systems that support sustainable food systems.
- Improved governance structures: Although smallholder farmers play an integral role in food security, little attention has been paid to improving their agricultural systems (Magidi et al., 2021). Supporting and incentivising smallholder farmers has the potential to enhance food security, improve rural livelihoods and reduce climate change vulnerabilities.

Implementation of Payment of Ecosystem Services

A study was conducted that tested the viability of payment of ecosystem services focusing on the Maloti-Drakensberg bioregion (Büscher, 2012). The study showed that land care management practices of landowners improved river flows and water quality and groundwater yields and increased baseflows during winter (dry seasons). Unchanged land use and well-managed land or fields have limited human-induced erosion and, therefore, produce fewer sediments that could end up in rivers, wetlands and dams. The study advocated for the implementation of ecosystem service payment (PES) as an approach to compensate landowners upstream for their effort to protect this strategic water source area. Their effort was perceived to deserve compensation because they were contributing to cost-effective water resource management. When clean water is available for use in KwaZulu-Natal (KZN) province and moved by interbasin transfers to Gauteng province and other provinces, the water utilities and municipalities earn income, which they share with the landowners taking care of the land (Wittmayer & Büscher, 2010).

Implementation of Ecological Infrastructure in the uMgeni Catchment

The use of nature-based solutions has increased recently due to the escalation of water treatment and other operations and the maintenance costs of grey infrastructure in the uMgeni basin (Van Niekerk et al., 2019). The Water Research Commission (WRC) and its partners investigated the effectiveness of using ecological infrastructure to aid grey infrastructure in enhancing water security in the uMgeni catchment in KZN (Jewitt et al., 2021). The study was part of national strategic integrated projects that focused on improving water security in South Africa. The project aimed to protect land use/cover, soil, wetlands and rivers and improve agricultural practices such as livestock grazing, which can contribute to reducing erosion and increasing the quantity and quality of water in the catchment. The ecological infrastructure is the networks of natural lands, working landscapes and other open spaces that are the substructure or underlying foundation on which the continuation or growth of ecosystem goods and services depends.

4.3 Barriers to Climate Change Adaptation in Rural Areas

Limitations and barriers to climate change adaptation and resilience-building initiatives are categorised into three distinct but interlinked groups that include (a) natural (ecological and physical),

(b) human and informational and (c) social (Rockström et al., 2023; Serote et al., 2023; Wilson et al., 2020):

(a) Natural barriers include ecological and physical factors associated with limitations emanating from natural environments that range from ecosystem boundaries to geographical and geological limitations (Wilson et al., 2020). Natural barriers to adaptations to climate change and limitations to the achievement of Sustainable Development Goals (SDG) require transformational change that controls exceeding the planetary boundaries of valuable resources (Rockström et al., 2023). As an example, an increase in sea level and temperature could cause some critical thresholds to exceed the adaptable conditions of some systems, such as mangrove and coral reef ecosystems. The challenge is that the changes could completely alter the functional state and integrity of the system of such ecosystems. A significant challenge in rural communities of Southern Africa is when ecosystems are no longer able to support and sustain livelihoods.

(b) Human and informational barriers to adaptation to climate change adaptation are related to knowledge, technological and economic limitations (Barnett et al., 2015). Included in this category are the spatio-temporal uncertainties related to weather forecasting, climate modelling and low levels of advocacy and data availability to stakeholders on the impacts of climate change (Wang et al., 2020). This category also comprises the lack of financial resources and assistance to facilitate adaptation interventions (Wang et al., 2020). There is a huge gap between developed and developing countries in terms of human and informational barriers to adaptation to climate change (Lee et al., 2022). Developed countries are privileged with a strong base and rich observational data and climate modelling projections to inform adaptation policies. However, developing countries such as Southern African countries have scant historical meteorological data and only rely on coarse-resolution modelling predictions, which do not provide enough information at the local scale (Lee et al., 2022).

(c) The social barriers to climate change adaptation comprise the psychological, behavioural and socio-institutional elements that guide societal reactions in the face of climate stress and change (Jones & Boyd, 2011). These social aspects of climate change adaptation are usually neglected within broader adaptation discourses, yet climate change impacts are felt more at the local level (Lee et al., 2022). Adaptation also occurs at the local level and, in most cases, without support from external interventions (Nhamo & Chilonda, 2012). As an example, South Africa initially invested in mitigation rather than adaptation since the policy perceived mitigation as having more substantial market incentives. However, this changed in 2011 as adaptation is now promoted prominently in the National Climate Change Response Policy (NDP, 2013).

The systematic identification of barriers to climate change adaptation and the intervention pathways at the local level form the basis for the successful implementation of climate change adaptation and policy formulation. This is particularly critical in rural communities in Southern Africa, which face limited adaptation capacity due to limited resources and technological development. A combination of these barriers and the increasing intensity and frequency of climate extremes restrict the ability of vulnerable groups to identify, assess and manage climate change risks (Kuran et al., 2020). The challenges pose the risk of dampening adaptation efficiency, slowing adaptation opportunities, increasing adaptation costs and restricting the formulation and implementation of adaptation policies, compounding the inability of poor communities to adopt efficient adaptive actions (Mpandeli et al., 2019).

4.4 Adaptation Strategies

The speed with which climate change manifests and intensifies is compounding existing vulnerability challenges in rural communities in developing countries, as they lack the resources to adapt. This has created new combinations of risks and vulnerabilities that may be costly to address if no immediate action is taken now. The challenges are exacerbated by widespread poverty, overdependence on natural resources for their livelihoods, reliance on rainfed agriculture, inequitable land distribution and limited access to financing and technology (Mabhaudhi et al., 2019b; Nhamo & Chilonda, 2012). There is an urgent need for rapid and extensive adoption of adaptation strategies at the current pace, which has placed many poor communities at risk (Alvar-Beltrán et al., 2021). An accelerated application of adaptation strategies is necessary to reduce vulnerability to future climate change extremes. Table 4.2 provides some of the adaptation strategies that can be adopted to reduce the risk and vulnerability of poor communities. An essential element needed to achieve the desired outcomes in sustainable and transformational development initiatives for agriculture is ensuring that investments in adaptation strategies are informed by robust evidence on past and future climate risks.

4.5 Recommendations

Climate change has emerged as one of the worst challenges facing humanity and threatens lives and livelihoods. Although various strategies have been proposed in the literature to mitigate, adapt and build resilience at the national and global levels, it is at the local level where the impacts of climate change are felt (Nhamo et al., 2020). Therefore, successful adaptation strategies are implemented at the local level, acknowledging that the impacts of climate change and adaptation differ from place to place. Therefore, the following recommendations are proposed to improve adaptation to climate change at the local level:

(a) Policy and decision-making must take a bottom-up approach when implementing climate change adaptation strategies and engaging local communities. Local communities have rich information and indigenous knowledge and have always been adapting to climate change or natural shocks by their means, in most cases without external support (Nhamo et al., 2019a). Indigenous knowledge must be integrated with novel approaches and climate-smart technologies to enhance adaptation and resilience building at the local level. Indigenous knowledge needs to be supported and improved, in addition to bringing entirely new approaches that can comprise cultural practices and face resistance (Cai et al., 2017; Chapman & Schott, 2020).
(b) Institutional support is a prerequisite to the success of climate change adaptation (Anguelovski & Carmin, 2011). There is a need to focus on microeconomic level actions, which are important indicators for the overall principles for effective adaptation that differ in space and time (Anguelovski & Carmin, 2011).
(c) Six guiding principles should be considered in adapting to climate change, as they span all policy domains. These include (IPCC, 2022) (i) ensuring the resilience of institutions that form the foundation that promotes rapid and inclusive development, (ii) creating a conducive environment that facilitates the adaptation of institutions and communities, (iii) promoting governance structures that facilitate equitable distribution of resources, (iv) enhancing community resilience and capacity to cope with and recover from shocks, (v) anticipating and managing risk by applying novel climate-smart technologies such as remote sensing and (vi) ensuring effective data collection and implementation of informed intervention strategies, improving preparedness and prioritising continuous monitoring.

These recommendations should be applied to start from the local level, responding to the local context, as each area is unique and, therefore, should be attended to in its specifics.

Table 4.2 Some adaptation strategies that can be adopted to reduce climate risk and vulnerability

Climate hazard	Practices	Adaptation benefits
Heatwaves	Grow heat-tolerant crops and focus on indigenous crops	Promote crops with a higher heat tolerance and optimal heat range Improve the productivity of areas where temperatures are projected to exceed heat thresholds that are harmful to existing cropping systems Indigenous crops are adapted to harsh local conditions Ensures food and nutritional security
	Short cycle varieties	Decrease the impact of heat stress at key phenological stages and improve yields Decrease the exposure of crops to heat as the growth cycle is shortened Enhances crop water productivity by reducing total water requirements during the growing season
Water stress (drought)	Intercropping	Short, vulnerable crops are protected from heat, and evapotranspiration is reduced as taller drought-tolerant crops cover them Irrigation provides water to many crops at once Practising hydroponics with recirculating water systems has the potential to reduce water losses Mulching, in combination with zero-tillage, reduces crop exposure to heat stress and increases soil moisture by reducing direct soil evaporation
	Irrigation scheduling and water-use-efficient irrigation types	Enhances water-use efficiency as sufficient water is supplied according to each crop's requirements Soil erosion is reduced and reduces macronutrient losses Reduce the risk of pests and diseases that occur under wet conditions
	Wastewater reuse	Adds nutrients to the plants and increases yields Promotes efficiency of water use and enhances sustainable freshwater use to address water scarcity Protects the environment by promoting a circular economy
	Adapted livestock breeds	Adopt livestock breeds with lower water intake to optimise water use Construct water reservoirs where livestock can drink Promote high livestock productivity by adopting ones that do well under heat and drought stress conditions
Land degradation and greenhouse gas emissions	Zero tillage	Reduces soil disturbances and retains organic content Vegetation facilitates higher infiltration Reduce soil erosion Evapotranspiration is reduced as the soil remains compact
	Site-specific nutrient and water management	Ensure water and food security Reduces the cost of inputs GHG emissions are reduced through the more efficient use of nitrogen
Pests and diseases	Biological control	Introduce natural enemies (insects) into the environment to feed on crop pests Appy beneficial entomopathogens to reduce chemical pesticides Reduces the use of chemical fertilisers, herbicides and pesticides that harm the environment
	Crop rotations	Controls the concentration of pests and diseases as each crop attracts different pathogens Reduce weed challenges and conserves soil moisture
Extreme weather events	Real-time weather forecasts	Adopt best harvesting practices, methods and timing to reduce losses Strengthen early warning systems to minimise the impact of weather extremes on crops and livestock Build climate-proof storehouses that meet sustainable structural requirements and standards Promote rainwater harvesting systems

Adapted from: Alvar-Beltrán et al. (2021)

4.6 Conclusions

Climate change adaptation begins with building resilience and taking bold decisions that include changing from the usual way of living and interacting with the environment. The challenges brought about by climate change require immediate action to guarantee the security of resources and, if ever, the SDG targets. This requires transformational change from all societies and communities, as the impacts of climate change do not recognise political boundaries, although resilience adaptation may differ. This chapter has highlighted the effectiveness of integrated management of socioecological and economic systems in enhancing adaptive capacity. Furthermore, the chapter has highlighted the importance of robust governance frameworks and adequate financial support to institutions and research in providing clear adaptation pathways that improve resilience-building initiatives. Therefore, a key component of successful adaptation is funding at all levels, including community-based adaptation initiatives. As adaptation involves changing from the usual way of living, it is faced with various barriers that need to be addressed quickly as the impacts of climate change worsen. The change is needed quickly, as climate change is manifesting rapidly, outpacing the initiatives being undertaken. Social barriers to climate change adaptation are the most challenging, as they contribute the most to enabling or preventing successful and logical adaptation from taking place. One major challenge of climate change adaptation is that the societal changes resulting from climatic regimes may promote discriminatory and exclusion and widen societal inequalities and marginalisation. These challenges can be solved by holistically addressing adaptation barriers. Policy is also key in adapting to climate change, as it provides pathways that facilitate the assessment of risk and vulnerability of the interconnected governance layers from local, national and global levels. Therefore, strategising and planning adaptation should consider the means to support adaptive capacity, minimising trade-offs like the creation of societal imbalances. Lastly, a key component of climate change adaptation is developing capacity and enhancing research and developing, responding to and addressing some of the social and other barriers.

References

Adelabu, D. B., Clark, V. R., & Bredenhand, E. (2020). Potential for sustainable mountain farming: Challenges and prospects for sustainable smallholder farming in the Maloti–Drakensberg mountains. *Mountain Research and Development, 40*, A1.

Alexandratos, N., & Bruinsma, J. (2012). *World agriculture towards 2030/2050: The 2012 revision* (Revision ed., p. 154). Food and Agriculture Organisation (FAO).

Alvar-Beltrán, J., Elbaroudi, I., Gialletti, A., Heureux, A., Neretin, L., & Soldan, R. (2021). *Climate resilient practices: Typology and guiding material for climate risk screening* (p. 36). Food and Agricultural Organisation of the United Nations (FAO).

Anguelovski, I., & Carmin, J. (2011). Something borrowed, everything new: Innovation and institutionalisation in urban climate governance. *Current Opinion in Environmental Sustainability, 3*, 169–175.

Ayanlade, A., Oluwaranti, A., Ayanlade, O. S., Borderon, M., Sterly, H., Sakdapolrak, P., Jegede, M. O., Weldemariam, L. F., & Ayinde, A. F. (2022). Extreme climate events in sub-Saharan Africa: A call for improving agricultural technology transfer to enhance adaptive capacity. *Climate Services, 27*, 100311.

Barnett, J., Evans, L. S., Gross, C., Kiem, A. S., Kingsford, R. T., Palutikof, J. P., Pickering, C. M., & Smithers, S. G. (2015). From barriers to limits to climate change adaptation: Path dependency and the speed of change. *Ecology and Society, 20*, 5.

Biagini, B., Bierbaum, R., Stults, M., Dobardzic, S., & McNeeley, S. M. (2014). A typology of adaptation actions: A global look at climate adaptation actions financed through the Global Environment Facility. *Global Environmental Change, 25*, 97–108.

Boretti, A., & Rosa, L. (2019). Reassessing the projections of the world water development report. *npj Clean Water, 2*, 1–6.

Büscher, B. (2012). Payments for ecosystem services as neoliberal conservation: (Reinterpreting) evidence from the Maloti-Drakensberg, South Africa. *Conservation and Society, 10*, 29–41.

Cai, X., Haile, A. T., Magidi, J., Mapedza, E., & Nhamo, L. (2017). Living with floods–household perception and satellite observations in the Barotse floodplain, Zambia. *Physics and Chemistry of the Earth, Parts A/B/C, 100*, 278–286.

Chapman, J., & Schott, S. (2020). Knowledge coevolution: Generating new understanding through bridging and strengthening distinct knowledge systems and empowering local knowledge holders. *Sustainability Science, 15*, 931–943.

Chiang, J. C., Kushnir, Y., & Giannini, A. (2002). Deconstructing Atlantic Intertropical Convergence Zone variability: Influence of the local cross-equatorial sea surface temperature gradient and remote forcing from the eastern equatorial Pacific. *Journal of Geophysical Research: Atmospheres, 107*, ACL 3-1-ACL 3-19.

Connolly-Boutin, L., & Smit, B. (2016). Climate change, food security, and livelihoods in sub-Saharan Africa. *Regional Environmental Change, 16*, 385–399.

Darjee, K. B., Neupane, P. R., & Köhl, M. (2023). Proactive adaptation responses by vulnerable communities to climate change impacts. *Sustainability, 15*, 10952.

Davis, C.L., Vincent, K., (2017) Climate risk and vulnerability: A handbook for Southern Africa, 2nd Council for Scientific and Industrial Research (CSIR), , p. 202.

Depietri, Y. (2020). The social–ecological dimension of vulnerability and risk to natural hazards. *Sustainability Science, 15*, 587–604.

Deranty, J.-P., & Corbin, T. (2022). Artificial intelligence and work: A critical review of recent research from the social sciences. *AI & Society*, 1–17. https://doi.org/10.48550/arXiv.2204.00419

Deutsch, S., Keller, R., Krug, C. B., & Michel, A. H. (2023). Transdisciplinary transformative change: An analysis of some best practices and barriers, and the potential of critical social science in getting us there. *Biodiversity and Conservation, 32*, 1–26.

Dwivedi, Y. K., Hughes, L., Kar, A. K., Baabdullah, A. M., Grover, P., Abbas, R., Andreini, D., Abumoghli, I., Barlette, Y., & Bunker, D. (2022). Climate change and COP26: Are digital technologies and information management part of the problem or the solution? An editorial reflection and call to action. *International Journal of Information Management, 63*, 102456.

FAO-IFAD-UNICEF-WFP-WHO. (2018). The state of food security and nutrition in the world 2018. In *Building climate resilience for food security and nutrition* (p. 202). Food and Agriculture Organisation of the United Nations.

Ferrant, G., Pesando, L. M., & Nowacka, K. (2014). *Unpaid Care Work: The missing link in the analysis of gender gaps in labour outcomes*. OECD Development Center.

Ghobakhloo, M., Iranmanesh, M., Mubarak, M. F., Mubarik, M., Rejeb, A., & Nilashi, M. (2022). Identifying industry 5.0 contributions to sustainable development: A strategy roadmap for delivering sustainability values. *Sustainable Production and Consumption, 33*, 716–737.

Guha-Sapir, D., Below, R., & Hoyois, P. (2023). The international disaster database. In *Centre for Research on the Epidemiology of Disasters (CRED): The CRED/OFDA international disaster database*. Université Catholique de Louvain.

IPCC. (2014). In O. Edenhofer, R. Pichs-Madruga, Y. Sokona, E. Farahani, S. Kadner, K. Seyboth, A. Adler, I. Baum, S. Brunner, & P. Eickemeier (Eds.), *Climate change 2014: Mitigation of climate change. Contribution of working group iii to the 5th assessment report of the Intergovernmental Panel on Climate Change (IPCC)* (p. 1454). Intergovernmental Panel on Climate Change (IPCC).

IPCC. (2022). Summary for policymakers. In H.-O. Pörtner, D. C. Roberts, E. S. Poloczanska, K. Mintenbeck, M. Tignor, A. Alegría, M. Craig, S. Langsdorf, S. Löschke, & V. Möller (Eds.), *Climate change 2022: Impacts, adaptation and vulnerability. contribution of working group II to the sixth assessment report of the intergovernmental panel on climate change. Intergovernmental panel on climate change*. Cambridge University Press.

Jewitt, G., Sutherland, C., Stuart-Hill, S., Taylor, J., Risko, S., Martell, P., & Browne, M. (2021). *Enhancing water security through restoration and maintenance of ecological infrastructure: Lessons from the uMngeni River Basin, South Africa* (p. 165). Water Research Commission (WRC).

Jones, L., & Boyd, E. (2011). Exploring social barriers to adaptation: Insights from Western Nepal. *Global Environmental Change, 21*, 1262–1274.

Kuran, C. H. A., Morsut, C., Kruke, B. I., Krüger, M., Segnestam, L., Orru, K., Nævestad, T. O., Airola, M., Keränen, J., & Gabel, F. (2020). Vulnerability and vulnerable groups from an intersectionality perspective. *International Journal of Disaster Risk Reduction, 50*, 101826.

Lee, S., Paavola, J., & Dessai, S. (2022). Towards a deeper understanding of barriers to national climate change adaptation policy: A systematic review. *Climate Risk Management, 35*, 100414.

Mabhaudhi, T., Chibarabada, T. P., Chimonyo, V. G. P., Murugani, V. G., Pereira, L. M., Sobratee, N., Govender, L., Slotow, R., & Modi, A. T. (2019a). Mainstreaming underutilised indigenous and traditional crops into food systems: A South African perspective. *Sustainability, 11*, 172.

Mabhaudhi, T., Nhamo, L., Mpandeli, S., Nhemachena, C., Senzanje, A., Sobratee, N., Chivenge, P. P., Slotow, R., Naidoo, D., & Liphadzi, S. (2019b). The water–energy–food nexus as a tool to transform rural livelihoods and well-being in Southern Africa. *International Journal of Environmental Research and Public Health, 16*, 2970.

Mabhaudhi, T., Haileslassie, A., Magidi, J., & Nhamo, L. (2022). *Irrigation suitability mapping examples from Zimbabwe, Zambia, Malawi and Kenya* (p. 52). International Water Management Institute (IWMI).

Magidi, J., van Koppen, B., Nhamo, L., Mpandeli, S., Slotow, R., & Mabhaudhi, T. (2021). Informing equitable water and food policies through accurate spatial information on irrigated areas in smallholder farming systems. *Water, 13*, 3627.

Mpandeli, S., Naidoo, D., Mabhaudhi, T., Nhemachena, C., Nhamo, L., Liphadzi, S., Hlahla, S., & Modi, A. (2018). Climate change adaptation through the water-energy-food nexus in southern Africa. *International Journal of Environmental Research and Public Health, 15*, 2306.

Mpandeli, S., Nhamo, L., Moeletsi, M., Masupha, T., Magidi, J., Tshikolomo, K., Liphadzi, S., Naidoo, D., & Mabhaudhi, T. (2019). Assessing climate change and adaptive capacity at local scale using observed and remotely sensed data. *Weather and Climate Extremes, 26*, 100240.

Mushore, T. D., Mhizha, T., Manjowe, M., Mashawi, L., Matandirotya, E., Mashonjowa, E., Mutasa, C., Gwenzi, J., & Mushambi, G. T. (2021). Climate change adaptation and mitigation strategies for smallholder farmers: A case of Nyanga District in Zimbabwe. *Frontiers in Climate, 3*, 676495.

NDP. (2013). *National development plan: Vision 2030* (p. 489). National Planning Commission (NDP).

Nelson, D. R., Adger, W. N., & Brown, K. (2007). Adaptation to environmental change: Contributions of a resilience framework. *Annual Review of Environment and Resources, 32*, 395–419.

Nesshöver, C., Assmuth, T., Irvine, K. N., Rusch, G. M., Waylen, K. A., Delbaere, B., Haase, D., Jones-Walters, L., Keune, H., & Kovacs, E. (2017). The science, policy and practice of nature-based solutions: An interdisciplinary perspective. *Science of the Total Environment, 579*, 1215–1227.

Nhamo, L., & Chilonda, P. (2012). Climate change risk and vulnerability mapping and profiling at local level using the Household Economy Approach (HEA). *Journal of Earth Science and Climate Change, 3*, 2.

Nhamo, L., & Ndlela, B. (2021). Nexus planning as a pathway towards sustainable environmental and human health post-Covid-19. *Environment Research, 192*, 110376.

Nhamo, L., Ndlela, B., Nhemachena, C., Mabhaudhi, T., Mpandeli, S., & Matchaya, G. (2018). The water-energy-food nexus: Climate risks and opportunities in southern Africa. *Water, 10*, 567.

Nhamo, L., Mabhaudhi, T., & Modi, A. (2019a). Preparedness or repeated short-term relief aid? Building drought resilience through early warning in Southern Africa. *Water SA, 45*, 75–85.

Nhamo, L., Matchaya, G., Mabhaudhi, T., Nhlengethwa, S., Nhemachena, C., & Mpandeli, S. (2019b). Cereal production trends under climate change: Impacts and adaptation strategies in Southern Africa. *Agriculture, 9*, 30.

Nhamo, L., Ndlela, B., Mpandeli, S., & Mabhaudhi, T. (2020). The water-energy-food nexus as an adaptation strategy for achieving sustainable livelihoods at a local level. *Sustainability, 12*, 8582.

Nhamo, L., Mpandeli, S., Liphadzi, S., & Mabhaudhi, T. (2022). Securing land and water for food production through sustainable land reform: A nexus planning perspective. *Land, 11*, 974.

Padmapriya, V., & Sujatha, D. (2023). Impact of blockchain technology on the urban environment and climate change. In *Green blockchain technology for sustainable smart cities* (pp. 97–117). Elsevier.

Peng, Y., Welden, N., & Renaud, F. G. (2023). A framework for integrating ecosystem services indicators into vulnerability and risk assessments of deltaic social-ecological systems. *Journal of Environmental Management, 326*, 116682.

Rauter, H. (2021). The role of Blockchain, big data, internet of things and artificial intelligence for transformative climate action. In *Planning climate smart and wise cities: A multidisciplinary approach* (pp. 99–137). Springer.

Rockström, J., Gupta, J., Qin, D., Lade, S. J., Abrams, J. F., Andersen, L. S., Armstrong McKay, D. I., Bai, X., Bala, G., & Bunn, S. E. (2023). Safe and just Earth system boundaries. *Nature*, 1–10. https://doi.org/10.1038/s41586-023-06083-8

Serote, B., Mokgehle, S., Senyolo, G., du Plooy, C., Hlophe-Ginindza, S., Mpandeli, S., Nhamo, L., & Araya, H. (2023). Exploring the barriers to the adoption of climate-smart irrigation technologies for sustainable crop productivity by smallholder farmers: Evidence from South Africa. *Agriculture, 13*, 246.

Van Niekerk, L., Adams, J. B., Allan, D. G., Taljaard, S., Weerts, S. P., Louw, D., Talanda, C., & Van Rooyen, P. (2019). Assessing and planning future estuarine resource use: A scenario-based regional-scale freshwater allocation approach. *Science of the Total Environment, 657*, 1000–1013.

Wang, W., Zhao, X., Cao, J., Li, H., & Zhang, Q. (2020). Barriers and requirements to climate change adaptation of mountainous rural communities in developing countries: The case of the eastern Qinghai-Tibetan Plateau of China. *Land Use Policy, 95*, 104354.

Williams, P. A., Crespo, O., Abu, M., & Simpson, N. P. (2018). A systematic review of how vulnerability of smallholder agricultural systems to changing climate is assessed in Africa. *Environmental Research Letters, 13*, 103004.

Wilmoth, J. R., Bas, D., Mukherjee, S., & Hanif, N. (2023). *World social report 2023: Leaving no one behind in an ageing world*. United Nations (UN).

Wilson, J. R., Foxcroft, L. C., Geerts, S., Hoffman, M. T., MacFadyen, S., Measey, J., Mills, A., Richardson, D. M., Robertson, M. P., & van Wilgen, B. W. (2020). The role of environmental factors in promoting and limiting biological invasions in South Africa. In *Biological invasions in South Africa* (pp. 355–385). Springer International Publishing.

Wittmayer, J. M., & Büscher, B. (2010). Conserving conflict? Transfrontier conservation, development discourses and local conflict between South Africa and Lesotho. *Human Ecology, 38*, 763–773.

Zunckel, K. (2003). Managing and conserving southern African grasslands with high endemism. *Mountain Research and Development, 23*, 113–118.

5 Climate Disaster Risk Reduction in Rural Southern Africa

David Chikodzi, Michael T. Mubvuma, and Elisha Mavodyo

5.1 Introduction

Climate change has been identified as one of the biggest threats to human development (see, e.g. World Bank, 2016; IPCC, 2021a, b, c; O'Neill et al., 2017). A report by the Intergovernmental Panel on Climate Change (IPCC) (2021a, b, c) concluded that the Southern African region is in the grip of a climate emergency. This highlights the importance of increased research on climate disaster risk reduction (DRR), particularly in rural areas of Southern Africa, where academic attention in this area is still limited. Given the apparent need for academic attention and the literature gap, this study explores mechanisms for reducing climate change disaster risk in rural areas of Southern Africa.

The Sixth Intergovernmental Panel on Climate Change (IPCC) Assessment Report, released in September 2021, observed that while climate disaster mitigation initiatives have become essential for the entire world, they are especially critical for Southern Africa, which is especially prone to the adverse impacts of climatic changes (IPCC, 2021a, b, c). The report also stated that the Southern African region was already rapidly experiencing the effects of climate change and more severe consequences compared to the global average. According to the report, climate risk in the Southern African region is exacerbated by a lack of adaptation capacity, as there is little funding available for investment in measures to protect communities and their livelihoods against current and predicted climate shocks. In Southern Africa, a large proportion of the population still has an extremely urgent need to build resilience to the shocks imposed by climate-related risks (Eckstein et al., 2021).

In the same spirit, several studies have reported adverse climatic conditions in the Southern African region in line with the IPCC (2021a, b, c), confirming the urgent need for climate action in the region. These include severe and recurring droughts (see, e.g. Mavodyo, 2023; Sousa et al., 2018), an increase in and adverse severity of heatwaves (van der Walt & Fitchett, 2022; Iyakaremye et al., 2022) and high-intensity tropical cyclones that affect parts of Mozambique and Zimbabwe (Nhamo et al., 2021; Nhamo & Chikodzi, 2021).

Although the narrative above highlights the growing recognition of the importance of improv-

D. Chikodzi (✉)
School of Natural Sciences, Great Zimbabwe University, Masvingo, Zimbabwe
e-mail: dchikodzi@gzu.ac.zw

M. T. Mubvuma
Gary Magadzire School of Agriculture and Engineering, Great Zimbabwe University, Masvingo, Zimbabwe
e-mail: mmubvuma@gzu.ac.zw

E. Mavodyo
Faculty of Commerce, University of Zimbabwe, Harare, Zimbabwe
e-mail: emavodyo@commerce.uz.ac.zw

M. Matsa et al. (eds.), *Climate Change Resilience in Rural Southern Africa*, Sustainable Development Goals Series, https://doi.org/10.1007/978-3-031-73600-1_5

ing disaster risk management in Southern Africa, as demonstrated by studies such as those by Balgah and Kimengsi (2022), Awu et al. and Sambo and Chikodzi (2020), among others, focusing on rural areas is more appropriate for a number of reasons. First, rural communities in Southern Africa rely heavily on subsistence agriculture and dryland, which is very sensitive to changes in climate. Climate stresses and shocks (such as droughts or floods) are felt acutely by rural poor people, who rely directly on the results of the dryland food system for their livelihoods, with severe consequences for the stability of their standard of living and welfare (see, e.g. Davies et al., 2009; Leal Filho et al., 2020). Second, as argued by Lee et al. (2022) and Mal et al. (2018), among others, climate change impacts have a direct link with high levels of inequality, poverty, poor education and health facilities, as well as other facets of the general population in low- and middle-income countries, leading to elevated climate change exposure and high levels of vulnerability and risk. Third, and most importantly, climate change has severe consequences on the enjoyment of human rights, and thus, rights holders have a moral obligation to take a human rights lens when responding to climate change by protecting vulnerable groups (Mayer, 2021). Thus, as a result of this valuable reasoning, the focus of our study is on rural communities in Southern Africa.

As the narrative above demonstrates, there is a critical need for a better understanding of the reduction of climate disaster risk in Southern Africa. To date, a small but growing body of literature has emerged in response to the region's urgent need for climate disaster risk management. Notable examples include studies by Nemakonde and Van Niekerk (2017) who developed a normative model for integrating organisations for disaster risk reduction and climate change adaptation within SADC member states; an earlier study by Vogel et al. (2010), which provided a detailed account of thoughts on drought risk management in Southern Africa; and, recently, a study by Leal Filho et al. (2020), which analysed the lived experiences from African pastoralist communities to manage climate change risks, hazards and extremes. Although these studies contribute significantly to the literature on reducing climate change disaster risk in the Southern African region, this study adds to the literature in several ways.

First, the existing literature on the subject needs to be more extensive and focus on other aspects of the reduction of climate disaster risk in the Southern African region. This study adds significantly to the literature by focusing on reducing the risk of climate disasters in the Southern African region, as such an understanding has the potential to be beneficial in guaranteeing resilient and sustainable development for the Southern African region in the midst of climate risk. Second, this study examines climate risk disaster reduction strategies in various countries in the Southern African region, providing opportunities for countries to share their best experiences on climate disaster risk reduction, which may improve resilience to shocks and stresses in rural communities in the region. For example, limited studies have shown how climate-smart conservation agriculture (CA) is and the potential it has to deliver on adaptation, mitigation and improved productivity by smallholder farmers in Southern Africa (Ngoma et al., 2021). Third, as will be discussed in greater detail in the next section of this chapter, this study used a mixed-methods approach, which has the well-known advantage of providing a more in-depth and comprehensive understanding of climate disaster risk reduction in rural areas in Southern African rural communities.

Thus, this study analyses the measures to reduce climate risk in selected rural communities in Southern Africa. The analysis in this chapter is guided by the Sendai Framework for Disaster Risk Reduction 2015–2030, which provides a global framework for DRR. The framework stipulates four priority areas that need attention to improve DRR. These include having a comprehensive understanding of disaster risk, strengthening disaster risk governance, building resilience to disasters through investing in DRR and enhancing disaster preparedness and building back better after disasters (UNISDR, 2015). Four themes guided the data collection and analysis in the study.

Thus, building on this introductory section, the rest of this chapter proceeds as follows. The next section reviews the methodology we relied on, followed by the presentation, analysis and discussion of the empirical results. Finally, the final section of this chapter concludes the study.

5.2 Materials and Methods

5.2.1 Study Area

Rural areas in Southern Africa face a cocktail of climate-related disasters. The most widespread and damaging are drought, floods, tornadoes, sea level rise, wildfires and cyclones (Sambo & Chikodzi, 2020). The study mainly focuses its content on rural spaces in Southern Africa, presenting examples of climate-related disasters in selected hotspot areas in the region. In the study, climate-related disasters are profiled for the entire Southern African region. However, the study also presents an in-depth study of DRR activities from four selected rural communities in South Africa and Zimbabwe. The communities studied are the Port Saint Jones and Mkhondo local municipalities in South Africa and the Chimanimani and Tsholotsho rural districts in Zimbabwe. The sites were selected for in-depth studies because they are viewed as localised hotspots for climate-related disasters in Southern Africa. The experiences of these study sites were then discussed with literature from other countries and places in the region. Figure 5.1 shows the focus area in Southern Africa and the communities selected for in-depth studies.

5.2.2 Methods

This study applied a mixed-methods design. Conventional content analysis was the primary source of data utilised in the chapter. Content analysis involves the systematic examination of text to highlight themes and patterns (Zembe et al., 2023). The sources of data used in the chapter include scientific articles, reports from DRR organisations, policy documents, relevant news articles and transcribed interviews. Three stages were followed in the conventional content analysis process. These are the following:

1. Defining the search criteria
2. Selection databases to search from
3. Search and extraction of relevant documents
4. Developing keywords and themes from the obtained content

The scientific databases searched included Web of Science, Research for Life, Google Scholar and the EM-DAT, CRED/UCLouvain. The inclusion criteria used the expressions disaster risk reduction AND rural Southern Africa, climate change AND rural disaster risk reduction in Southern Africa, climate disaster preparedness AND rural Southern Africa and climate disaster risk management AND rural Southern Africa. The search for scientific databases yielded 425 journal articles and book chapters. These had to be further narrowed down to those that address climate-related hazards, rural communities and building community resilience. This reduced the relevant articles to 95, of which 45 were finally used in the investigation after further screening using the abstract of the articles. The EM-DAT, CRED/UCLouvain database yielded 3095 disasters from the African continent from the year 1900 to 2023. This database had to be further filtered to Southern Africa and then to weather-related hazards. This left the relevant disasters being 557, and these included tropical cyclones, heatwaves, cold waves, convective storms, riverine floods, flash floods, tornadoes, drought, wildfires, lightning and landslides.

Primary data to support the one collected from conventional content analysis were gathered through questionnaire surveys and key informant interviews conducted in South Africa and Zimbabwe. Data were collected between October 2019 and June 2022. These were periods after each of these places had been hit by climate-related disasters. Table 5.1 summarises the number of interviews with key informants and household questionnaires completed during the data collection process. Not all questions from the household questionnaires were compulsory to

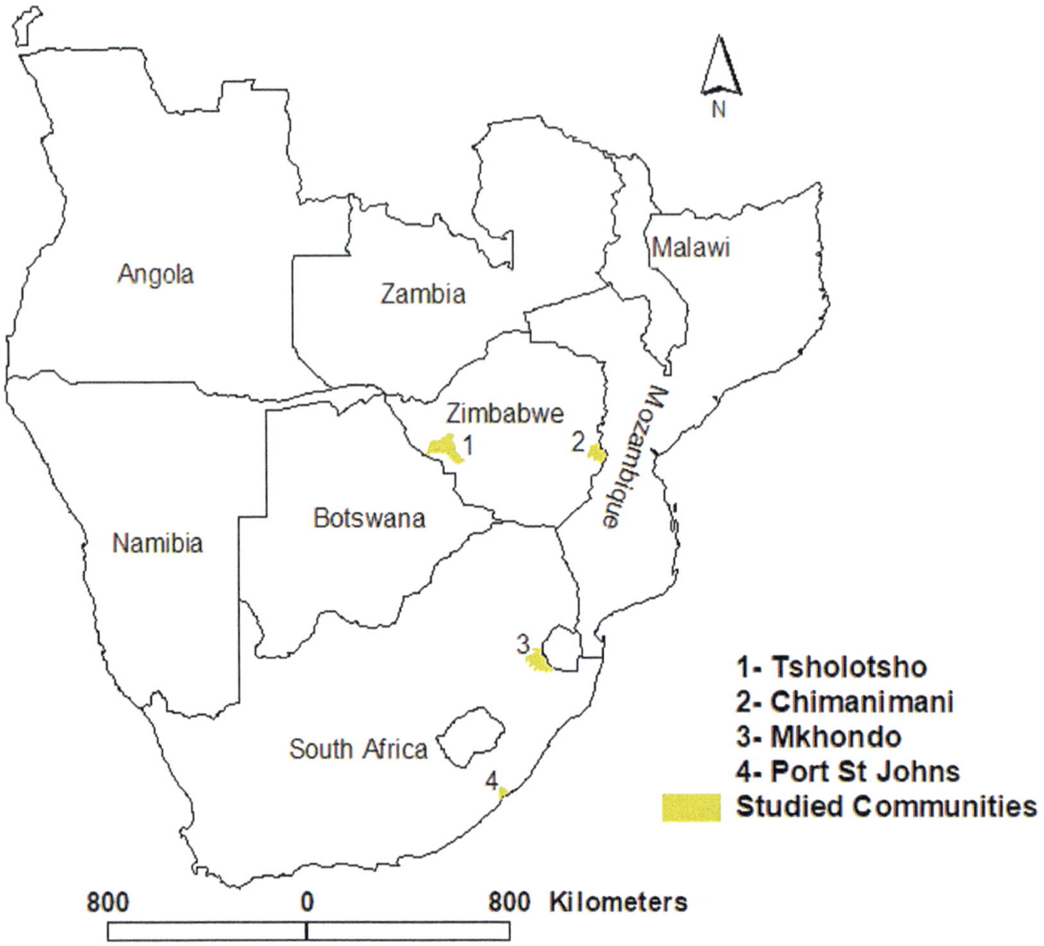

Fig. 5.1 Selected study sites. (Source: Produced by the authors)

Table 5.1 Key informant interviews and household questionnaires used

Place	Household questionnaires completed (N)	Key informant interviews were done
Chimanimani	368	20
Port Saint Johns	249	10
Tsholotsho	60	5
Mkondo	–	10

Source: Produced by authors

answer. Hence, N (sample size) is sometimes presented as lower than the maximum number of participants from the area. N will represent the number of participants who will have responded to that question.

The questionnaires targeted communities that had been affected by climate-related hazards such as tropical cyclones (Chimanimani and Tsholotsho), flooding (Port Saint Johns, Tsholotsho, Chimanimani) and tornadoes (Mkondo). The questionnaire survey targeted the head of household or any adult over the age of 18 years whenever the head was not available. QuestionPro analytics software was used to capture the questionnaire data. QuestionPro Analytics can analyse both qualitative and quantitative data using word clouds, cross tabulations, text analysis and descriptive statistics, which were used in the study. Quantitative data was presented in the form of descriptive statistics (frequencies) and figures. Some of the questions in the questionnaire allowed multiple responses as long as the answer was relevant, and hence, some of the frequencies may have totals above 100%. Each figure of the questionnaire survey will show both the frequency and N (sample size).

Key informant interviews were conducted with relevant stakeholders in DRR circles at the local and district levels in all study sites. These included employees from local municipalities such as municipal managers, district development coordinators, members of the civil protection units, environmental managers, human settlement planning officers, disaster management team members, local traditional leadership and local farmers and managers from forestry plantation companies in the areas. The interviews lasted for about an hour on average, and issues discussed included impacts, trends, projections, losses, costs, mitigation, management, anticipatory action and resilience building on climate-related disasters. The interview data was analysed in multiple stages. The interview data was first transcribed and then cleaned. This was then followed by grouping the data into different themes to facilitate fine-grain analysis as guided by the objectives of the study. The qualitative data were then presented as text discussions and word clouds.

5.3 Presentation of the Findings

The findings are organised around the identified themes to help achieve a better understanding of the reduction of the risk of climate disasters in rural Southern Africa. The identified themes were extracted from the four priority areas prescribed by the Sendai Framework to guide the effective implementation of the DRR processes. The presentation of the findings from selected communities for each theme was done in detail only for sites that showed interesting findings, and comparisons were made to see if there were significant variations between all study sites.

5.3.1 Climate Disaster Risk Profile for Rural Southern Africa

From 1900 to 2023, there has been a significant increase in the occurrence of extreme climate-related events in Southern Africa. Floods, violent storms and droughts have been by far the most frequent and damaging (Fig. 5.2) occurrences. These disasters have occurred in Southern Africa more than 556 times since the year 1900 (EM-DAT, CRED/UCLouvain database, 2023). However, it is in the last 30 years that the region has experienced an exponential increase in these extreme events, with flooding being the most recurrent and costly.

The hardest hit places in terms of the numbers affected and the value of property damaged by climate-related disasters in Southern Africa are urban settlements. This is mainly due to the high population densities in these areas. However, the number of individual rural spaces affected by climate-related disasters in Southern Africa is very high (Fig. 5.3). From the EM-DAT, CRED/UCLouvain database, the burden of extreme hydroclimatic events falls mainly on the eastern parts of Southern Africa, which include Mozambique and Madagascar, eastern parts of Zimbabwe and South Africa and the southern parts of Malawi.

Findings from the EM-DAT database align with the regional literature on climate-related hazards. Otto et al. (2022) and Langa et al. (2023) highlight the increased frequency of extreme weather events in Mozambique, with four cyclones and two storms in 3 years. Rural communities in Mozambique, Madagascar and Malawi struggle to recover between consecutive events. Van Der Walt and Fitchett (2022) note the increased occurrence, intensity and duration of warm temperatures in South Africa. Dube and Chikodzi (2021) find an increased impact of tornadoes and flooding in Mpumalanga province, affecting socio-economic development. Chisale et al. (2021) identified drought, strong winds, heat and flooding as threats to local livelihoods near Mchinji and Phirilongwe forest areas in Malawi.

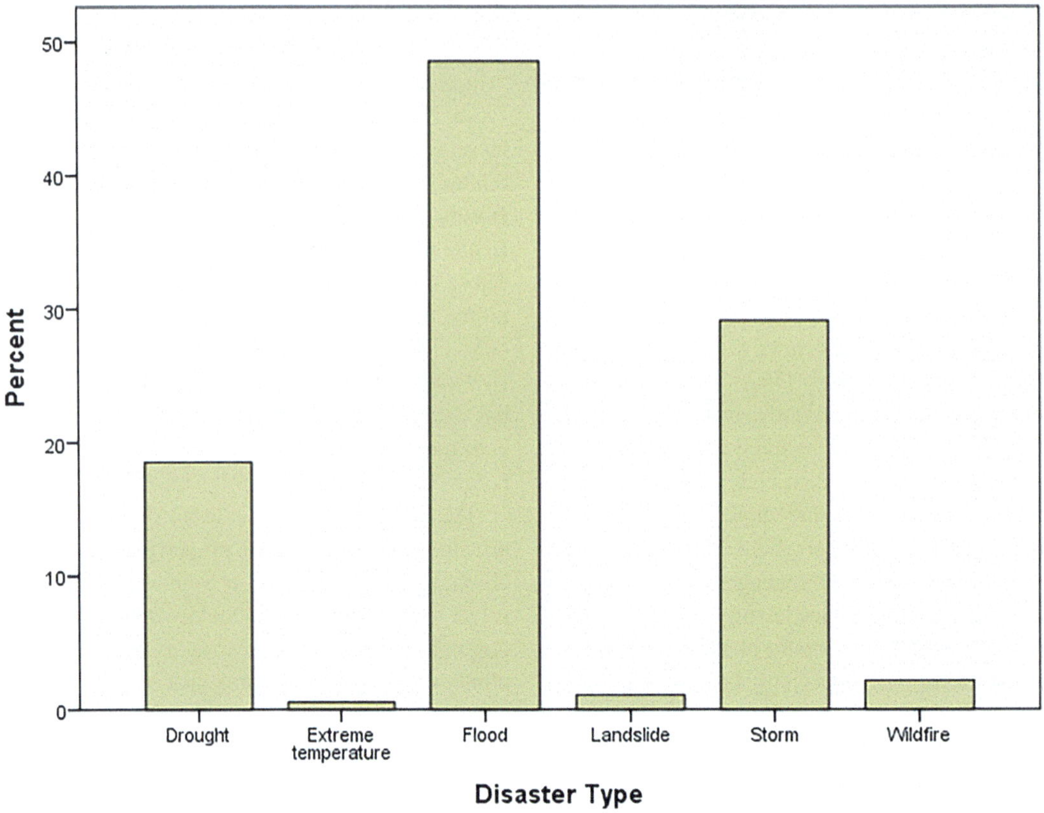

Fig. 5.2 Occurrence of climate-related disasters in Southern Africa. (Source: Authors, data from EM-DAT, CRED/UCLouvain database, 2023)

The questionnaire survey conducted during the 2019 cyclone Idai in Chimanimani, Zimbabwe, revealed that the productive assets that supported local livelihoods were the most affected. During a questionnaire survey, 32% of the participants were found to have lost their productive farms and crops that were close to harvest (Fig. 5.4). According to key informants, this damage to harvest had the negative effect of damaging the agricultural value chain of the area, which includes both the informal and sectors that are mainly dependent on agricultural produce to function. Furthermore, due to extensive damage to roads and bridges, most of the Chimanimani rural district became inaccessible, further affecting the speed with which affected communities could recover their livelihoods from the shock. This also impacted the livelihoods anchored in the transport sector. Key informants also reviewed that the loss of productive rural assets during disasters tends to erode the ability of affected communities to build resilience to future shocks and leaves them vulnerable to even minor exposure.

The same pattern was observed at the other studied sites. Key informants in Port Saint Johns highlighted that during the April 2019 floods, some places became inaccessible soon after the disaster, hindering rescue efforts, and productive agricultural land was damaged. In Tsholotsho, key informants highlighted that some of the flood-affected households had to be permanently relocated from their flood-prone area. One of the largest displacements occurred during the Dineo-induced floods of 2017, during which approximately 859 people were displaced from the Mahlaba, Mahlosi, Mbanyana, Tshino and Mapili areas. In Port Saint Johns, relocation is also being considered for settlements vulnerable to flooding, especially those close to the Mzimvubu River and Green Farm.

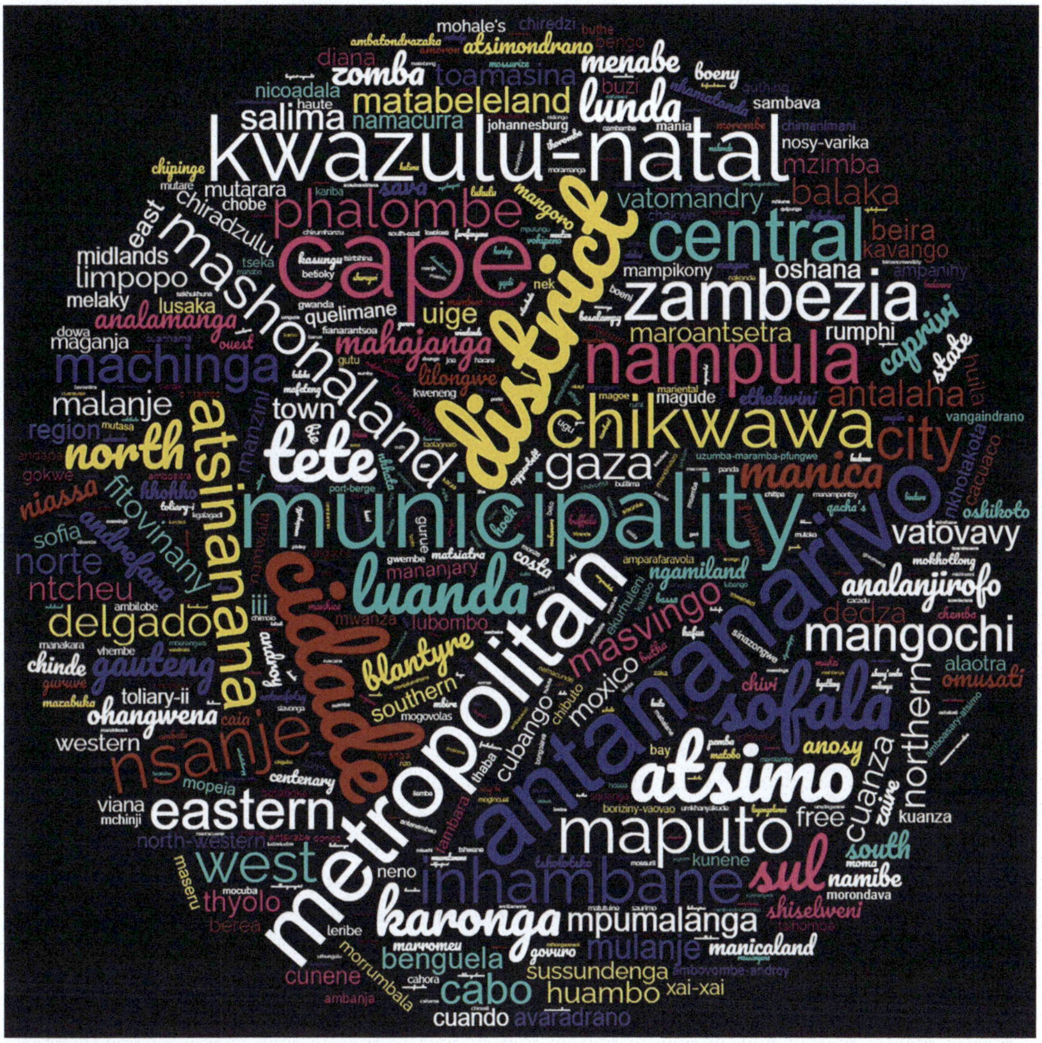

Fig. 5.3 Hotspot of climate-related disasters in Southern Africa. (Source: Authors, data from EM-DAT, CRED/UCLouvain database, 2023)

5.3.2 Disaster Risk Governance in Rural Southern Africa

Disaster risk governance in rural Southern Africa involves multiple actors at different levels and timescales. According to Tall et al. (2013), it includes disaster prevention, response and recovery frameworks, evacuation planning, training drills, coordination, information exchange and financial resource mobilisation. The process is complex, involving state and non-state actors, as well as formal and informal institutions. Jordhus-Lier (2019) advocates for DRR institutions that engage communities in planning and execution to ensure that decisions are locally relevant and responsive to needs.

Uneven power relations among DRR actors in rural Southern Africa can lead to the dominance of certain groups, marginalising less powerful ones in decision-making. Gillard et al. (2016) argue that a limited understanding of power dynamics has contributed to the failure of many DRR initiatives in the region. The Sendai Framework for DRR emphasises that at-risk communities should be central to DRR processes and involved in decision-making across all levels.

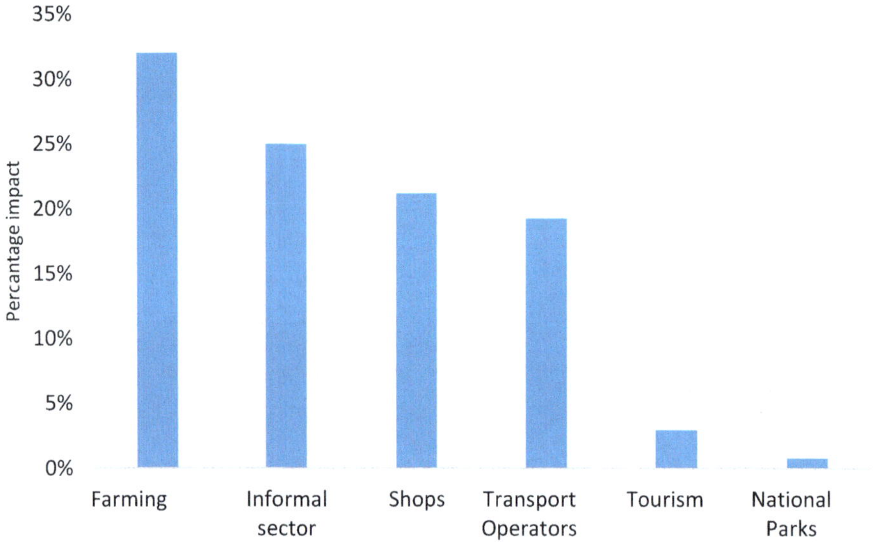

Fig. 5.4 Economic activities affected during the Idai cyclone in Chimanimani, $N = 368$. (Source: Produced by authors)

Local knowledge, which has evolved through lived experiences, is crucial for effective DRR (Trogrlić et al., 2022). However, in many DRR structures in rural Southern Africa, this local knowledge needs to be utilised more.

In the Tsholotsho Rural District of Zimbabwe, frequently hit by floods, the survey participants perceived that only formal institutions were active in the DRR, neglecting local and informal institutions (Fig. 5.5). Key informants identified Mapili, Mbamba, Mahlosi, Mahlaba, Mbanyana and Tamuhla as flood-prone communities affected by the Gwayi River and its tributaries. The government, NGOs and the Civil Protection Unit dominate DRR activities, sidelining informal structures such as traditional leadership, which have valuable local knowledge.

In Chimanimani, local actors such as religious leaders, traditional leaders and community organisations were the first responders to Tropical Cyclone Idai in 2019. Due to the damaged infrastructure and bad weather, external responders were unable to access the area. Before external aid arrived, affected communities in Kopa and Ngangu showed cohesion and unity, clearing roads, burying the dead and sharing resources. The Chimanimani hotel owners provided food and shelter to people without housing without demanding payment. However, the arrival of external help disrupted community cohesion, leading to demands for payment in US dollars, even for removing obstacles from personal properties.

In the South African context, the Disaster Management Act provides a provision for all interested stakeholders to contribute to the DRR process. The interested stakeholders may include traditional leadership, local communities, the private sector and volunteers (Zweig, 2017). Local municipalities are obliged to involve local actors in developing local disaster risk profiles and to improve understanding and participation in DRR issues by these communities (Van Niekerk et al., 2018). Despite this responsibility of the local authorities, in-depth interviews reviewed that local municipality actors viewed DRR as a preserve of mainly the district and provincial governments and, therefore, did not set aside significant funding for the process. A questionnaire survey conducted in Port Saint Johns (PSJ) after the 2019 floods showed that the most affected received assistance from the local municipality, friends and relatives (Fig. 5.6). These were the first responders to the crises, yet they were the most handicapped to respond due to limited budgeting. Of

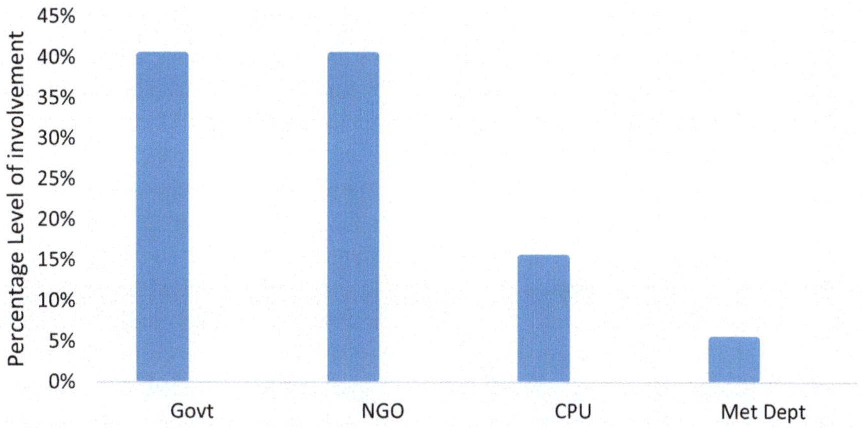

Fig. 5.5 Institutions involved in DRR in Tsholotsho, Zimbabwe. $N = 60$. (Source: Produced by the authors)

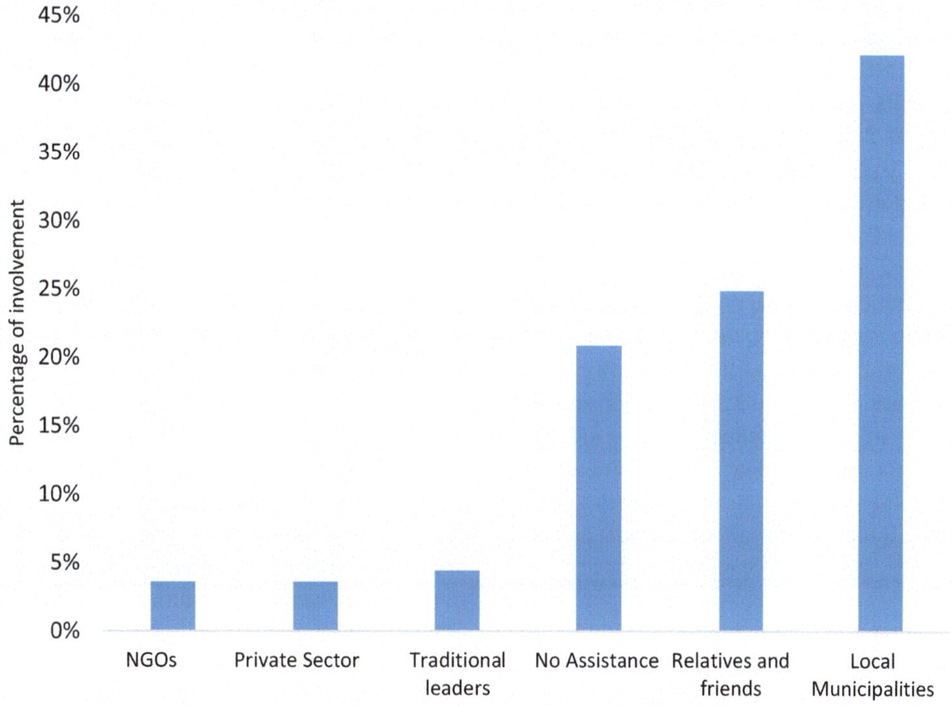

Fig. 5.6 DRR actors in Port Saint Johns, $N = 249$. (Source: Produced by authors)

note is that in PSJ, NGOs played a minimal role in DRR compared to Tsholotsho and Chimanimani in Zimbabwe. The local private actors played an important role in providing temporary accommodation, food and water to the affected population during flood disasters. This was completely missing Tsholotsho and very limited in Chimanimani.

In the Mkhondo local municipality, where tornadoes and related flooding are increasingly becoming a regular occurrence, key informants strongly believed that the traditional local leadership was being undermined in terms of the role they play in DRR. They noted that a culture of disrespect for traditional authority has led to an increase in climate-related disasters. They gave examples of the most affected populations being those who refused to be subjected to their culture and traditional leadership. This then leads them to construction on vulnerable and yet sacred places, which angers their ancestors, hence the increase in incidences of extreme events. The participants also advocated for greater inclusion of local actors and traditional leadership in the DRR processes to avoid poor adaptation to climate change-related impacts.

Trogrlić et al. (2022) argue that community participation in DRR activities serves to create an inclusive and robust process given the predicted intensification of extreme climate-related events. In Malawi and Mozambique, DRR policies recognise local actors as essential participants in the successful implementation of DRR. Their national DRR policies require community-level plans, drills and training programmes to help local actors understand and implement the DRR process. Despite these progressive provisions in the DRR policies, these structures are noted to be largely handicapped by limited funding for their operationalisation (Chikodzi & Kenny, 2021). Key informants from Mkondo argue for the need to develop community-based DRR systems that view the process as a partnership between local communities and mandated agencies and their partners. In Chimanimani, key informants observed the need to engage the private sector to work in tandem with local communities to enable them to respond proactively to disasters and build climate-resilient rural communities.

5.3.3 Financing Disaster Risk Reduction in Rural Southern Africa

In most rural parts of Southern Africa, DRR activities could be better funded and compete for resources with other national needs (Chikodzi & Kenny, 2021). Most DRR actors and institutions are financially limited in fulfilling their mandate (Ibid.). The burden of funding DRR activities falls largely on national governments and donor agencies. However, it is supposed to be a shared responsibility of the state and non-state actors to be effective (Gumbo et al., 2021). Risk assessments for the season inform fiscal distributions of DRR funds and are usually inadequate for risk mitigation (Mavhura & Mapuva, 2022). Therefore, innovative ways of investing in DRR in rural Southern Africa are needed. Serious engagements with the private sector and exploring different insurance mechanisms can be the key to unlocking investments in DRR (Bhavnani et al., 2008).

In Port Saint Johns and Chimanimani, 35% and 34%, respectively, of the questionnaire survey participants' households highlighted the unavailability of financing options from banks, insurance and other financial institutions as the biggest drawback to their households in recovering and rebuilding better after climate disasters. Twenty-four per cent of PSJ participants and 20.6% of Chimanimani also observed limited financial muscle at the household level as limiting their recovery after disasters. The savings and other funds to which these households had access were significantly less than what was required to adequately respond to perceived threats imposed on them by climate change. Without external interventions from the government and their development partners, these communities highlighted that they would still be inadequately prepared for the following disasters and their ability to cope was eroded by each disaster that occurred. The key informants of PSJ and Chimanimani advocated for dedicated funding that could support the replenishment of damaged livelihood assets and also for programmes that spearhead livelihood protection. At Mkondo, just like many municipalities, participants observed that DRR financing is a key competence of the district municipality, which is the Gert Sibande District Municipality. However, the provincial and central government, through the National Disaster Management Centre, could take over DRR recovery efforts when the impacts of the disaster are such that local and district authorities have limited capacity to cope and recover from the shock.

In Tsholotsho, key informants observed that the government of Zimbabwe, together with its development partners, was leading the process of financing climate-related DRR activities, including building back better damaged livelihoods and relocation of households settled in high-risk areas. They observed that NGOs such as Plan International had produced an anticipatory action framework to weather- and climate-related hazards for the sake of protecting humanitarian and development gains. According to key informants, the produced anticipatory action plan came with prearranged financing and long listing of pre-agreed actions that may be implemented when trigger levels for drought or floods are reached. Key informants deemed long listing to be crucial because, during emergencies, there is usually limited time for responders to convene and deliberate on appropriate actions when trigger points are reached, especially for rapid-onset events such as floods that have very short lead times.

Chikodzi and Kenny (2021) argued that the scaling up of insurance coverage would assist the recovery process post-disasters in rural Southern Africa. This is against the background of this financing model playing a significant role in post-disaster recovery in communities of the developed world (Sheehan et al., 2023). Since 2020, Zimbabwe has been part of the Africa Disaster Risk Financing Capacity of the African Development Bank. This project aims to increase Zimbabwe's ability to cope with climate-related risks such as droughts, cyclones and floods. The project also promotes the adoption of DRR financing mechanisms, such as sovereign risk insurance, as a vehicle to build climate-conscious rural communities (African Development Bank, 2023). Insurance coverage targets at-risk communities such as smallholder farmers and vulnerable rural communities who will receive assistance from insurance payouts in the event of droughts (Ibid.). There is a need to upgrade such financing models to rural Southern Africa to lessen the burden of DRR financing on donors and governments.

5.3.4 Enhancing Disaster Preparedness in Rural Southern Africa

Disaster preparedness is part of a community's ability to respond adequately to perceived threats. Communication with at-risk populations is critical to enhancing their preparedness for risk. According to Sambo and Chikodzi (2020), timely warnings to at-risk communities of the pending threats of disaster give them time to take evasive action and protect their properties and livelihoods. A good early warning system, however, needs to balance a tangle of conflicting requirements, such as the cost and reliability of the forecasting method used. Across Southern Africa, the skills to forecast and respond to perceived climatic extremes are unique for all countries.

Questionnaire surveys in Tsholotsho and PSJ showed that the communication of warnings by local government officials, traditional leaders, churches and family members was shown by the participants to be the most effective method of preparing the at-risk populations for the imminent disaster (Fig. 5.7). Therefore, this calls for the establishment of effective community-level DRR structures, given that the warnings they issue have greater penetration within communities. Notable at the study sites is the impact of text messages sent directly to the mobile phones of the at-risk population. According to key informants, these were usually sent via bulk messaging by the mobile network operators. Other modes of communication are received differently in the two study areas of Tsholotsho and Port Saint Johns. For example, social media reaches more people in Tsholotsho than in PSJ, and television has more viewership in PSJ than almost none in Tsholotsho. This brings about the need to diversify the means of communication with the at-risk population when issuing warnings.

Furthermore, communication should be transmitted via locally relevant platforms. This can only be done when prior engagement with the community on DRR issues has been done. In both PSJ and Tsholotsho, there is a significant number of people who need access to any form of early warning system or communication about disasters. These are usually the already marginalised members of society who lack access to any form of media. This group of people is also often the most impacted by disasters, yet they receive the least assistance if any at all. Care must be taken to involve this group of people in the DRR processes.

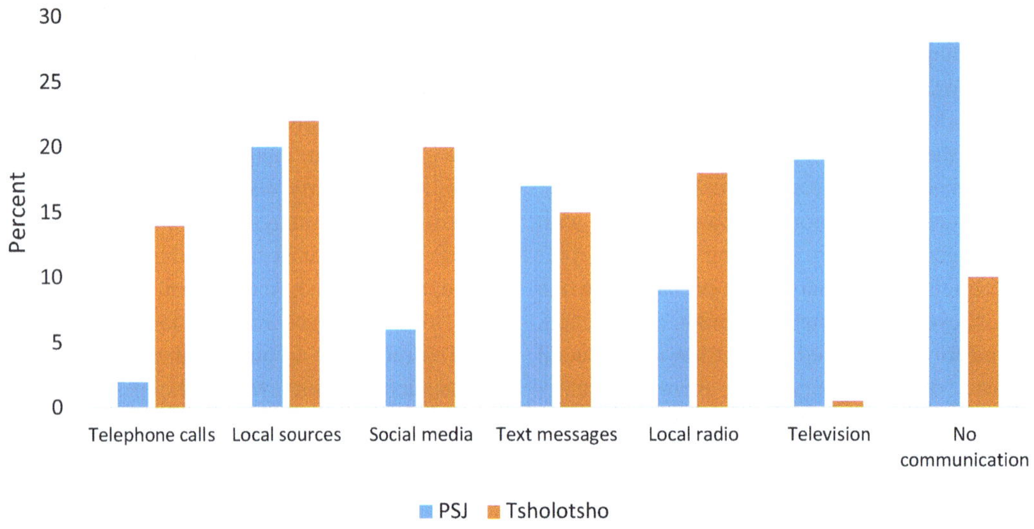

Fig. 5.7 Communication methods used in PSJ (2019) and Tsholotsho (2017 and 2018) during flood disasters. (Source: Produced by the authors)

Key informants noted that social protection programmes are crucial in improving the resilience of rural communities to climate-related disasters such as droughts and flooding. The main objective of these programmes is to provide safety nets to chronically poor households through strategies such as cash or voucher assistance, recovery loans and labour market programmes such as food for work and unemployment benefits. These social protection programmes have been found to stimulate the rapid recovery of productive livelihood assets lost during disasters. Chimanimani key informants highlighted that Zimbabwe had a long history of cash transfers to communities affected by disasters to alleviate the resulting shock. Where appropriate, the government has supported this initiative through the Cash Working Group in the Food Security and Livelihoods Cluster. The biggest challenge to cash-based interventions in Zimbabwe during the writing of this work was the hyperinflationary situation; hence, there was a need to put measures in place to ensure that the transfer value was retained. Participants observed that cash transfers provide more flexibility to both aid providers and affected communities compared to actual food distribution during drought or other disasters. They highlighted that cash transfers and vouchers ensure that the specific needs of the targeted recipients, such as food, clothing and productive assets, are addressed. This option also stimulates local markets and improves the dignity of households affected by disasters. However, the feasibility of cash transfers must be analysed for every community being considered. Participants from communities affected by cyclone Idai in Chimanimani reviewed that cash transfers provided flexibility, choice and self-respect for recipients. Some participants reviewed that cash programmes had resulted in disharmony and, in some cases, gender-based violence within households, with the main contestation being how best to use the received cash.

Conservation agriculture (CA) has been shown to help farmers cope with the impacts of drought and climate change (Mavesere & Dzawanda, 2023). CA is based on three principles: minimum soil disturbance, crop residue retention and crop rotation. Thus, the application of CA has been observed to avert problems of land degradation, increase crop productivity, reduce soil erosion, decrease carbon footprint, minimise energy use, maintain soil physicochemicals and improve food security (Giller et al., 2009; Marongwe et al., 2012; Brouder & Gomez-Macpherson, 2014; Tanyanyiwa et al., 2022;

Mulimbi et al., 2023a, b). Kumawat et al. (2023) reported that compared to conventional crop production practices, CA-based soil and crop management practices had a positive effect of increasing water productivity by between 18% and 66%, soil organic carbon by 12% and 93%, crop productivity by 3.8% and 76.2% and energy use efficiency by 8.9% and 40.2%. In addition, Mugandani and Mafongoya (2019) noted that among farmers practising CA in the Chivi, Murehwa and Mutoko districts of Zimbabwe, close to 82% of them used CA to manage soil erosion, while 72% used CA for land restoration potential. In Zimbabwe, conservation agriculture is practised by more than 300,000 smallholder farmers drawn from a total of 1500 wards. Successful stories of CA have been reported in dryland regions of the country, such as Mwenezi, Chiredzi, Gwanda and Manicaland region and Mashonaland Central (Marongwe et al., 2012; Brouder & Gomez-Macpherson, 2014; Tanyanyiwa et al., 2022). In South Africa, interesting reports on CA application were reported in the Eastern Cape region (Mulimbi et al., 2023a), while in the Democratic Republic of Congo, CA practice has registered successful stories in Maniema province (Mulimbi et al., 2023b). In the application of CA in livestock farming in Zambia, production has also been reported to increase crop productivity and improve the adaptation of rural farmers to climate-change-related shocks such as drought (Giller et al., 2009).

Anticipatory action (AA) is also known as anticipatory humanitarian action, forecast-based financing, forecast-based action, early action, early warning, early action or risk-informed early action, depending on the context and actor (IFRC, 2023). AA can be seen as an umbrella term for a cocktail of disaster forecast-based approaches that anticipate hazards and their impacts and facilitate proactive action (Hansen et al., 2022). AA should not be seen as replacing investment in DRR but as improving disaster preparedness, linking disaster resilience building and disaster response. It aims to shift the mentality of DRR from reactive to anticipatory. Globally, there has been a drive for international humanitarian players to shift from a primarily responsive approach to an AA approach, which has outcomes that are more efficient and dignified and protects hard-won development gains (WFP, 2022). AA is made up of three core elements, which are forecasting model or trigger with a clear decision-making protocol to guide decision-making; preallocated financing, which is available within the window of opportunity; and pre-agreed actions that deliver the impact reduction mechanisms for the forecasted risks (REAP, 2022).

AA has been used in many parts of rural Southern Africa. In Mozambique, AA has been activated for tropical cyclones in 2020 and 2021 (Anticipatory Hub, 2023). This led to the reinforcement and protection of housing, schools and infrastructure, and it also led to the distribution of sandbags, insecticide-treated mosquito nets and water purification kits such as filters, chlorine and tablets. Volunteers were sent to communities to strengthen early warnings and warnings (de la Poterie et al., 2023; Anticipatory Hub, 2023). In Zambia, AA was activated during the July 2021 floods. This led to community awareness of the early warning issued and the early action that needed to be taken to prepare for the risk. In the process, evacuation routes, disaster-free zones and temporary accommodation areas were identified. AA also led to the prepositioning and distribution of WASH items and the mobilisation of communities to construct culverts to direct the waterway from key facilities (de la Poterie et al., 2023). In Zimbabwe, a number of organisations have taken the lead in developing and implementing AA. The World Food Programme (WPP) developed an AA for drought in the Mudzi district. The AA aims to reduce the impact of droughts in the district through action based on the forecast of drought categorised as mild, moderate and severe through the use of the standardised precipitation index (Nobre & Bonifacio, 2022). The first activation took place in Mudzi in September 2021. District-level forecasts for the rain months of November and January indicated that the moderate drought trigger was reached. WFP AA provided access to adequate water supply to prevent possible production losses in areas with the highest risk of drought by installing solar-powered boreholes (WFP, 2022). They also provided critical climate information together with actionable agriculture advisories for a potential below-average rainfall season. Given its relative success, AA is now being implemented in four districts: Mudzi, Mbire, Binga and Matobo (WFP, 2022).

5.3.5 Obstacles to DRR in Rural Southern Africa

Chimanimani key informants noted significant challenges in information sharing and data availability within DRR circles, preventing effective climate action. Although DRR policies in Zimbabwe provide a framework for state and non-state actors to collaborate during disasters, timely information sharing with coordinating agencies is often lacking, creating critical gaps in the DRR process.

In South Africa, the Disaster Management Act of 2002 (amended in 2015) promotes collaboration between the government, civil society and the private sector to implement effective DRR (Van Niekerk et al., 2018). The Act emphasises the need for consistency among diverse actors. For example, those who forecast hazards may need more authority to issue early warnings. In contrast, those who can issue warnings may need more capacity to respond, highlighting the need for coordinated efforts (Chikodzi & Kenny, 2021).

In Mkondo, key informants noted that miscommunication, inter-agency rivalries and lack of trust among DRR actors hinder the process. A robust AA system faces challenges such as accessing quality hazard assessments, vulnerability and impact data necessary to develop accurate triggers and standard procedures (IFRC, 2023). The available data are often inaccurate, disjoint and of course resolution, making it difficult to produce triggers for various locations (WFP, 2022).

Most DRR structures in Southern Africa operate under decentralised structures. Limited funding has always been a handicap for such decentralised structures. In Zimbabwe, Malawi and Mozambique, such structures cascade down to the ward level but usually need more operational budgets to activate their structures in the absence of pending disaster and mostly rely on erratic donor funding (de la Poterie et al., 2023). Community-level DRR structures at the community level need to be adequately equipped and staffed. Limited resources imply constrained engagement with local communities in DRR issues. Hence, exposure to identified hazards remains high (Zembe et al., 2023). Therefore, the decentralisation of DRR activities in most parts of rural Southern Africa is just on paper, thus not making the desired impact. In some instances, the community-based DRR agent works on a voluntary basis with no budget, which limits their capacity to engage the communities, hence introducing both horizontal and vertical disconnections (Weingärtner & Spencer, 2019).

During disasters in most parts of Southern Africa, there are always challenges with bureaucracy, procurement and logistics. These challenges occur more during fast-onset climatic hazards such as cyclones and flooding. According to Roots (2021), for international DRR agencies, the delay in receiving from abroad and the need for strict adherence to organisational procedures for the procurement and distribution of DRR aid were familiar sources of delay. In most cases, funding available for the prepositioning of key requirements usually needs to be improved for the scale of the disaster.

There is also a general shortage of most of the skills required for the entire DRR process. Limited forecasting skills are considered a significant gap in the DRR process in Southern Africa (Sambo & Chikodzi, 2020). Season forecasting, for example, usually needs to provide more details on the timing, distribution and quantity of the hazard of interest to support critical decisions on livelihood protection and preposition essential requirements (de la Poterie et al., 2023). To give an example, the uncertainty that characterises tropical cyclone paths and slow-onset events such as droughts can be compounded by poor forecasting skills in making critical decisions during disaster periods. Given the shortages of expertise in the DRR field, training and refresher courses are needed for all critical actors in the field. This helps the DRR actors to familiarise themselves with the trending innovations in the area (Zembe et al., 2023).

5.4 Conclusions

This study explored the reduction of the risk of climate disasters in rural Southern Africa. Our study used a mixed-methods approach that involved collecting information through content analysis and primary data sources. The content analysis collected data from online scientific databases using the defined inclusion criteria. Primary data sources included household questionnaire surveys and key informant interviews from South Africa's Port Saint Jones and Mkhondo local municipalities, as well as Zimbabwe's Chimanimani and Tsholotsho rural districts. As a result, several conclusions were drawn from this study.

Our findings show that rural Southern Africa is vulnerable to numerous climate-related hazards, including droughts, floods, cyclones and tornadoes, with increasing frequency. These disasters often result in the loss of productive assets since agriculture is the primary income. Various policies, organisations and institutions have been established for DRR (disaster risk reduction), placing at-risk communities at the centre and devolving DRR structures to the lowest community levels. Local institutions, including traditional leadership, are crucial as first responders to rapid-onset events but are often underresourced and not prominently included in DRR efforts, creating significant gaps. The financial burden of DRR falls mainly on governments and donor agencies, which are often financially restricted. Effective communication is key to preparing at-risk communities, and local sources are the most effective for early warnings. Anticipatory action for DRR is partially practiced and needs upscaling for better impact. The main challenges to DRR initiatives in rural Southern Africa include limited funding, technical skills and logistical issues.

Given the findings, some recommendations can be offered. There is a need to integrate DRR in all policies and sectors, from national to local governments. This integration needs to come with a mandatory percentage of the total budgets invested in DRR. Integration must occur in tandem with the integration of DRR into development planning at all levels of society. The capacity development in DRR activities is needed to enable the strengthening of knowledge in the area with different spatial coverage. Such capacity development must take a cross-sectoral approach. The DRR institutions must be strengthened by incorporating more local actors in rural Southern Africa.

References

African Development Bank. (2023). *Zimbabwe – Africa Disaster Risk Financing (ADRiFi) capacity-building project*. Retrieved from: https://projectsportal.afdb.org/dataportal/VProject/show/P-ZW-AA0-018#:~:text=The%20project%20will%20enhance%20Zimbabwe's,solution%20for%20addressing%20climate%20risks. Accessed 22 Mar 2024.

Anticipatory Hub. (2023). *Forecast-based Financing: Closing the gap between disaster preparedness and emergency relief in Southern Africa – Mozambique*. Retrieved from: https://www.anticipation-hub.org/experience/anticipatory-action-in-the-world/mozambique/forecast-based-financing-closing-the-gap-between-disaster-preparedness-and-emergency-relief-in-southern-africa-mozambique. Accessed 22 Mar 2024.

Bhavnani, R., Vordzorgbe, S., Owor, M., & Bousquet, F. (2008). *Disaster risk reduction in the sub-Saharan Africa region*. Washington, DC.

Balgah, R. A., & Kimengsi, J. N. (Eds.). (2022). *Disaster management in sub-Saharan Africa: Policies, institutions and processes*. Emerald Publishing Limited.

Brouder, S. M., & Gomez-Macpherson, H. (2014). The impact of conservation agriculture on smallholder agricultural yields: A scoping review of the evidence. *Agriculture, Ecosystems & Environment, 187*, 11–32.

Chisale, H. L., Chirwa, P. W., Babalola, F. D., & Manda, S. O. (2021). Perceived effects of climate change and extreme weather events on forests and forest-based livelihoods in Malawi. *Sustainability, 13*(21), 11748.

Chikodzi, D., & Kenny, L. R. (2021). Disaster risk reduction governance in southern Africa: Focus on budgets and institutional competencies. In *Cyclones in southern Africa: Volume 2: Foundational and fundamental topics* (pp. 227–241).

Davies, M., Guenther, B., Leavy, J., Mitchell, T., & Tanner, T. (2009). Climate change adaptation, disaster risk reduction and social protection: Complementary roles in agriculture and rural growth? *IDS Working Papers, 2009*(320), 01–37.

de la Poterie, A. T., Castro, E., Jr., Rahaman, H., Heinrich, D., Clatworthy, Y., & Mundorega, L. (2023). Anticipatory action to manage climate risks: Lessons from the red cross red crescent in southern Africa, Bangladesh, and beyond. *Climate Risk Management, 39*, 100476.

Dube, K., & Chikodzi, D. (2021). Characterisation and impacts of tornado-induced flooding and windstorms in Mpumalanga Province, South Africa. In *The increasing risk of floods and Tornadoes in Southern Africa* (pp. 257–269). Springer.

Eckstein, D., Künzel, V., & Schäfer, L. (2021). *The global climate risk index 2021*. Germanwatch.

EM-DAT, CRED/UCLouvain. (2023). *EM-DAT, cred international disaster database*. Universite Catholique de Louvain. Available on www.emdat.be

Gillard, R., Gouldson, A., Paavola, J., & Van Alstine, J. (2016). Transformational responses to climate change: Beyond a systems perspective of social change in mitigation and adaptation. Wiley *Interdisciplinary Reviews: Climate Change, 7*(2), 251–265.

Giller, K. E., Witter, E., Corbeels, M., & Tittonell, P. (2009). Conservation agriculture and smallholder farming in Africa: The heretics' view. *Field Crops Research, 114*(1), 23–34.

Gumbo, T., Ingwani, E., Mufungizi, A. A., & Dumba, S. (2021). Planning for climate and cyclone resilient transport infrastructures in southern Africa: Lessons from Zimbabwe. In *Cyclones in Southern Africa: Volume 3: Implications for the sustainable development goals* (pp. 159–176).

Hansen, J., Hellin, J., Rosenstock, T., Fisher, E., Cairns, J., Stirling, C., Lamanna, C., van Etten, J., Rose, A., & Campbell, B. (2022). Climate risk management and rural poverty reduction. *Agricultural Systems, 172*, 28–46.

IFRC. (2023). *Forecast-based action by the dref– Overview of early action protocols*. International Federation of the red Cross red Crescent. https://public.tableau.com/app/profile/ifrcgo/viz/NEWFbADashboard/Dashboard. Accessed 29, 30 Oct 2023.

IPCC. (2021a). *The intergovernmental panel on climate Change's (IPCC) 6th assessment report: Physical science basis*. https://www.ipcc.ch/report/sixth-assessment-report-working-group-i/#:~:text=AUTHORS-,Report,first%20part%20of%20the%20Sixth%20Assessment%20Report%2C%20Climate%20Change%202021%3A%20The,-Physical%20Science%20Basis

IPCC. (2021b). Climate change 2021-the physical science basis. *Interactions, 49*(4), 44–45.

IPCC. (2021c) *Climate change has already hit southern Africa*. Here's how we know Published: October 24, 2021 11.07am SAST. https://theconversation.com/climate-change-has-already-hit-southern-africa-heres-how-weknow169062#:~:text=The%20Intergovernmental%20Panel,climate%20change%20projections

Iyakaremye, V., Zeng, G., Ullah, I., Gahigi, A., Mumo, R., & Ayugi, B. (2022). Recent observed changes in extreme high-temperature events and associated meteorological conditions over Africa. *International Journal of Climatology, 42*(9), 4522–4537.

Jordhus-Lier, D., Saaghus, A., Scott, D., & Ziervogel, G. (2019). Adaptation to flooding, pathway to housing or 'wasteful expenditure'? Governance configurations and local policy subversion in a flood-prone informal settlement in Cape Town. *Geoforum, 98*, 55–65.

Kumawat, A., Yadav, D., Srivastava, P., Babu, S., Kumar, D., Singh, D., Vishwakarma, D. K., Sharma, V. K., & Madhu, M. (2023). Restoration of agroecosystems with conservation agriculture for food security to achieve sustainable development goals. *Land Degradation & Development, 34*, 3079.

Langa, J. M. D. R. C., Bruna, N., Monjane, B., do Amaral, G., da Amélia Fé, E. A., Rafael, B. P., et al. (2023). Extreme climatic events and climate change policies: A call for climate justice action in Mozambique. In *Climate justice in the majority world* (pp. 42–60). Routledge.

Leal Filho, W., Taddese, H., Balehegn, M., Nzengya, D., Debela, N., Abayineh, A., et al. (2020). Introducing experiences from African pastoralist communities to cope with climate change risks, hazards and extremes: Fostering poverty reduction. *International Journal of Disaster Risk Reduction, 50*, 101738.

Lee, C. C., Yuan, Z., Lee, C. C., & Chang, Y. F. (2022). The impact of renewable energy technology innovation on energy poverty: Does climate risk matter? *Energy Economics, 116*, 106427.

Mal, S., Singh, R. B., Huggel, C., & Grover, A. (2018). Introducing linkages between climate change, extreme events, and disaster risk reduction. In *Climate change, extreme events and disaster risk reduction: Towards sustainable development goals* (pp. 1–14).

Marongwe, L. S., Kwazira, K., Jenrich, M., Thierfelder, C., Kassam, A., & Friedrich, T. (2012). An African success: The case of conservation agriculture in Zimbabwe. In *Sustainable intensification* (Vol. 9, pp. 153–161). Routledge.

Mavhura, E., & Mapuva, J. (2022). Barriers for local authorities to invest in disaster risk reduction: Evidence from Zimbabwe. *South African Geographical Journal, 104*(1), 122–136.

Mayer, B. (2021). Climate change mitigation as an obligation under human rights treaties? *American Journal of International Law, 115*(3), 409–451.

Mavesere, F., & Dzawanda, B. (2023). Effectiveness of Pfumvudza as a resilient strategy against drought impacts in rural communities of Zimbabwe. *GeoJournal, 88*(3), 3455–3470.

Mavodyo, E. (2023). The impact of climate change on food insecurity in the southern African development community. *JDE (Journal of Developing Economies), 8*(1), 162–183. https://doi.org/10.20473/jde.v8i1.43534

Mugandani, R., & Mafongoya, P. (2019). Behaviour of smallholder farmers towards adoption of conservation agriculture in Zimbabwe. *Soil Use and Management, 35*(4), 561–575.

Mulimbi, W., Nalley, L. L., Strauss, J., & Ala-Kokko, K. (2023a). Economic and environmental comparison of conventional and conservation agriculture in South African wheat production. *Agrekon, 62*, 1–19.

Mulimbi, W., Nalley, L., Nayga, R. M., Jr., & Gaduh, A. (2023b). Are consumers willing to pay for conservation agriculture? The case of white maise in The Democratic Republic of the Congo. In *Natural resources forum* (Vol. 47, No. 1, pp. 22–41). Blackwell Publishing Ltd.

Ngoma, H., Angelsen, A., Jayne, T. S., & Chapoto, A. (2021). Understanding adoption and impacts of conservation agriculture in eastern and southern Africa: A review. *Frontiers in Agronomy, 3*, 671690.

Nhamo, G., Dube, K., & Saurombe, T. (2021). Impact of tropical Cyclone Idai on tourism attractions and related infrastructure in Chimanimani, Zimbabwe. In *Cyclones in Southern Africa: Volume 3: Implications for the sustainable development goals* (pp. 245–264).

Nhamo, G., & Chikodzi, D. (2021). Cyclones in southern Africa: Volume 1: Interfacing the catastrophic impact of cyclone Idai with SDGs in Zimbabwe. Springer Nature.

Nobre, G. G., & Bonifacio, R. (2022). *Drought Forecast-based Financing: Lessons learned in building a trigger menu for anticipatory action in Mozambique and Zimbabwe*. EGU General Assembly.

Otto, F. E., Zachariah, M., Wolski, P., Pinto, I., Nhamtumbo, B., Bonnet, R., …, & Harrington, L. J. (2022). *Climate change increased rainfall associated with tropical cyclones hitting highly vulnerable communities in Madagascar, Mozambique & Malawi*. Retrieved from: http://www.indiaenvironmentportal.org.in/files/file/climate_change_rainfall_madagascar.pdf. Accessed 21 Mar 2024.

Sheehan, B., Mullins, M., Shannon, D., & McCullagh, O. (2023). On the benefits of insurance and disaster risk management integration for improved climate-related natural catastrophe resilience. *Environment Systems and Decisions, 43*(4), 639–648.

Sousa, P. M., Blamey, R. C., Reason, C. J., Ramos, A. M., & Trigo, R. M. (2018). The 'Day Zero' Cape Town drought and the poleward migration of moisture corridors. *Environmental Research Letters, 13*(12), 124025.

Tall, A., Patt, A. G., & Fritz, S. (2013). Reducing vulnerability to hydro-meteorological extremes in Africa. A qualitative assessment of national climate disaster management policies: Accounting for heterogeneity. *Weather and Climate Extremes, 1*, 4–16.

Tanyanyiwa, V. I., Kanyepi, T., & Katanha, A. (2022). Zimbabwe's Pfumvudza agriculture programme—Reality or rhetoric? In *Sustainable agriculture and food security* (pp. 327–347). Springer International Publishing.

Trogrlić, R. Š., Duncan, M., Wright, G., van den Homberg, M., Adeloye, A., & Mwale, F. (2022). Why does community-based disaster risk reduction fail to learn from local knowledge? Experiences from Malawi. *International Journal of Disaster Risk Reduction, 83*, 103405.

REAP. (2022). *Early action: The state of play 2021*. Risk Informed Early Action Partnership.

Roots, R. (2021). Review of the pilot phase of the Forecast-based Action by the Disaster Relief Emergency Fund (FbA by the DREF) 2018–2020. Internal Report for IFRC, Geneva.

Sambo, P., & Chikodzi, D. (2020). Deciphering adoption and adaption of ICTs in humanitarian services: The southern African context. In *Information and communication technologies for humanitarian services*. The Institution of Engineering and Technology. https://doi.org/10.1049/PBTE089E_ch15

UNISDR. (2015). Global assessment report on disaster risk reduction. In *Making development sustainable: The future of disaster risk management*. United Nations Office for Disaster Risk Reduction (UNISDR).

Nemakonde, L. D., & Van Niekerk, D. (2017). A normative model for integrating organisations for disaster risk reduction and climate change adaptation within SADC member states. *Disaster Prevention and Management: An International Journal, 26*(3), 361–376.

O'Neill, B. C., Oppenheimer, M., Warren, R., Hallegatte, S., Kopp, R. E., Pörtner, H. O., & Yohe, G. (2017). IPCC reasons for concern regarding climate change risks. *Nature Climate Change, 7*(1), 28–37.

Van Der Walt, A. J., & Fitchett, J. M. (2022). Extreme temperature events (ETEs) in South Africa: A review. *South African Geographical Journal, 104*(1), 70–88. https://doi.org/10.1080/03736245.2021.1907219

Van Niekerk, D., Wentink, G. J., & Shoroma, L. B. (2018). Natural hazards governance in South Africa. In *Oxford research encyclopedia of natural hazard science*. https://doi.org/10.1093/acrefore/9780199389407.013.246

Vogel, C., Koch, I., & Van Zyl, K. (2010). "A persistent truth"—Reflections on drought risk management in southern Africa. *Weather, Climate, and Society, 2*(1), 9–22.

World Bank. (2016). *High and dry: Climate change, water, and the economy*. The World Bank.

Zweig, P. J. (2017). Collaborative risk governance in informal urban areas: The case of Wallacedene temporary relocation area. *Jàmbá: Journal of Disaster Risk Studies, 9*(1), 386.

Weingärtner, L., & Spencer, A. (2019). Analysing gaps in the humanitarian and disaster risk financing landscape. *ODI and Start Network, London.*

World Food Programme. (2022). *A system to anticipate and address the impacts of drought in Zimbabwe: An overview of partnerships, capacity strengthening and activations*. Retrieved from: https://docs.wfp.org/api/documents/WFP-0000142536/download/?_ga=2.69597524.55942037.1711129122-895266523.1711129122. Accessed 22 Mar 2023.

Zembe, A., Nemakonde, L. D., & Chipangura, P. (2023). A policy coherence framework for food security, climate change adaptation and disaster risk reduction in South Africa. *International Journal of Disaster Risk Reduction, 95*, 103877.

6 Climate Governance and Resilience Action: The Southern African Context

Leonard Chitongo, Vincent Chenzi, and Caitlin Blaser Mapitsa

6.1 Introduction and Background

Climate change has ravaged the globe by intensifying and even triggering disasters in the form of heatwaves, floods, droughts, cyclones, desertification and the rise in sea levels. Climate change must be managed through climate governance management systems. According to Jagers and Stripple (2003), climate governance is the assortment of procedures and methods that are devised and employed to guide societal systems in the prevention, mitigation and or adaptation to hazards associated with climate change. In a general sense, the term 'governance' encompasses the processes that define standards for the allocation of power, the development of policy goals and the involvement of individuals and other relevant stakeholders in the process of making decisions. Mugambiwa (2021) asserts that there is a prevailing consensus within the scientific community about the existence of climate change, emphasising the urgency for communities to address and mitigate its impacts proactively. The incorporation of adaptation measures has been recognised as a significant component of government strategies in addressing climate change since the inception of climate-related deliberations. Although the precise role of adaptation remains elusive, it is evident that the implementation of comprehensive and sustainable adaptation strategies will be imperative to mitigate the impacts of climate change effectively.

In Zimbabwe, an average of two cyclones affect the country each year, especially in the Manicaland and Mashonaland East provinces (World Data, 2023). However, due to climate change, the frequency of cyclones that affect Zimbabwe has increased to an annual average of five cyclones (Africanews, 2022). This basically means that Zimbabwe and its Southern African neighbours must prepare for more devastating disasters. The point is that Southern Africa is vulnerable to droughts due to the high frequency and severity of extreme weather events, including droughts and floods. Therefore, a robust disaster governance and resilience mechanism is urgently required for the region to cope with the increasing frequency and effects of disasters and other extreme weather events.

The most devastating tropical cyclone that affected Zimbabwe was Cyclone Idai. It was one

L. Chitongo (✉)
School of Social Sciences, College of Humanities, University of KwaZulu-Natal, Durban, South Africa

V. Chenzi
Department of Peace, Security and Society, Faculty of Arts and Humanities, University of Zimbabwe, Harare, Zimbabwe

C. Blaser Mapitsa
Wits School of Governance, University of the Witwatersrand, Johannesburg, South Africa
e-mail: Caitlin.Mapitsa@wits.ac.za

M. Matsa et al. (eds.), *Climate Change Resilience in Rural Southern Africa*, Sustainable Development Goals Series, https://doi.org/10.1007/978-3-031-73600-1_6

of the worst tropical cyclones on record to affect Africa and the Southern Hemisphere between the 4th and 21st of March 2019 (Chapungu, 2020). The long-lived storm caused catastrophic damage and a humanitarian crisis in Mozambique, Madagascar, Zimbabwe and Malawi, leaving more than 1500 people dead and many more missing while displacing more than two million inhabitants within the region (Chapungu, 2020). In Chimanimani, agriculture, being the main economic activity, was greatly affected. Many farms, including support infrastructure, were destroyed, resulting in the loss of arable land, livestock and stored cereals. The destroyed livestock and infrastructure included forty dip tanks, tobacco barns, hectares of grazing land, more than 300 cattle and thousands of chickens, sheep and goats in Chimanimani and Chipinge. This destruction directly affected livelihoods, especially for community members who relied on agriculture, as it destroyed beans, tomatoes, maize fields, banana plantations, onions, cotton and sugar cane. Community members, especially women and girls, failed to adapt to the challenges of destroyed livelihoods and embraced harmful coping mechanisms, especially prostitution, which exposed them to gender-based violence (GBV), sexually transmitted infections (STIs) and unwanted pregnancies. Several stakeholders, including local authorities, non-government organisations (NGOs) and foreign governments, were heavily involved in addressing the effects of the disaster.

Matopos, on the other hand, is one of the high-altitude catchment areas for the Limpopo River Basin. The ecosystem in the Matopos area affects all downstream water resources. There have been several recent droughts in Matopos that have threatened livelihoods in the area. Unlike cyclones, which are short in scale but intensive, droughts are slower in onset but can have long-standing implications; given that rangelands require time to recover after a drought, there is bound to be increased human-wildlife conflict due to shared water sources.

From exploring these two cases in depth, the chapter argues that governance, when communities are recovering from disaster or near tipping points, is fragmented. The capacities of stakeholders responsible for climate management to improve community development and cushion the effects of shocks and stressors in the aftermath of disasters also proved to be ineffective. There is an urgent need to remodel current disaster governance and resilience efforts by focusing on specific core gaps. These include a need to clarify the mandates, roles and responsibilities of all stakeholders and to ensure a match between the scope of the mandate and the capacity of each role player. A second consideration is a mismatch between the scope of the problem and the mandates of those working for solutions. Finally, there is a greater need to align incentives towards integrated planning to strengthen investment in resilient infrastructure.

The main aim of the chapter is to understand better resilience in rural Zimbabwe and the implications this has for climate governance using the case study of Cyclone Idai in March 2019 and recurrent droughts in Matopos. In order to address these issues, the chapter is divided into five sections, excluding the introduction. The first part examines the geographical landscape of the two case study areas, Chimanimani and Matopos. The conceptual framework forms the second part, while the methodology constitutes the third segment. The results section forms the fourth phase, and the discussion makes up the fifth and final portion of the chapter.

6.2 Geographical Landscape of Manicaland and Matabeleland South

The two case studies discussed in this chapter are both located in Zimbabwe. Cyclone Idai mainly affected parts of Chimanimani and Chipinge, both located in the Manicaland province of Zimbabwe (Chapungu, 2020). Manicaland province is a province that covers primarily the eastern highlands and the southeastern plateau of Zimbabwe. It is one of the ten provinces of Zimbabwe. The topography of Manicaland deviates from the areas below 915 meters and soars to

Nyanga Mountain, the highest mountain in Zimbabwe at 2.592 meters. The most significant part of the Manicaland province is covered in giant mountains and thick indigenous forests (Risiro, 2019). The annual average high temperature is approximately 23 °C. The annual average low temperature is 13.9 °C. The average annual precipitation is usually around 1098 mm. The northern part of Manicaland is drained to the north by tributaries of the Zambezi River, such as the Gairezi River and the Nyangombe River. The central and southern parts of the province are drained to the south by tributaries such as the Save and Oji rivers. Some areas along the eastern edge of the state, such as Honde Valley, Burma Valley and the eastern Chimanimani and Chipinge districts, drain eastward into the Pungwe and Buzi rivers. It is made up of seven districts, namely, Buhera, Chimanimani, Chipinge, Makoni, Mutare, Mutasa and Nyanga (Jerie & Ndabaningi, 2011). The two districts of Chimanimani and Chipinge were the ones that were mainly affected by Cyclone Idai in March 2019.

On the other hand, Matopos is located in Matabeleland, a southern province situated in southwestern Zimbabwe. Major economic activities include subsistence agriculture and livestock farming. However, the province is located on the edge of the Kalahari Desert, so it is arid and very dry (Maviza & Ahmed, 2020). The arid climate is not hospitable to agriculture. Hence, droughts and lack of economic opportunities have caused widespread poverty and migration out of the province. It is 949.86 meters (3116.34 feet) above sea level and has a subtropical steppe climate. The province's annual temperature is 23.57 °C (74.43 °F) and is 0.84% higher than the average of Zimbabwe (Mguni, 2021). Matabeleland South typically receives about 58.84 millimetres (2.32 inches) of precipitation and has 74.45 rainy days (20.4% of the time) annually. Like Manicaland province, Matabeleland South province has seven districts and these are Beitbridge, Bulilima, Gwanda, Insiza, Mangwe, Matobo/Matopos and Umzingwane (Mguni, 2021).

6.3 Resilience Framework

Disaster resilience has been conceptualised in different contexts of social, ecological and environmental interactions. The term came into light after 2005 when 168 countries agreed to adopt the Hyogo Action Framework 2005–2015 to build the resilience of nations and communities to disasters. The Sendai Framework for Disaster Risk Reduction 2015–2030 was endorsed in 2015 to strengthen resilience by implementing integrated and inclusive measures for disaster risk reduction as a successor instrument to the Hyogo Framework. In the same year, almost all nations agreed on the 2030 Agenda for Sustainable Development (Pradhan & Chauhan, 2020). Resilience is the ability to withstand or successfully cope with adversity (Werner & Smith, 2001). Conner (1993) views it as the ability to demonstrate both strength and flexibility during the change process while displaying minimal dysfunctional behaviour. For this study, resilience is the ability to anticipate, endure, recuperate from and overcome disaster-induced distress (Acosta et al., 2017). Resilience theory argues that it is not the nature of adversity that is important but how we deal with it (Ackerman & Maslin-Ostrowski, 2002). When we face adversity, misfortune or frustration, resilience helps us bounce back. It helps us survive, recover and even thrive in the face and wake of misfortune. When faced with natural disasters, it is essential to know how well people adapt to the dynamic environment. Resilience analysis looks at whether people and institutions can continue living and working generally without having to adapt. It may also mean that people are capable of adapting to disaster by being able to maintain or quickly recover their well-being or their normal functioning. Resilient systems show high levels of diversity, which can be realised by creating strong institutions that enhance community cohesion (Bahadur et al., 2010). Therefore, resilient systems call for the building of durable community structures and knowledge and jointly exploring existing and new ways to cope with potential uncertainty.

The resilience framework examines the interaction between the actions of communities and the harsh environment, as well as subsequent actions and reactions. In this framework, the exercise of power, institutional support, the inherent knowledgeability of communities and their resolve to overcome the effects of climate change impact different levels of society. This collectivism implies that for long-term sustainable resilience, all aspects of society should be involved. The framework recognises the complementary roles of short-term and long-term interventions in building resilient communities. The framework reflects the culture of the study population and the interaction with the community that represents the larger societal structures. The framework is contextually and culturally appropriate and presents a multifaceted view of resilience while providing a broad spectrum of options (Jack & Kholeif, 2007).

6.4 Methods and Research Design

The study adopted a multi-method qualitative research approach with a comparative case study research design. The main objective of the study was to understand better resilience in rural Zimbabwe and the implications this has for climate governance. The design of case study research was vital for this type of study because it not only enabled research to analyse the relationship between phenomena, context and people but also provided great flexibility in data collection through various means while capturing the context and lived experiences of the participants (Tetnowski, 2015). Three research questions guided the research. What is the nature of resilience in rural Zimbabwe? What are the barriers to implementing resilience-building efforts? What are possible strategies that can be adopted to strengthen climate governance?

The authors used two different approaches at each study site. In Chimanimani, 24 key informants were purposively sampled based on their experience with the cyclone and its aftermath. The reason why Chimanimani was selected is because the region in Zimbabwe was the worst affected by Cyclone Idai in terms of death toll and infrastructure losses. The study did not cover all of Chipinge district but focused on parts of the district that were mainly affected by the cyclone, especially Ward 21 (Munsaka et al., 2021). Furthermore, the manner in which disaster governance and resilience exercises were conducted in Chimanimani in the aftermath of Cyclone Idai also made it a good case study with ample lessons for future governance and resilience building in Zimbabwe and the Southern Africa region in general.

Likewise, Matopos was also selected as the second case study because it is prone to droughts due to poor rainfall, arid climate and high temperatures in the Matabeleland south province. To identify critical informants for the study, the researchers used a snowball sampling technique (Parker et al., 2019). Initially, the researchers had three key informants from the Chimanimani community who were able to identify and introduce the researchers to other key informants. These key informants included a traditional leader (headmen), a ward councillor and a religious leader. The community comprised survivors and witnesses of the cyclone. These informants were critical in providing a balanced perspective on governance and resilience-building efforts from both an administrative and benefactor's perspective.

The study conducted in-depth interviews with the key informants that lasted 2 hours per session. These interviews were conducted between May 1, 2023, and May 26, 2023, by the University of Zimbabwe disaster management team accompanied by students and the researcher(s). The university students helped to collect data by interviewing members of the Rusitu and Kopa communities. However, authorisation was first obtained from the Chimanimani District Administrator's Office (DA) before entering the district for the research. However, a sample population of 24 informants may appear too small for a study of this size; since the research was qualitative, a small sample population was more appropriate for providing quality data while reducing the challenge of data saturation and redundancy (Kerr et al., 2010). In terms of data analysis, the study applied the thematic approach since it is suitable for qualitative data as it enabled analysis according to themes (Clarke et al., 2015). These thematic approaches focused on governance and resilience-building measures from a national, regional, communal and family level.

The study also used a focus group discussion (FGD) as a data collection tool (Nyumba et al., 2018). The FGD consisted of 12 participants, including community leaders (councillor, village head and headman). The community leaders only participated in the FDGs to triangulate the data that was gathered from the community members and through personal observations to make comparisons between the official standpoint and the community viewpoint. FGDs and interviews were both conducted because preliminary observations indicated direct contradiction between what community leaders said and what community members stated about the cyclone, governance and resilience efforts. Hence, to properly triangulate the collected data, the FGD only included community members (witnesses and survivors) except the community heads. This was meant to get a clear official opinion from the governing authorities on disaster governance and resilience measures that were implemented after the cyclone. The interviews with community members were meant to corroborate the authenticity of the information provided by community leaders.

In Matopos, eight key informant interviews were conducted with village leaders, and these were complemented with a participant-coded narrative approach, which is a qualitative approach that allows research participants to be involved in the coding of their data, strengthening reliability as well as participation in the research process (Blaser Mapitsa, 2023). Respondents were randomly sampled in Wards 16 and 17 to give a broadly representative view of that population. The participants themselves then identified the themes that were used to analyse the data.

The researchers adhered to all ethical protocols, including seeking ethical clearance from the Ethics Review Board of Marondera University of Agricultural Sciences and Technology. All informants for the study were first asked to agree to participate in the study and were informed about the study. The researchers also protected the identity of key informants who requested to remain anonymous using pseudonyms in Chimanimani and by storing data with redacted identifying information in Matopos.

6.5 Presentation of the Findings

6.5.1 Stakeholder Roles and Mandates

It was clear at both sites that there was a divergence in understanding the roles and mandates of different individuals and organisations in the climate governance space, including functions from planning to project implementation around adaptation or disaster response. To contextualise the stakeholder response, nearly half of the Matopos respondents reported that they did not have someone they could turn to for support to respond to climate shocks or stressors, and the half that did have sources of support turned mainly to family and neighbours. However, local government, NGOs and other organisations also played an important role in the response related to climate stress.

One of the critical roles of local authorities in collaboration with the Government of Zimbabwe (GoZ) in building resilience was the construction of houses for survivors and those affected by the cyclone. During the FGD, the primary respondents included the Ward 21 councillor and village head. Ward 21 lost more than 248 residents, including entire families. Various stakeholder groups, including the Air Force of Zimbabwe (AFZ), Civil Protection Unit (CPU) and the Zimbabwe National Army (ZNA), were heavily involved in assisting the affected, especially in terms of providing emergency aid to survivors, search and rescue and distributing emergency aid. The GoZ also provided presidential input (seeds, fertilisers and agricultural implements) to revitalise community agricultural activities.

Women and people living with disabilities were shortlisted as priority groups during community resilience-building efforts, as they were considered the most affected. However, girls were left out of priority groups. According to Councillor Dhliwayo, 400 women were allocated houses despite the fact that only 55 had small deeds prior to the cyclone. Chimanimani Ward 21 had residential properties that were allocated by the Chimanimani council, which had provided title deeds to the owners of the residential houses. These houses were mainly owned by civil servants who operated in the community, especially teachers, agricultural extension officers, police and health workers. However, most of these houses were destroyed by the cyclone, and the Chimanimani council had the records of the property owners. The selection of these women was based on the head of the household. Therefore, all female-headed households were prioritised both in facto and de jure female-headed households.

Currently, there is a move away from a needs-based approach to a human rights-based approach (HRBA) in climate change programming. The HRBA recognises and empowers the voices and knowledge of all people, with an emphasis on vulnerable groups (Gabel, 2016), to change their situation of vulnerability. In this case, people's right to shelter, food security, access to health, water, employment and dignity were drastically affected by the anthropogenic climate change-driven disaster (Gatenio Gabel & Gabel, 2016). The state and nonstate actors, the duty bearers, have a legal and moral obligation to fulfil the human rights of those who are vulnerable to the impacts of climate change. Hence, to address the rights of the victims, women were given houses because they were considered vulnerable because most of these women had lost their husbands to the cyclone and were left to fend for their families. Other women were organised into groups to embark on projects operated at the Ngorima clinic. These groups received sawing machines to help them start sustainable livelihood projects. However, there was much disempowering of women during these processes. In a properly executed HRBA, the main goal is to empower those in vulnerable situations, including women, to effectively participate at each stage of the climate change programming process. However, these women were not consulted, and what they were assisted with in order to address their issues was a far cry compared to what they had lost.

Furthermore, in situations where the rights of vulnerable groups are violated, there should be room for redress mechanisms. However, such mechanisms were neither established nor considered. Men who were left vulnerable, like their female counterparts, were also assisted with empowerment projects to help them adapt to the post-disaster phase. They were given grinding mills and photocopiers to start small businesses. Furthermore, other local men were offered employment by various organisations that were embarking on reconstruction projects to financially empower them and their families.

On a rather sad note, some female survivors claimed that they had to offer sexual services to the local and traditional leadership in exchange for food hampers. Mrs. Chidawaredza (pseudonym) stated that the headmen requested sexual favours with unmarried women and that those who refused were denied aid. However, it was not only the local and traditional leadership that was alleged to have abused their authority. Personnel from various NGOs operating in the area were also accused of ill-treating survivors and eschewing funds and resources. These actions only worsened the trauma of the cyclone-affected persons. State operatives, especially soldiers, were also accused of serious abuses including misappropriation of resources, beating, raping and impregnating affected persons. These malpractices resulted in the breakup of marriages, the spread of sexually transmitted infections (STIs) and a general breakdown in the disaster response effort. Although the authors did not manage to speak with NGOs to authenticate these allegations, they managed to speak with serving members of the ZNA, which was actively involved and its members were implicated in these abuses. It was revealed that the ZNA fired several of its employees for misconduct during Chimanimani's interventions (military personnel 2023).

The local authorities established an early warning system to protect against future disasters. Megaphones were given to community members for convenience in disseminating information during disasters. District development coordinators (DDCs) in the area received special radios that broadcast early warning communication about emerging disasters. The DDCs then relayed the information to the entire community. In addition, three evacuation centres were established in the area at Rusitu High, Dzingire and Ndakopa Primary School.

Several non-governmental organisations (NGOs) provided food aid, psychosocial support and emergency resources. These included Mercy Corps, World Vision, Plan International, Family Aids Caring Trust (FACT), Jairos Jiri, World Food Programme (WFP), Angel of Hope, International Red Cross (IRC) and Red Crescent Movement (RCM), among others. These organisations provided aid by airlifting the resources since all roads had been destroyed. Ward 21 has five schools that serve as temporary rehabilitation centres. Priority was given to cyclone victims with chronic illnesses such as high blood pressure, high blood sugar and asthma. These victims were assisted with drugs and food hampers that were differentiated according to the effect of the cyclone on the victims and the nature of their chronic illness. Emergency assistance was also extended to community members as they were considered vulnerable since there were no stores in the area shortly after the cyclone. Religious organisations, including the Roman Catholic, Seventh Day Adventist and Anglican churches, provided food aid, blankets, clothes and toiletries for the affected persons and survivors.

In terms of resilience building, the multi-stakeholder interventions were mainly short-term measures that only addressed the immediate challenges but did not address the long-term needs of the community. This section was guided by the research question: How effective were the multi-stakeholder governance and resilience-building efforts in the aftermath of Cyclone Idai? Data from this section were mainly drawn from recipients of disaster response aid, and they differed significantly from the official point of view highlighted above. The study results generally indicated that the disaster response intervention was labelled by most affected people and survivors as ineffective. Community members stated that they only received food supplies during the first few months of the disaster and afterwards received nothing. Also, the clothes they received, were either too big or too small since the leaders and their relatives appropriated the best clothes.

Furthermore, one key informant posited: 'The cyclone destroyed our businesses, fields, homes, and property... It created a massive gap in terms of recovering to our initial production levels. Currently, we are only living from hand to mouth until we can get back on our feet'. This scenario was worsened by the handling of government-sponsored resilience and governance initiatives, which the people claimed were politicised and discriminatory. The presidential input scheme, for example, failed to benefit all the community members since only supporters of the ruling party were given the inputs, while other community members were denied access.

One of the challenges that affected the resilience-building efforts was the allocation and distribution of new housing and residential stands to affected persons and survivors of the cyclone. Survivors of the cyclone highlighted that the new houses and residential stands were not distributed relatively. One community member also stated that she spent 2 years constructing a two-room house without any assistance. To make matters worse, she had to stop her two children from attending school to raise the funds to build the house. These observations are a direct contradiction to the councillor's claim that all affected persons and survivors were allocated stands. At Kopa, only former landowners received new stands and houses, while their tenants received nothing. A servivor of the cyclone stated that, 'We were forced to leave the relief centres with nowhere to go before we were allocated stands, and they promised us stands, but up to now, we have yet to receive anything'. The logic for evicting cyclone-affected people from the relief centres was to conceal the fact that the stands they had been allocated had already been allocated to undeserving people.

Furthermore, contrary to the official claim that schools and roads were reconstructed, it was observed that these interventions were not completed due to a lack of resources. For example, reconstruction at the Dzingire primary was halted due to a lack of resources. Road construction is still incomplete due to resource constraints. Furthermore, in addition to food donations, no livelihood projects were introduced to help people with disabilities (PLWD).

Community leaders agreed that the occurrence of the cyclone, though alarming, was also a blessing in disguise. They argued that before the cyclone, most of the roads in the area were impassable; however, the reconstruction efforts, though incomplete, managed to improve the area from its initial condition before the cyclone. On a similar note, some community members also saw the cyclone as an opportunity for local business owners. Webster Dengende, a shopkeeper, also argued that the cyclone benefited his business because it led to the further discovery of alluvial gold deposits that attracted gold panners whose activities created an increased local demand for essential goods and services.

6.5.2 Matching the Scale of the Problem and Response

One of the challenges of resilience building in Zimbabwe is that the scope and mandate of organisations that can build resilience and respond to natural shocks and stressors do not always have the required level of work to respond to the problem. As demonstrated in Chimanimani, while a number of NGOs and government departments have a broader mandate, the front line of disaster response was local organisations with a presence in the community. Although this helped reach households who needed assistance, often, the nature of the problems was not local. For example, people with chronic conditions had difficulties accessing medication due to interrupted supply chains. Although responders could occasionally source medicine from far-away locations, repairing supply chain issues is outside the scope of their work. Similarly, in Matopos, drought had a number of complex causes. However, few organisations responding to the context of hunger or acute shortage were able to influence the climactic patterns elsewhere in the region.

When considering the complex system of the Limpopo River Basin, where upstream activities in Matopos or Chimanimani affect ecosystems in Mozambique, it is even more apparent that collaboration needs to occur across borders and at multiple scales of work. However, we found little evidence that these systems are being built or given the capacity required for effective, inclusive, polycentric governance. The implications of this are discussed in the following section, where institutional incentives for planning remain fragmented and isolated.

6.5.3 Stronger Incentive Systems for Integrated Planning

It is apparent from both systems in crisis, post-cyclone in Chimanimani, as well as those nearing tipping points, such as in Matopos, that both emergency responses and ongoing developmental support are not taking place in a consistently or systematically integrated fashion. As an example, there is a lack of integration between land use planning, service provision and disaster response. Several affected persons and survivors of the cyclone claimed that they were allocated stands by the Chimanimani council within the convergence zone. This was a direct breach of a regulation that stipulates that buildings should be constructed at least 50 meters away from rivers and water bodies. In Kopa, however, houses, shops and local administrative offices were constructed in the midst of the convergence zone, where more than 200 lives were lost by being swept away by the flooded rivers, including houses, offices, vehicles and property. According to Kazambara (pseudonym) a local resident and survivor of the cyclone:

> 'There were more than 280 missing people in Kopa and, to avoid exposure, the authorities did not recognise these victims because they were responsible for settling them in a hazardous location. This was done to avoid litigation. Two communities in Chimanimani were mainly affected by the cyclone, including Kopa and Rusitu villages. In Rusitu vil-

lage, residents settled in a safer region than in Kopa. Therefore, to protect itself from public scrutiny, the state media emphasised the effects of the cyclone in Rusitu village, where far fewer victims were affected since they were settled in a proper settlement region. Consequently, disaster response efforts were focused more on Rusitu, where little damage had been done….'

Buildings, including a police station, veterinary offices, shops and teachers' and agricultural extension officers' houses, were situated in the high-risk area and were all swept away with the occupants. The authorities suppressed the information by preventing journalists and reporters from accessing the affected area. These efforts concealed the fact that people were once settled at the epicentre of the cyclone, thus ultimately denying them a chance to access help from stakeholders. Therefore, the disaster could have been avoided if the stands had been allocated away from the confluence of major rivers (Rusitu, Nyahode and Chipita) in the Rusitu Valley. Furthermore, the establishment of the three evacuation centres in Dzingire, Ndakopa and Rusitu was not an adequate measure of disaster mitigation and resilience. Most community members claimed that they did not know that those three schools were designated local evacuation centres. Access to information is a fundamental human rights issue. The Paris Agreement recognises the importance of access to information when responding to climate change.

Furthermore, community members also dismissed the use of megaphones as an early warning system. Instead, they stated that the only early warning system available to them is the Econet unsolicited short-text message service (SMS) through mobile phones. However, these mobile alerts are affected by poor network connectivity and lack of electricity in the community.

6.6 Discussion of the Findings

One of the challenges of resilience building in Zimbabwe is that the scope and mandate of organisations that build resilience and respond to natural shocks and stressors do not have the required level of competence and resources. The majority of the population in Southern Africa, approximately 56%, lives in rural settlements and is at least partially dependent on the natural environment for their livelihoods (Helen Suzman Foundation, 2024). Mugambiwa (2021) states that climate change and its associated stressors have a significant impact on human development by supporting or destroying the livelihood systems of the most impoverished and vulnerable populations. The livelihood strategies of farmers and the issues they face in relation to climate change exhibit significant variation on a global scale (Nicholls et al., 2007). One approach employed by farmers to manage the impacts of both climatic and non-climatic shocks is the diversification of household incomes (Brown et al., 2006).

6.6.1 Resilience Strategies

When considering the complex system of the Limpopo River Basin, where upstream activities in Matopos or Chimanimani affect ecosystems in Mozambique, it is even more apparent that collaboration needs to occur across borders. Active participation of stakeholders in decision-making processes related to climate resilience is fundamental. This participatory approach ensures that the strategies developed are context-specific and reflect the needs and concerns of people living in rural communities. The participation of affected and interested parties ensures a sustainable, holistic, integrated approach (Jagers & Stripple, 2003). Participation provides an opportunity for revisiting social partnerships and developing shared goals among citizens, civil society, the state and the private sector. This will promote mutual accountability between stakeholders.

6.7 Challenges Encountered in Building Resilience

6.7.1 Ecological and Environmental Barriers

Recurring cyclones in Chimanimani have exacerbated vulnerability in the district. Between January 2019 and March 2023, six cyclones affected the region (World Data, 2023). The occurrence of cyclones is a major issue that restricts the adoption of climate-resilient agricultural techniques and endangers livelihoods depending on their magnitude (Munsaka et al., 2021). Dryland regions, as is the case of Matopos (Greenberg et al., 2022), are characterised by severe climatic conditions, such as elevated temperatures and erratic precipitation patterns. The impact of climate change has intensified preexisting extreme weather phenomena, resulting in increased levels of food insecurity and a persistent decrease in agricultural productivity.

According to a report by the United Nations Development Programme (UNDP) in 2021, it is projected that if water stress patterns persist in some African countries, a significant number of individuals, ranging from 75 to 250 million, could be affected by water stress by 2030. Furthermore, there is a possibility of a substantial decline, up to 50%, in rainfed agricultural production (World Food Programme, 2021). In regions characterised by elevated temperatures and limited and unpredictable precipitation, small-scale farmers will face more significant challenges in sustaining their livelihoods. These changes have been recognised by the UNDP (2021) as exacerbating the vulnerability of their economic activities. Consequently, there is a need for more effort to adapt to and mitigate the consequences of droughts and temperature fluctuations.

6.7.2 Finance and Infrastructure Deficits

Building climate resilience in rural areas such as the Matatiele Local Municipality in South Africa often requires substantial financial resources and infrastructure development. Unfortunately, this has remained a hindrance to building improvements (Alfred Nzo District Municipality, 2015; Kruger & Shongwe, 2004). Ziervogel and Parnell (2012) attest to the absence of specific funding allocations to address climate change at the community level. In South Africa, there is a scarcity of local financial resources allocated for supporting climate change adaptation projects at the national, provincial and municipal levels (Sibiya et al., 2023). Zimbabwe, like any other country in the world, has been affected by the COVID-19 pandemic (Chitongo et al., 2020). It further impedes the ability to fund the Civil Protection Unit (CPU), which oversees disaster management.

6.7.3 Information and Awareness Gaps

The lack of current and readily available climate information has been a significant hindrance to informed decision-making in South Africa (Mubaya et al., 2012). Lack of information on the combined impacts of sea storms and rising sea levels has affected the potential adaptation strategies of small farmers and consequently limited their responses. Studies have indicated that farmers who have access to weather data and engage in community-based monitoring tend to be more aware of climate changes and are more likely to make the necessary adjustments. Cultural and cognitive barriers have been identified as discouraging factors for the use of external information (Shackleton et al., 2015; Nielsen & Reenberg, 2010). An examination of livestock producers in the Eastern Cape region of South Africa shows that more than half of the smallholder farmers' population have not implemented measures aimed at adapting to climate change with underlying factors such as inadequate access to climate-related information and insufficient knowledge regarding adaptive options and modern farming techniques (Motsa, 2011).

The challenges of climate governance emerging from the study show the low levels of adaptive ability in Chimanimani and Matobo districts, resulting from a multitude of socio-economic and environmental issues, which will continue to increase the intensity of the consequences of climate change encounters. Although there is a clear need for measures to promote climate resilience, identified challenges must be overcome through sustainable land management and adaptive strategies that can be implemented through participatory approaches. Adapting to climate change in the agricultural sector requires a comprehensive approach to climate protection. Of particular importance is the generation and dissemination of climate-resilient technologies and practices by investing in research and advisory services (such as heat-, flood- and drought-resistant crops and stress-tolerant livestock breeds). This requires investments in infrastructure and capacity building of global, regional, national and subnational research and advisory systems.

6.8 Conclusions

Governance mechanisms to build climate resilience in Zimbabwe are fragmented and, as a result, incomplete in their support to communities that are adapting to the effects of climate change. While there are structures that are theoretically in place for planning in a climate-responsive way, the practice is not reliably resulting in more resilient communities. A lack of communication and integration between the identification of floodplains or other disaster-prone ecologies and critical-related functions such as building permits and land use planning is preventing some of the sound policy directives from translating into more resilient communities. This is compounded by a mismatch between the mandate and capacity of some organisations that becomes apparent in times of crisis, such as time lags at Home Affairs when people are missing from a disaster. There is a mismatch between the scope and mandate of organisations responding to crises and the scope of the problem. Climate change is caused by global forces that are primarily outside the control of national crisis response organisations. As a result, central activities for integrated planning and prevention are challenging to coordinate and lie with transboundary and multilateral organisations whose timelines are often ill-equipped to deal with disasters. The contest for emergency response and resilience building in both Chimanimani and Matopos has underscored the need for greater alignment of incentives among different actors, from individual steps taken at the household level towards climate adaptation to government support and response. This is best addressed through the strengthening of integrated planning.

References

Ackerman, R. H., & Maslin-Ostrowski, P. (2002). *The wounded leader: How real leadership emerges in times of crisis*. Jossey-Bass.

Acosta, J. D., Chandra, A., & Madrigano, J. (2017). An agenda to advance integrative resilience research and practice: Key themes from a resilience roundtable. *Rand Health Quarterly, 7*(1), 5.

Africanews. (2022). *Zimbabwe ready for six more cyclones*. https://www.africanews.com/2022/02/10/zimbabwe-ready-for-six-more-cyclones//. Accessed May 18, 2023.

Alfred Nzo District Municipality. (2015). Climate change response strategy. Technical report. *Environmental Resource Management Department*. Matatiele, Eastern Cape, South Africa.

Bahadur, A. V., Ibrahim, M., & Tanner, T. (2010). *The resilience renaissance? Unpacking of resilience for tackling climate change and disasters. Strengthening Climate Resilience Discussion Paper 1*. Institute of Development Studies.

Blaser Mapitsa, C. (2023). Methods to make sense of resilience: Lessons from participant coded micronarratives. *Canadian Journal of Program Evaluation. 38*(1), 99–114.

Brown, D. R., Stephens, E. C., Ouma, J. O., Murithi, F., & Barrett, C. B. (2006). Livelihood strategies in the rural Kenyan highlands (December 1, 2006). *African Journal of Agricultural and Resource Economics*. Available at SSRN: https://ssrn.com/abstract=1846849

Chapungu, L. (2020). *Mitigating the impact of cyclone disasters: Lessons from Cyclone Idai*. Briefing paper, South African Institute of International Affairs (SAIIA).

Chitongo, L., Chikunya, P., & Marango, T. (2020). Do economic blueprints work? Evaluating the prospects and challenges of Zimbabwe's transitional stabilisation programme. *African Journal of Governance and Development, 9*(1), 7–20.

Clarke, V., Braun, V., & Hayfield, N. (2015). Thematic analysis. In J. A. Smith (Ed.), *Qualitative psychology: A practical guide to research methods* (pp. 222–248). SAGE Publications.

Conner, D. R. (1993). *Managing at the speed of change: How resilient managers succeed and prosper where others fail*. Villard Books.

Gatenio Gabel, S., & Gabel, S. G. (2016). Understanding human rights and social policy. In *A rights-based approach to social policy analysis* (pp. 17–38). Springer.

Gabel, S. G. (2016). A rights-based approach to social policy analysis. Springer International Publishing.

Greenberg, S., Drimie, S., Losch, B., & Jila, N. (2022). Agroecological initiatives in Matatiele and Umzimvubu, Eastern Cape. TAFS phase 2, final site report.

Helen Suzman Foundation. (2024). *The Southern African Development Community I – Population*. https://hsf.org.za/publications/hsf-briefs/the-southern-african-development-community-i-population

Jack, L., & Kholeif, A. (2007). Introducing strong structuration theory for informing qualitative case studies in organisation, management and accounting research. *Qualitative Research in Organizations and Management: An International Journal, 2*(3), 208–225.

Jagers, S., & Stripple, J. (2003). Climate governance beyond the state. *Global Governance, 9*(3), 385–399.

Jerie, S., & Ndabaningi, T. (2011). The impact of rainfall variability on rain-fed tobacco in Manicaland province of Zimbabwe. *Journal of Sustainable Development in Africa, 13*(1), 241–1250.

Kerr, C., Nixon, A., & Wild, D. (2010). Assessing and demonstrating data saturation in qualitative inquiry supporting patient-reported outcomes research. *Expert Review of Pharmacoeconomics & Outcomes Research, 10*(3), 269–281. https://doi.org/10.1586/erp.10.30

Kruger, A. C., & Shongwe, S. (2004). Temperature trends in South Africa: 1960-2003. *International Journal of Climatology, 24*(15), 1929–1945.

Maviza, A., & Ahmed, F. (2020). Analysis of past and future multi-temporal land use and land cover changes in the semi-arid Upper-Mzingwane sub-catchment in the Matabeleland south province of Zimbabwe. *International Journal of Remote Sensing, 41*(14), 5206–5227.

Mguni, N. L. (2021). *Water, sanitation, and hygiene in health care facilities: Status in rural areas of Matabeleland South Province of Zimbabwe*. University of Johannesburg (South Africa).

Motsa, B. (2011). Financial barriers to the implementation of climate change adaptation measures. In L. Masters & L. Duff (Eds.), *Overcoming barriers to climate change adaptation implementation in Southern Africa* (pp. 141–161). Africa Institute of South Africa.

Mubaya, C. P., Njuki, J., Mutsvangwa, E. P., Mugabe, F. T., & Nanja, D. (2012). Climate variability and change or multiple stressors? Farmer perceptions regarding threats to livelihoods in Zimbabwe and Zambia. *Journal of environmental management, 102*, 9–17.

Mugambiwa, S. S. (2021). Multilevel approach to climate change adaptation in rural Zimbabwe: A well-articulated framework devoid of effective implementation. *African Journal of Development Studies (AJDS), 11*(4), 165–181.

Munsaka, E., Mudavanhu, C., Sakala, L., & Manjeru, P. (2021). & Matsvange, D. (2021). When disaster risk management systems fail: The case of cyclone Idai in Chimanimani District, Zimbabwe. *International Journal of Disaster Risk Science, 12*, 689–699. https://doi.org/10.1007/s13753-021-00370-6

Nicholls, R. J., Wong, P. P., Burkett, V. R., Codignotto, J. O., Hay, J. E., McLean, R. F., Ragoonaden, S., & Woodroffe, C. D. (2007). Coastal systems and low-lying areas. Climate change 2007: Impacts, adaptation and vulnerability. In M. L. Parry, O. F. Canziani, J. P. Palutikof, P. J. van der Linden, & C. E. Hanson (Eds.), *Contribution of working group II to the fourth assessment report of the intergovernmental panel on climate change* (pp. 315–356). Cambridge University Press.

Nielsen, J. Ø., & Reenberg, A. (2010). Cultural barriers to climate change adaptation: A case study from Northern Burkina Faso. *Global Environmental Change, 20*(1), 142–152.

Nyumba, O. T., Wilson, K., Derrick, C. J., & Mukherjee, N. (2018). The use of focus group discussion methodology: Insights from two decades of application in conservation. *Methods in Ecology and Evolution, 9*(1), 20–32. https://doi.org/10.1111/2041-210X.12860

Parker, C., Scott, S., & Geddes, A. (2019). *Snowball sampling*. SAGE research methods foundations. https://doi.org/10.4135/9781526421036831710

Pradhan, C., & Chauhan, R. (2020). Disaster resilience for sustainable development in Nepal. In I. Pal, J. von Meding, S. Shrestha, I. Ahmed, & T. Gajendran (Eds.), *An interdisciplinary approach for disaster resilience and sustainability. MRDRRE 2017. Disaster risk reduction (methods, approaches and practices)*. Springer.

Risiro, J. (2019). The challenges of integrating indigenous knowledge in the teaching of weather and climate in geography in Manicaland province of Zimbabwe. *The Journal of Geography Education in Africa, 2*, 30–46.

Shackleton, S., Ziervogel, G., Sallu, S., Gill, T., & Tschakert, P. (2015). Why is socially-just climate change adaptation in sub-Saharan Africa so challenging? A review of barriers identified from empirical cases. *Wiley Interdisciplinary Reviews: Climate Change, 6*(3), 321–344.

Sibiya, A., van der Westhuizen, J., & Sibiya, B. (2023). Challenges Experienced by SMMEs and Interventions by the South African National and Provincial Government: A Literature Review. *African Journal of Inter/Multidisciplinary Studies , 5*(1), 1–11.

Tetnowski, J. (2015). Qualitative case study research design. Perspectives on fluency and fluency disorders, *25*(1), 39–45. https://doi.org/10.1044/ffd25.1.39

World Data. (2023). *Cyclones in Zimbabwe*. https://www.worlddata.info/africa/zimbabwe/cyclones.php. Accessed May 18, 2023.

World Food Programme (WFP). (2021). *Climate Change in Southern Africa*. https://executiveboard.wfp.org/document_download/WFP0000129015#:~:text=Impacts%20of%20changes%20in%20temperatures%20and%20rainfall%20patterns%20are%20already,intensity%20of%20droughts%20and%20floods

Ziervogel, G., & Parnell, S. (2012). South African coastal cities: governance responses to climate change adaptation. In Climate change at the city scale (pp. 223–243). Routledge.

7 The State of Climate Resilience in Rural Communities in Southern Africa: Case Study of Dryland Regions of Zimbabwe

Michael T. Mubvuma, Dorcas Chapungu, and David Chikodzi

7.1 Introduction

Climate change refers to a change in climate over a long period of years (30–35 years). Climate in Southern Africa is changing (Dahwa et al., 2022), and the impacts are mostly felt particularly in rural communities because they rely on rain-fed agriculture as the main livelihood activity (Dibakoane et al., 2022), limited adaptive capacities and high sensitivities. The climate of Southern Africa exhibits high spatial and temporal variability. Over the years, the frequency and severity of drought, floods and cyclones have increased as a result of changing climate. The duration of the growing season is shorter for most staple food crops due to climate change (Mubvuma, 2013; Ngoma et al., 2023). Indeed, the Southern African region is bearing the brunt of climate change due to high exposure to intensified climate-related natural disasters, precipitation anomalies and temperature extremes (Tim et al., 2023; Chivangulula et al., 2023; Mashao et al., 2023).

In addition, close to one-third of droughts in the world have been recorded in Southern Africa. Climate change is responsible for more than 1000 deaths (Addaney et al., 2023), while 13 million people have been seriously affected by climate change in Southern Africa (Menashe-Oren et al., 2023). The increase in the frequency of drought, floods and other climate extremes is compromising food, water, housing, health and nutritional security in numerous households with grave concerns about outbreaks of diseases (Chivangulula et al., 2023; Simanjuntak et al., 2023). Furthermore, climate change has worsened economic challenges in Southern Africa due to the over-reliance on dryland agriculture as the primary source of livelihood in Southern Africa (John et al., 2023). The poor economic situation, particularly the record of skyrocketing inflation in some countries (Udeagha & Breitenbach, 2023), and poverty have worsened the situation.

The climate-induced challenges have caused market instability and price distortions and also resulted in many people failing to purchase food and other livelihood measures (Udeagha & Breitenbach, 2023). The havoc caused by climate change has raised serious questions about whether the region is resilient enough to face the full wrath of climate change. Limited studies have reported on the resilience status of rural

M. T. Mubvuma (✉)
Gary Magadzire School of Agriculture and Engineering, Great Zimbabwe University, Masvingo, Zimbabwe
e-mail: mmubvuma@gzu.ac.zw

D. Chapungu
Faculty of Arts, Theology, Culture and Heritage Studies, Reformed Church University, Masvingo, Zimbabwe

D. Chikodzi
School of Natural Sciences, Great Zimbabwe University, Masvingo, Zimbabwe

M. Matsa et al. (eds.), *Climate Change Resilience in Rural Southern Africa*, Sustainable Development Goals Series, https://doi.org/10.1007/978-3-031-73600-1_7

communities to climate change in Southern Africa. In fact, hardly any study has reported on the resilience of rural communities in Southern Africa. Where attempts were made to report on the resilience of communities in the region (Theron, 2013; Bahta, 2022; Nhemachena et al., 2020; Muchaku et al., 2023; Nyamwanza et al., 2023), the studies did not exhaustively evaluate the importance of the resilience index and its components in the region. Furthermore, the relationship between the resilience index and important livelihood activities such as access to informal safety nets, access to markets, use of early warning systems and livestock and cash crop sales at the household level has rarely been reported in the literature (Chivangulula et al., 2023; Simanjuntak et al., 2023).

Resilience is "the ability of at-risk individuals, households, communities, and systems to anticipate, cushion, adapt, bounce back better, and move on from the effects of shocks and hazards in a manner that protects livelihoods and recovery gains and supports sustainable transformation" (UNDP, 2019a, b page 12). Over the years, studies have recommended a resilience approach to development in communities, where the focus is placed on understanding the shocks and stresses experienced by communities. Shocks are simply high-impact and short time-bound events that are usually of limited duration and characterised by sudden onset (Carlson et al., 2012; UNDP, 2019a, b). Stress refers to slow-onset events, changes and longer-term dynamics that are more extended disruptions and can have high impacts (UNDP, 2019a, b). The resilience capacity of rural communities in Southern Africa is adversely affected by climatic and socio-economic shocks such as climate change, drought, hyperinflation, diseases and market instability (Carlson et al., 2012).

The urgent need to evaluate the level of resilience in rural communities in Southern Africa cannot be overly emphasised. The literature on the state of components of the resilience index, such as absorptive capacities, adaptive capacities and transformative capacity index of rural communities in Southern Africa, is still limited (Chivangulula et al., 2023; Simanjuntak et al., 2023). The resilience capacity of rural communities, when exposed to climate change, depends on the type of climate change-related shocks, the prevailing economic conditions of the country and a set of policies to enhance the resilience of communities to climate change-related shocks (Mentges et al., 2023). Hardly any study has systematically reviewed the state of resilience in rural communities in Southern Africa against climate and socio-economic shocks (van Aswegen et al., 2020; Bahta, 2022). Moreover, relationships between resilience capacity index parameters and socio-economic shocks have not been fully exhausted to understand options to improve the livelihood of people in rural communities (Bahta, 2022).

Furthermore, there is a lack of information on the relationship between food crops and livestock production and the resilience capacity index of rural communities in Southern Africa when subjected to climate change-related shocks (van Aswegen et al., 2020; Bahta, 2022). Therefore, there is a need for studies that have a deep understanding of the resilience state in rural communities in Southern Africa against climate change and socio-economic shocks. Furthermore, it is necessary to unpack the relationship between resilience index parameters and climate change-related shocks in Southern Africa.

The main objective of this study was to determine the state of the resilience of communities in rural areas of drylands areas of Zimbabwe in the Southern African region and to determine the relationship between components of the resilience capacity index and the total resilience index in rural areas when subjected to climate change-related shocks. The research aims to assess the current state and various components of resilience in rural communities located in the drylands of Zimbabwe. It aims to evaluate the contribution of food crops and livestock to the resilience capacity index of rural communities in Southern Africa when faced with climate change-related shocks. Furthermore, the study will explore the relationship between food crops and livestock production and the resilience capacity index of rural communities in the drylands of Zimbabwe under similar climate change-related shocks.

7.2 Literature Review

7.2.1 Components of Resilience in Southern Africa

The resilience capacity index is made up of three major components, namely, absorptive capacity, adaptive capacity and transformative capacity index. Absorptive capacity is an important key driver of the resilience capacity index (Beynon, 2023) and measures the level of communities that drive resilience and the ability to absorb shocks and survive beyond the impacts. The absorptive capacity of a rural community is defined in terms of the ability of rural communities to minimise exposure to expected future shocks and stresses and how the community can recover quickly when subjected to shocks and stresses. Literature on the state of components of the resilience index, such as absorptive capacities, adaptive capacities and transformative capacity index of rural communities in Southern Africa, is still limited (van Aswegen et al., 2020; Bahta, 2022).

The adaptive capacity of rural communities measures the ability of communities to adapt their way of life and survive after being exposed to shocks and stress. The adaptive capacity of the rural community is built on alternative livelihood strategies based on changing conditions. Research, innovations and cultural interventions are important key drivers to improving the adaptive capacity of rural communities. In the study, the adaptive capacity index was calculated from seven components: access to informal safety nets, social capital bonding, cash savings, access to remittances, access to humanitarian assistance, asset ownership score and shock preparedness. ZRBF defined access to informal safety nets as the total number of community organisations that are active in providing safety nets to households within communities. Bonding social capital was defined as a measure of whether the household could seek help from relatives or non-relatives within their communities when faced with shock or stress that need money or food. The cash saving component was taken as a binary variable, and the likelihood of having a person with cash savings (i) at home or no cash (0) was measured.

Similarly, access to remittances was taken as a binary variable and measured the likelihood that a household would receive remittances on a regular (1) basis or not regularly (0). The component of access to humanitarian assistance was defined as a measure of a household having received assistance either from the government or NGOs within 12 months. Asset ownership was calculated from 9 livestock assets and 25 productive assets. Thus, asset ownership was constructed from 34 assets, and each asset contributed a single score. Shock preparedness and mitigation were defined as a measure of response to three important questions: How prepared is the household or community to have programmes that provide emergency food or cash assistance when exposed to the shock? The second question was evaluated to determine whether the community has a clear disaster management/response programme in place. The third question evaluated whether the community had an action adaption planning or a resilience committee.

The transformation capacity index measures the ability of a rural community to transform and change in a positive way to improve its resilience capacity. It involves policies/regulations, governance mechanisms, community networks and infrastructure and covers both formal and informal social protection mechanisms that constitute an enabling environment for systemic change. The transformative capacity index was calculated using seven components, namely, access to markets, access to formal safety nets, access to essential services, access to infrastructure, access to agricultural services, gender norms and collective action. Access to infrastructure and access to essential services were calculated to calculate the total number of infrastructure or essential services, respectively, available to households in communities; the gender norm component was constructed based on 13 individual women decision-making power questions related to buying, selling, spending money and participation in groups. The collective action component measures the benefits achieved from a community working as a group.

7.2.2 Resilience Framework

The study looked at the state of resilience in Southern Africa, based on the resilience framework. Important questions about resilience, such as "Resilience for whom? What is resilience for? Resilience to what? Resilience through what? Moreover, resilience and so what?", were answered in the study. The first question on resilience was answered by identifying rural communities in Southern Africa as our target group.

The question is: What resilience was affected by defining the study context in terms of social groups? The study looked at rural communities in the context of their main livelihood activity, agriculture and other related activities. The question: Resilience to what was solved by identifying a basket of shocks and stresses that are disturbing rural communities in their quest to achieve resilience. The identified shocks include natural hazards such as drought, low market prices of crops and livestock products.

7.3 Materials and Methods

7.3.1 Study Site

The study was carried out in dryland parts of Zimbabwe and in Southern African countries that include Angola, Malawi, South Africa, Zambia and Zimbabwe (Fig. 7.1). The study area is found in the tropics and subtropics. It is facing the negative effects of climate change in the form of increased/extreme temperatures, frequent drought, intense floods and cyclones. The study area also shares similarities in adaptability and resilience levels, as well as livelihoods dominated by agriculture.

7.3.2 Study Design

The study design used a combination of two methods. The first method purportedly analysed primary data collected by the Zimbabwe Resilience Building Fund (ZRBF), from field surveys in dryland parts of Zimbabwe during the years 2019–2021. The analysed data, together with published papers from ZRBF, were used to determine the different components of resilience capacity in dryland parts of Zimbabwe and ultimately to determine the contribution of food crops and livestock in dry rural communities of Zimbabwe on the resilience capacity index when subjected to climate change-related shocks and to determine the state of resilience in dryland parts of Zimbabwe. Zimbabwe was used as an example case study of countries in Southern Africa. The second method used was a systematic review of articles to determine the state of resilience in rural communities in Southern Africa when they are subjected to climate change-related shocks. The ZRBF survey and the systematic method of reviewing the literature are presented in the following sections.

7.3.2.1 Zimbabwe Resilience Building Fund (ZRBF) Survey

The study used a mixed-methods approach using both qualitative and quantitative methods of data collection. Quantitative data was used to determine the state of resilience of rural communities in Zimbabwe (as a case study of a Southern African country); therefore, data was collected from three output monitoring surveys conducted during the years 2015–2021 by the Zimbabwe Resilience Building Fund (ZRBF), while a systematic review of published data, drawn from Scopus database, was used to determine the state of resilience in other Southern African countries.

The ZRBF conducted three rounds of surveys at different times in the years 2019–2021. The first survey round (n = 3440) was conducted in 2019. The second survey round (n = 3353) was carried out in 2020, while the third and last survey round (n = 3353) was carried out in 2021. In all three surveys, FGDs and KII were used to triangulate the data collected at the household level (UNDP, 2020a, b).

7.3.2.2 Systematic Review Method

The systematic review registers were independently identified according to the updated guidelines of PRISMA (2020) (Page, 2020), performing a database search during the period from 1 July 2023 to 30 August 2023 (Fig. 7.2). The search process used inclusion criteria such as "Southern

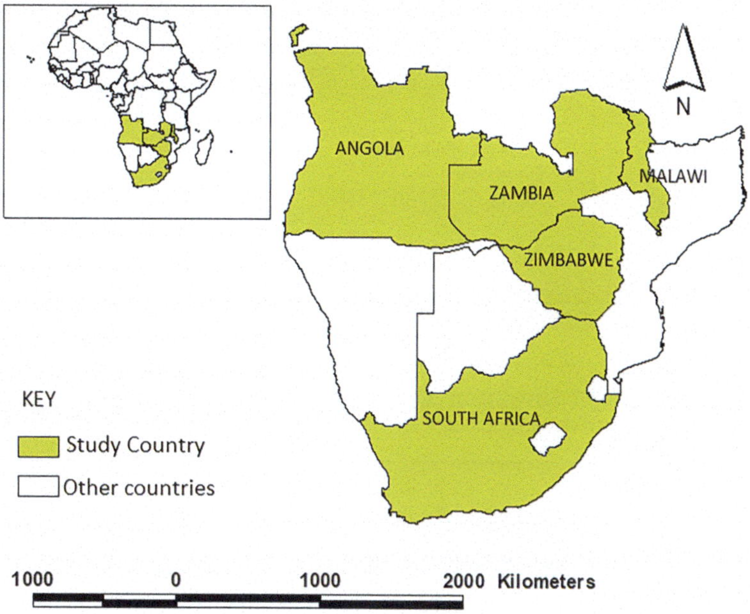

Fig. 7.1 Map of the study area

African rural community resilience and climate change", "climate shocks and rural resilience in Southern Africa" and "livelihood shocks and resilience capacity in rural Southern Africa". Thus, registration was defined as a process of searching relevant published peer-reviewed articles from the Scopus database. The Scopus database was chosen for the study based on the authors' availability. Figure 7.2 summarises how the process of identifying studies/research papers via databases and registers was conducted. The study identified reports/published papers through the Scopus database. Thus, a total of 1484 reports were selected for evaluation. The articles were screened, removing 144 duplicate reports. In addition, 1254 reports were removed because they were not relevant to the study area. Thirty-four reports were removed because they needed to address the general objectives of the study. Thus, 52 reports were successfully used for the review study.

7.3.3 Data Sources and Analysis

Data collected from ZRBF surveys included crop types, crop yield, crop sales, livestock numbers, livestock types, access to formal and informal safety nets, bonding social capital, cash savings, access to remittances, productive assets, access to humanitarian assistance, shock preparedness and mitigation, social capital, human capital, bridging social capital, linking social capital, livelihood diversification, access to financial services, exposure to information, access to markets, access to agricultural services, access to basic services, access to infrastructure, gender norm and collective action.

A regression analysis between the resilience capacity index and the resilience components (absorptive capacity, adaptive capacity and transformative capacity) was performed using SPSS software (version 24) with the objective of determining the relationship and contribution of each resilience component in building resilience of the rural community in the dry areas of Zimbabwe. The collected data from ZRBF was subjected to multivariate analysis and regression analysis. The contribution of food crops and livestock sales in rural communities in the dryland parts of Zimbabwe to the resilience capacity index when subjected to climate change-related shocks was determined by performing a Mann-Kendall correlation test.

Data on crops and livestock (numbers, sales) were linearly regressed against the resilience

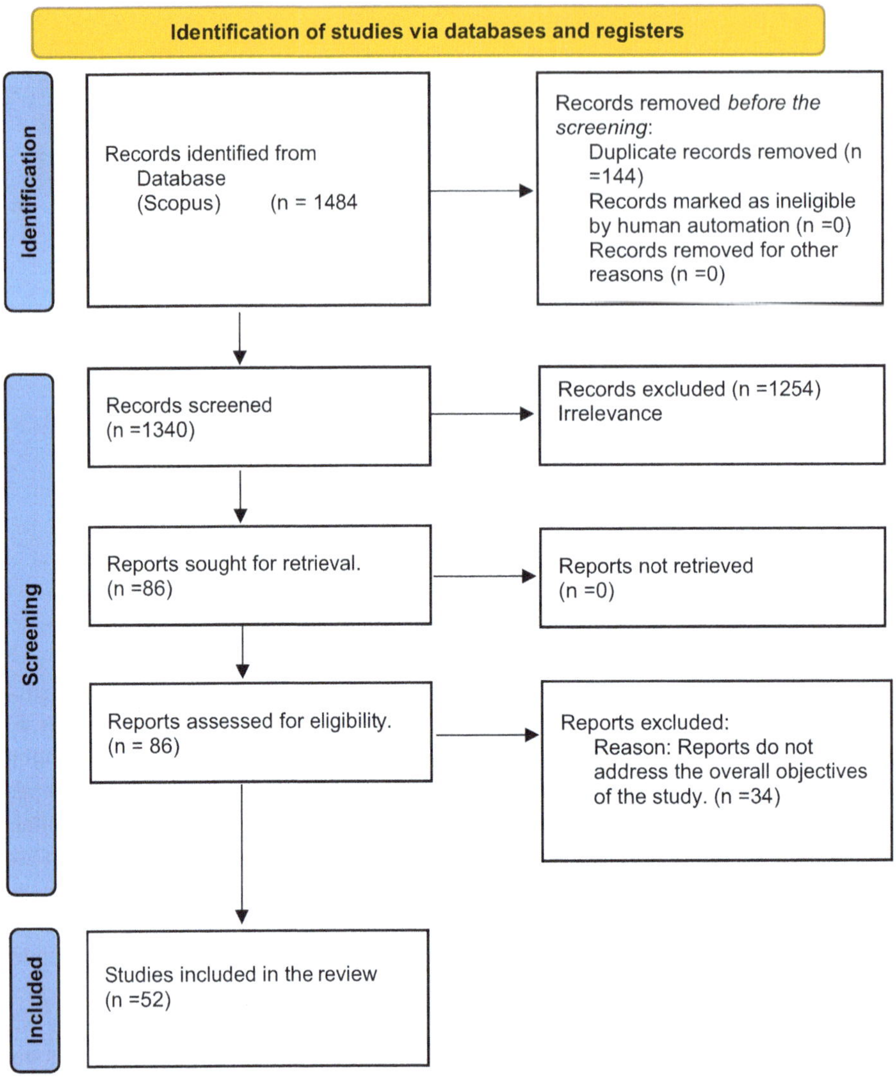

Fig. 7.2 PRISMA 2020 flow diagram: Adapted and modified (Page et al., 2021)

index (RI) and weighted with crop and livestock diseases, livestock theft, drought and low market prices as covariates (Berghof et al., 2021). Analysis of the results also included regression coefficient (R^2) and the probability of accepting at a 95% confidence interval. Linear regression and partial correlation analysis were used to evaluate the relationship between food crops and the resilience index in rural communities. Assessment of the existence of strategies that can increase the resilience of crops, markets or livestock-based livelihoods when communities are faced with environmental and socio-economic shocks was conducted using a nonparametric Mann-Kendall test and Pearson product-moment correlation test. Statistical correlation in food crops and livestock practices/strategies was evaluated using Kendall's tau, where a value $> +0.6$ was considered to have a strong positive relationship, while < -0.4 was considered to have a weak relationship (Hennemuth et al., 2013). The MK test was selected for use in the study following recommendations from previous studies: where Ngongondo et al. (2011), Mubvuma (2018) reported that the MK test has the advantage of being simple, robust and less sensitive to outliers and missing data. The MK test is less sensitive to abrupt changes in data trends that may be due to nonhomogeneity or linearity in the data set (Hennemuth et al., 2013; Noroozi et al., 2023; Pandey et al., 2023). The MK test is computed as follows:

$$Z_{MK} = \begin{cases} \dfrac{s-1}{\sqrt{\operatorname{var} s}} & \text{if } s > 0 \\ 0 & \text{if } s = 0 \\ \dfrac{s+1}{\sqrt{\operatorname{var}(s)}} & \text{if } s < 0 \end{cases} \quad (7.1)$$

S is defined as follows:

$$S = \sum_{i-1}^{n-1} \sum_{j=i+1}^{n} sign\left(x_j - x_i\right) \quad (7.2)$$

where x_j and x_i are sequential data values (Kendal, 1975; Mann, 1945; Noroozi et al., 2023; Pandey et al., 2023).

SPSS version 24 software was used to perform the MK test. The null hypothesis was tested at a confidence level of 95% for all parameters under evaluation (Gouda, 2015).

7.4 Results and Discussion

7.4.1 Components of the Resilience Capacity Index in Dry Areas of Zimbabwe

Table 7.1 shows the relationship between the resilience capacity index and different components of the resilience capacity of rural communities in the dryland parts of Zimbabwe. Data were measured by the ZRBF in Zimbabwe during the period from March 2019 to July 2021 of the three outcome surveys. All components of resilience capacity showed a positive significant regression relationship ($P < 0.01$) against the resilience capacity index, with adaptive capacity, absorptive capacity and transformative capacity index showing regression coefficients of 0.6720, 0.5831 and 0.6589, respectively. The results clearly suggest that a change in adaptive capacity, absorptive capacity or transformative capacity index would significantly change the resilience capacity of rural communities against climate change-related shocks.

Both the results of the ZRBF survey and 30% of the articles searched under systematic review (Bassongui & Younes, 2023; Bag et al., 2023; Tsaurai, 2023; Ismail et al., 2023; Zarepour Moshizi et al., 2023; Nyamwanza et al., 2023; Muchaku et al., 2023; Mthembu & Hlophe, 2020; Nkonya et al., 2023; Sema et al., 2023; Slayi et al., 2023; Twine, 2013) showed that the resilience capacity of people in the dryland parts of Zimbabwe was a function of absorptive capacity index, adaptive capacity index and transformative capacity index. Components of absorptive capacity included access to informal safety nets, livestock assets and the level of shock preparedness. Similarly, the components of adaptive capacity were human capital, access to financial services, exposure to information, livestock assets and production assets. In addition, the components for transformative capacity included bridging social capital, linking social capital, access to agricultural services, access to formal safety nets, access to markets, access to basic services, access to infrastructure and use of collective action.

The positive contribution of different components of resilience capacity clearly suggests that improving financial services, exposure to information, livestock assets and production assets, among other components, will indeed improve the resilience of rural communities (Saeed et al., 2023).

7.4.2 Effect of Crop, Livestock and Climate-Related Shocks on Resilience Index

Table 7.2 shows the contribution of food crops and livestock in rural communities in dryland parts of Zimbabwe to the resilience capacity index when subjected to climate change-related shocks. The results are from three surveys conducted by ZRBF during 2019, 2020 and 2021. Climate shocks in the form of drought, excessive rains and floods had a significant negative effect ($P < 0.02$) on the resilience of rural communities

Table 7.1 Relationship between the resilience capacity index and its components

Resilience components	Regression coefficient	P-value
Absorptive capacity index	0.6720	$P < 0.01$
Adaptive capacity index	0.5831	$P < 0.01$
Transformative capacity index	0.6589	$P < 0.01$
Combined resilience capacity index	0.6273	$P < 0.01$

in Zimbabwe during the period 2019–2021 (Table 7.2). This clearly suggests that when rural communities face climate shocks, their ability to anticipate, cushion, adapt, bounce back better and move on from the effects of shocks in a way that protects their livelihoods and recovery gains is compromised. This low resilience to climate shocks could be due to the over-reliance of communities on rain-fed agriculture, which is directly affected by climate hazards (Bahta, 2022). In addition, low resilience could also be due to inadequate early warning systems, limited information flow or unavailability of alternative livelihood activities. Indeed, several studies in Southern Africa have recommended the need for the use of early warning systems to capacitate and improve the resilience of rural communities against natural hazards related to climate change (Theron, 2013; Bahta et al., 2016; Kamara et al., 2018; Nhamo et al., 2019; van Aswegen et al., 2020; Bahta, 2021, 2022; Maganga et al., 2021; Bahta & Lombard, 2023).

The results of the study also showed that crops and livestock pests as shocks could affect the ability of rural communities to achieve resilience. The resilience of rural communities was significantly ($P < 0.02$) affected by the incidence of crop diseases or pests. Similar reports were noted by Nhemachena et al. (2020). Furthermore, the low prices for crops and livestock on the market had a significant effect on the resilience of rural communities. The death of livestock was amplified by diseases, and several cattle were killed. Low prices incapacitated farmers in their purchasing power to buy food, farming inputs and livestock drugs. Interventions to improve market access to rural communities may improve their resilience capacity. However, the effect of crop damage caused by wildlife or livestock theft did not affect the resilience of rural communities.

7.4.3 The Relationship Between Livestock and the Resilience Index in Rural Communities

Table 7.3 presents partial correlation results between different livestock and the resilience index in rural communities.

Rural communities that owned sheep and goats, cattle and donkeys showed greater resilience ($P < 0.05$) compared to communities that had pigs and poultry. Therefore, the ownership of pigs and poultry proved to be not helpful for rural communities in achieving resilience capacity. Indeed, the results clearly showed that livestock plays an important role in building resilience in rural communities. However, the type of livestock is also vital to determine the level of resilience of rural communities. Donkeys are important for tillage and transportation activities, particularly in short distances or in areas where the road network is very poor. In addition, the ownership of cattle helps rural communities with food (milk and meat) and provides transport and tillage. Cattle work as a store of value and may be sold to generate income to buy food in the event of an economic shock. Several studies have reported the importance of livestock in building resilience in rural communities in Zimbabwe (UNDP, 2020b; Flower & Panetta, 2010).

7.4.4 Correlation Between Food Crops and the Resilience Index in Rural Communities Under Drought Conditions

Table 7.4 shows the results of partial correlations between different food crops and the resilience index in rural communities under drought conditions. Rural communities that grow sorghum and pearl millet in drought-hit areas showed more

Table 7.2 Effect of crop, livestock and climate-related shocks on resilience index

	2019		2020		2021	
Climate and socio-economic shocks	Adjusted R^2	*P*-value	Adjusted R^2	*P*-value	Adjusted R^2	*P*-value
Climate shocks						
Drought	−0.482	0.001*	−0.326	0.018*	−0.471	0.002*
Excessive rains	−0.334	0.007*	−0.406	0.003	−0.343	0.000*
Floods	−0.303	0.000*	−0.437	0.000*	−0.371	0.000*
Crops and livestock shocks						
Crop diseases/pest	0.050	0.000*	0.042	0.009*	0.047	0.000*
Crop damage by wildlife	0.048	0.105	0.035	0.301	0.032	0.218
Livestock theft	−0.005	0.654	0.020	0.124	0.028	0.343
Low crop and livestock prices	0.024	0.294	0.007	0.231	0.013	0.065
Livestock death from diseases	0.141	0.213	0.423	0.004*	0.478	0.003*
Economic shocks						
Sharp price increases	0.337	0.000*	0.294	0.000*	0.421	0.000*
Low prices of agricultural products	0.347	0.006*	0.528	0.005*	0.363	0.0014*

*Significance at 0.05 level

Table 7.3 Correlation between livestock and the resilience index in rural communities under drought conditions

Control variables			Donkeys	Pigs	Sheep and goats	Cattle	Poultry
Resilience capacity index	Donkeys	Correlation	1.000				
		Significance (two-tailed)	.				
	Pigs	Correlation	0.005	1.000			
		Significance (two-tailed)	0.643	.			
	Sheep and goats	Correlation	0.569	0.438	1.000		
		Significance (two-tailed)	0.004	0.241	.		
	Cattle	Correlation	0.607	0.327	0.362	1.000	
		Significance (two-tailed)	0.003	0.355	0.212	.	
	Poultry	Correlation	0.238	0.439	0.132	0.378	1.000
		Significance (two-tailed)	0.429	0.724	0.581	0.286	.

resilience ($P < 0.05$) compared to communities that grow maize and groundnuts (Table 7.4). Thus, the cultivation of groundnuts and maize when faced with drought shock proved to be not helpful for rural communities in achieving resilience capacity. The results suggest that the cultivation of food crops plays an important role in the building of resilience in rural communities. However, the type of food crops is essential in determining the level of resilience of rural communities. Sorghum and pearl millet are drought-tolerant crops and thus perform well even when grown under water-limiting conditions. Several studies have reported the importance of millets in building resilience of rural communities in drylands of Zimbabwe (UNDP, 2020b). The growing of drought-escaping crops or varieties that are tolerant to drought has been recommended in the literature (Mubvuma, 2013).

7.4.5 Summary of the State of Resilience in Selected Southern African Countries

Of the 52 reports in the systematic review, four reported that rural communities in Zimbabwe are more prone to drought shocks (UNDP, 2019a, 2020b; Matenge et al., 2023; Zhu & Shang, 2023). Their low capacity to cope with drought

Table 7.4 Correlation between food crops and the resilience index in rural communities

Control variables			Sorghum	Maize	Groundnuts	Pearl millet
Resilience capacity index	Sorghum	Correlation	1.000			
		Significance (two-tailed)	.			
	Maize	Correlation	0.003	1.000		
		Significance (two-tailed)	0.0612	.		
	Groundnuts	Correlation	0.237	0.397	1.000	
		Significance (two-tailed)	0.002	0.429	.	
	Pearl millet	Correlation	0.674	0.327	0.427	1.000
		Significance (two-tailed)	0.004	0.00.5	0.001	.
		Correlation	0.528	0.684	0.676	0.532
		Significance (two-tailed)	0.001	0.006	0.004	0.003

stemmed from limited education levels, poor agronomic practices, poor market access facilitates and inadequate water facilities for animal and human drinking. According to the ND-Gain matrix index (G index score) (Cevik & Miryugin, 2023), Zimbabwe was ranked the 44th most vulnerable country and was the 188th readiness country in 2021. The high vulnerability rate coupled with low readiness clearly suggests the country's low capacity to respond to climate shocks. This could be the reason why the country has a low resilience capacity.

In addition, three publications (Cevik & Miryugin, 2023; Matenge et al., 2023; Zhu & Shang, 2023) highlighted the challenge of the inaccessibility of disaster recovery policies as a major cause of low resilience of rural communities to drought. However, two publications (Fowler & Panetta, 2010; UNDP, 2019b) reported that significant effort was being made at the policy level to develop irrigation infrastructure that would assist in agricultural inputs and promote climate-smart agriculture techniques so as to ensure food security even under drought shocks.

Similarly, in Zambia, two publications by Ndoma and Siachiwena (2023) reported that frequent and intense drought was the most prevalent shock affecting rural communities. Ndoma and Siachiwena (2023) reported that at the national level, rural communities were 24% resilient and 76% vulnerable to drought and climate change shocks. The cause of this vulnerability was the inadequacy of the resources of rural communities to withstand drought shocks, which accounted for 63% of the population. Better than most African countries, Zambia has a lower G index score of 132, a vulnerability rate of 0.48 and a readiness of 0.324.

In South Africa, 4 out of the 21 publications (Bahta et al., 2016; Theron, 2013; Kamara et al., 2018; Khoza et al., 2019) reported that the level of resilience capacity of rural communities was low due to low education level, inadequate training information, lack of land tenure, limited agro-processing training, limited access to the market and low income outside of the farm. South Africa was ranked 111th most vulnerable country and was the 120th readiness country in 2021. This ranking was based on the country's low G index, low vulnerability rate (0.39) and low readiness of 0.356.

In Malawi, 4 out of the 21 publications reported flooding as the most important climate shock. Mwalwimba et al. (2023) reported the lack of a flood vulnerability assessment framework in most rural communities and stressed that this was reducing the resilience capacity of rural communities. Furthermore, in a survey reported by Ferreira et al. (2021), it was observed that the number of poor people in rural Malawian communities (10.2 million) was 6.7 times greater than that of people in urban areas. These views were supported by Benson and De Weerdt (2023), who reported 30% rural poverty rates compared to 11% in urban areas. In addition, in 2021, Malawi had a high G index of 161, a high vulnerability rate of 0.542 and a low readiness of 0.294. This low readiness clearly suggests Malawi's low resilience capacity in terms of its ability to respond to climate change and other related disasters. However, despite the low resilience among rural communities in Malawi, significant efforts are being made to improve the situation. For example, Kayamba-Phiri et al. (2023) reported several key strategies, such as the development of early warning systems, the use of climate-smart agriculture and the construction of dams and mitre drains. In addition, recent studies have reported the development of a variable common flood resilience framework in rural communities in Malawi with the objective of improving their resilience capacity (Dewa et al., 2023).

In Angola, drought was identified as the leading climate shock that affects the resilience of rural communities. The country has a low resilience capacity to drought and climate change. In 2021, Angola had a high G index of 159, a high vulnerability rate of 0.510 and a low readiness of 0.268. Extreme drought conditions are affecting livestock farmers by causing water shortages, hunger, famines and the death of livestock. Serrat Capdevila et al. (2023) reported that rural communities in Angola were not yet resilient to drought and went on to present four ways in which communities were vulnerable to drought. The categories are (1) lack of adequate knowledge of the available water sources in the country, (2) insufficient investment in water resource infrastructure at the community level, (3) lack of efficient mechanisms to repair, maintain and guarantee the functionality of infrastructure at the community level and limited institutional capacity at various levels to prepare for, mitigate and respond to droughts.

7.5 Conclusions and Recommendations

This study examined the resilience status of communities in the rural drylands of Zimbabwe. It determined the relationship between the components of the climate resilience capacity index and the total resilience index. The study also systematically reviewed the resilience state in selected countries in Southern Africa. The findings show that rural communities in dryland parts of Zimbabwe are exposed to frequent and intense droughts that compromise their resilience to climate shocks. In rural communities, resilience was significantly affected by the incidence of crop diseases or pests. Farming activities such as crop and livestock production in terms of types and numbers were shown to affect the resilience of rural communities significantly. Owning cattle helps rural communities improve resilience. Communities that relied on indigenous food crops such as sorghum and pearl millet exhibited more resilience compared to those that only grew maize. Given the socio-economic and environmental similarities across the dryland regions of Southern Africa, these findings can be generalised for the region and the main factors influencing resilience, including education level, general training, access to information, land tenure, agro-processing training, distance to market and income diversification. The study recommends the enhancement of climate information systems and the mobilisation of resources to address the underlying causes of climate vulnerability from a policy and strategic point of view. However, as a limitation, not all Southern African countries were evaluated in the study. For future studies, we recommend evaluating dryland parts of all southern African countries.

References

Addaney, M., Kwadwo Dumenu, W., & Jarbandhan, D. B. (2023). Introduction to climate change, adaptation policy and resilient development in Africa. In *Climate change in Africa: Adaptation, resilience, and policy innovations* (pp. 1–9). Springer Nature Switzerland.

Bahta, Y. T., Jordaan, A., & Muyambo, F. (2016). Communal farmers' perception of drought in South Africa: Policy implication for drought risk reduction. *International Journal of Disaster Risk Reduction, 20*, 39–50.

Bahta, Y. T. (2021). Perception of agricultural drought resilience in South Africa: A case of smallholder livestock farmers. *Jàmbá: Journal of Disaster Risk Studies, 13*(1), 1–11.

Bahta, Y. T. (2022). Nexus between coping strategies and households' agricultural drought resilience to food insecurity in South Africa. *Land, 11*(6), 893.

Bahta, Y. T., & Lombard, W. A. (2023). Nexus between social vulnerability and resilience to agricultural drought amongst South African smallholder livestock households. *Atmosphere, 14*(5), 900.

Bag, S., Dhamija, P., Singh, R. K., Rahman, M. S., & Sreedharan, V. R. (2023). Big data analytics and artificial intelligence technologies based collaborative platform empowering absorptive capacity in health care supply chain: An empirical study. *Journal of Business Research, 154*, 113315.

Bassongui, N., & Ben Younes, H. (2023). Does absorptive capacity matter in the impact of immigration on innovation in Africa? Evidence from panel quantile regression with nonadditive fixed effects. *International Regional Science Review, 46*(2), 210–232.

Benson, T., & De Weerdt, J. (2023). *Employment options and challenges for rural households in Malawi: An agriculture and rural employment analysis of the fifth Malawi integrated household survey, 2019/10* (Vol. 40). International Food Policy Research Institute.

Berghoff, A. S., Sessa, C., Yang, J. H., Tsourti, Z., Tsang, J., Tabernero, J., Peters, S., Linardou, H., Letsch, A., Haanen, J., & Garralda, E. (2021). Female leadership in oncology—has progress stalled? Data from the ESMO W4O authorship and monitoring studies. *ESMO Open, 6*(6), 100281.

Beynon, J. (2023). *Who should pay? Climate finance fair shares* (No. 311). Center for Global Development.

Carlson, J. L., Haffenden, R. A., Bassett, G. W., Buehring, W. A., Collins III, M. J., Folga, S. M., Petit, F. D., Phillips, J. A., Verner, D. R., & Whitfield, R. G. (2012). *Resilience: Theory and Application* (No. ANL/DIS-12-1). Argonne National Lab.(ANL), Argonne, IL (United States).

Cevik, S., & Miryugin, F. (2023). Rogue waves: Climate change and firm performance. *Comparative Economic Studies, 65*(1), 29–59.

Chivangulula, F. M., Amraoui, M., & Pereira, M. G. (2023). The drought regime in Southern Africa: A systematic review. *Climate, 11*(7), 147.

Dahwa, E., Mudzengi, C. P., Mubvuma, M., Maravanyika, T., Chapungu, L., & Chikodza, E. (2022). Optimizing productivity in semi-arid dryland agriculture for developing countries: insights from zimbabwe. In *Climate change adaptations in dryland agriculture in semi-arid areas* (pp. 233–249). Singapore: Springer Nature Singapore.

Dewa, O., Makoka, D., & Ayo-Yusuf, O. A. (2023). Measuring community flood resilience and associated factors in rural Malawi. *Journal of Flood Risk Management, 16*(1), e12874.

Dibakoane, S., Siyongwana, P., & Shabalala, A. N. (2022). Vulnerability, impact and adaptation strategies of female farmers to climate variability. *Jàmbá: Journal of Disaster Risk Studies, 14*(1), 1302.

Ferreira, I. A., Salvucci, V., & Tarp, F. (2021). Poverty and vulnerability transitions in Myanmar: An analysis using synthetic panels. *Review of Development Economics, 25*(4), 1919–1944.

Fowler, B., & Panetta, D. (2010). *CARE international Zimbabwe: Improving small-holder farmers' access to basic financial services and agricultural input and output markets*. AKF Savings Groups Learning Initiative.

Gouda, M. A. (2015). Common pitfalls in reporting the use of SPSS software. *Medical Principles and Practice, 24*(3), 300–300.

Hennemuth, B., Bender, S., Bülow, K., Dreier, N., Keup-Thiel, E., Krüger, O., Mudersbach, C., Radermacher, C., & Schoetter, R. (2013). *Statistical methods for the analysis of simulated and observed climate data: applied in projects and institutions dealing with climate change impact and adaptation*. CSC.

Ismail, I. J., Amani, D., & Changalima, I. A. (2023). Strategic green marketing orientation and environmental sustainability in sub-Saharan Africa: Does green absorptive capacity moderate? Evidence from Tanzania. *Heliyon, 9*(7), e18373.

John, T. W., Sušnik, J., Masia, S., & Jewitt, G. (2023). Towards realisation of nexus-doing at the grassroots level: Water-energy-food governance assessment in the Songwe River Basin (Tanzania and Malawi). *Environmental Science & Policy, 150*, 103596.

Kamara, J. K., Akombi, B. J., Agho, K., & Renzaho, A. M. (2018). Resilience to climate-induced disasters and its overall relationship to well-being in southern Africa: A mixed-methods systematic review. *International Journal of Environmental Research and Public Health, 15*(11), 2375.

Kayamba-Phiri, F., Khulumbo, B., & Nkhulembe, M. (2023). *From response to preparedness: Enhancing community-led disaster risk management in Malawi*. International Food Policy Research Institute.

Kendal, A. P. (1975). A comparison of "influenza C" with prototype myxoviruses: Receptor-destroying activity (neuraminidase) and structural polypeptides. *Virology, 65*(1), 87–99.

Khoza, T. M., Senyolo, G. M., Mmbengwa, V. M., & Soundy, P. (2019). Socio-economic factors influencing smallholder farmers' decision to participate in agro-processing industry in Gauteng province, South Africa. *Cogent Social Sciences, 5*(1), 1664193.

Kobayashi, L., Coimbra, R., Goes Jr, A. M., Reva, V., Santorelli, J., Moore, E. E., Galante, J., Abu-Zidan, F., Peitzman, A. B., Ordonez, C., & Maier, R. V. (2020). AAST-WSES guidelines on diagnosis and management of peripheral vascular injuries. *The Journal of Trauma and Acute Care Surgery, 89*(6), 1183–1196.

Maganga, A. M., Chiwaula, L., & Kambewa, P. (2021). Climate-induced vulnerability to poverty among smallholder farmers: Evidence from Malawi. *World Development Perspectives, 21*, 100273.

Mann, H. B. (1945). Nonparametric tests against trend. *Econometrica: Journal of the Econometric Society*, 245–259..

Mashao, F. M., Mothapo, M. C., Munyai, R. B., Letsoalo, J. M., Mbokodo, I. L., Muofhe, T. P., Matsane, W., & Chikoore, H. (2023). Extreme rainfall and flood risk prediction over the East Coast of South Africa. *Water, 15*(1), 50.

Matenge, R. G., Parida, B. P., Letshwenyo, M. W., & Ditalelo, G. (2023). Impact of climate variability on rainfall characteristics in the semi-arid Shashe catchment (Botswana) from 1981–2050. *Earth, 4*(2), 398–441.

Menashe-Oren, A., Bocquier, P., Ginsburg, C., Compaoré, Y., & Collinson, M. (2023). The dynamic role of household structure on under-5 mortality in southern and eastern sub-Saharan Africa. *Demographic Research, 49*, 249–294.

Mentges, A., Halekotte, L., Schneider, M., Demmer, T., & Lichte, D. (2023). A resilience glossary shaped by context: Reviewing resilience-related terms for critical infrastructures. *International Journal of Disaster Risk Reduction, 96*, 103893.

Mubvuma, M. T. (2013). Climate change: matching growing season length with maize crop varietal life cycles in semi-arid regions of Zimbabwe. *Greener Journal of Agricultural Sciences, 3*(12), 809–816.

Mubvuma, M. T. (2018). Planting date as an adaptive strategy to improve yield of Chickpea (Cicer arietinum) under under climate change condition in Southern Africa https://univendspace.univen.ac.za/bitstream/handle/11602/1218/Thesis%20-%20Mubvuma,%20m.%20t.-.pdf?sequence=1

Muchaku, S., Magaiza, G., & Hamandawana, H. (2023). Translating indigenous knowledge into actionable climate-change adaption strategies: A case study of Maluti-a-Phofung Local Municipality, Free State Province, South Africa. *Sustainability, 15*(2), 1558.

Mthembu, A., & Hlophe, S. (2020). Building resilience to climate change in vulnerable communities: A case study of uMkhanyakude district municipality. *Town and Regional Planning, 77*, 42–56.

Mwalwimba, I. K., Manda, M. Z., & Ngongondo, C. (2023). *Towards a framework for flood vulnerability assessment for urban and rural areas in Malawi*. https://dx.doi.org/10.21203/rs.3.rs-2757046/v1

Ndoma, S., & Siachiwena, H. (2023). South Africans dissatisfied with government performance on child well-being.

Ngongondo, C. S., Xu, C. Y., Tallaksen, L. M., Alemaw, B., & Chirwa, T. (2011). Regional frequency analysis of rainfall extremes in Southern Malawi using the index rainfall and L-moments approaches. *Stochastic Environmental Research and Risk Assessment, 25*, 939–955.

Ngoma, H., Finn, A., & Kabisa, M. (2023). Climate shocks, vulnerability, resilience and livelihoods in rural Zambia. *Climate and Development, 16*, 1–12.

Nhamo, L., Mabhaudhi, T., & Modi, A. T. (2019). Preparedness or repeated short-term relief aid? Building drought resilience through early warning in southern Africa. *Water SA, 45*(1), 75–85.

Nhemachena, C., Nhamo, L., Matchaya, G., Nhemachena, C. R., Muchara, B., Karuaihe, S. T., & Mpandeli, S. (2020). Climate change impacts on water and agriculture sectors in southern Africa: Threats and opportunities for sustainable development. *Water, 12*(10), 2673.

Nkonya, E., Kato, E., Msimanga, M., & Nyathi, N. (2023). Climate shock response and resilience of smallholder farmers in the drylands of South-Eastern Zimbabwe. *Frontiers in Climate, 5*, 890465.

Noroozi, A. A., Homaei, M., & Torkaman, M. (2023). Analysis of the trends of climate parameters using Mann Kendal test (TFPW-MK) in Khuzestan province. *Journal of Environmental Science Studies, 8*(2), 6510–6521.

Nyamwanza, A., Bhatasara, S., Chimhowu, A., Jacobs, P., Macheka, L., Nwosu, C., Nezomba, H., Miller, M., Mafunga, T., & Maila, M. (2023) *Climate adaptation and sustainable rural health outcomes in Southern Africa*. University of Zimbabwe Publications.

Page, M. J., McKenzie, J. E., Bossuyt, P. M., Boutron, I., Hoffmann, T. C., Mulrow, C. D., Shamseer, L., Tetzlaff, J. M., Akl, E. A., Brennan, S. E., & Chou, R. (2021). The PRISMA 2020 statement: an updated guideline for reporting systematic reviews. *BMJ*, 372.

Pandey, V., Pandey, P. K., Chakma, B., & Ranjan, P. (2023). Influence of short-and long-term persistence on identification of rainfall temporal trends using different versions of the Mann-Kendall test in Mizoram, Northeast India. *Environmental Science and Pollution Research, 31*, 1–20.

Saeed, S., Makhdum, M. S. A., Anwar, S., & Yaseen, M. R. (2023). Climate change vulnerability, adaptation, and feedback hypothesis: A comparison of lower-middle, upper-middle, and high-income countries. *Sustainability, 15*(5), 4145.

Sema, R. P., Matelele, L. A., Maluleke, N. L., Phora, G., Mokoena, M. L., Netnou-Nkoana, N. C., Bairu, M. W., Gerrano, A. S., Khoza, L., Shargie, N., & Vernooy, R. (2023). *Strengthening farmer-managed seed systems in South Africa*. Bioversity International; Pretoria (South Africa): ARC; DARLLD.

Serrat Capdevila, A., Limones Rodriguez, N., Marzo Artigas, J., Wijnen, M. M. P., & Petrucci, B. (2023). *Resiliência à Seca e Segurança Hídrica no sul de Angola (Portuguese)*. World Bank Group.

Simanjuntak, C., Gaiser, T., Ahrends, H. E., Ceglar, A., Singh, M., Ewert, F., & Srivastava, A. K. (2023). Impact of climate extreme events and their causality on maise yield in South Africa. *Scientific Reports, 13*(1), 12462.

Slayi, M., Zhou, L., & Jaja, I. F. (2023). Exploring farmers' perceptions and willingness to tackle drought-related issues in small-holder cattle production systems: A case of rural communities in the eastern cape, South Africa. *Applied Sciences, 13*(13), 7524.

Theron, L. C., & Theron, A. (2013). Positive adjustment to poverty: How family communities encourage resilience in traditional African contexts. *Culture & Psychology, 19*(3), 391–413.

Tim, N., Hünicke, B., & Zorita, E. (2023). Simulated rainfall extremes over southern Africa over the 20th and 21st centuries. *Natural Hazards and Earth System Sciences Discussions, 2023*, 1–22.

Tsaurai, K. (2023). Foreign direct investment-growth nexus in BRICS: How relevant are the absorption capacities? *Scientific Annals of Economics and Business, 70*(1), 41–55.

Twine, W. (2013). Multiple strategies for resilient livelihoods in communal areas of South Africa. *African Journal of Range & Forage Science, 30*(1–2), 39–43.

Udeagha, M. C., & Breitenbach, M. C. (2023). The role of financial development in climate change mitigation: Fresh policy insights from South Africa. *Biophysical Economics and Sustainability, 8*(1), 1.

UNDP. (2019a). ZRBF Resilience Knowledge Hub: Jolicoeur, D., Rex, E., Kutukwa, N., Sundsmo, A., Stack, J., Musandu, N., & Nembaware, N. (2019). Foundations in Resilience Curriculum.

UNDP. (2019b). ZRBF Resilience Knowledge Hub: Mock, N., Stack, J., & Sundsmo, A. (2019). Assessment of the Zimbabwe Resilience Building Fund Crisis Modifier Mechanism.

UNDP. (2020a). ZRBF Resilience Knowledge Hub: Langworthy, M., Fox, K., Martin, S., Woodson, L., Aziz, T., Al-haddad, R., Cuellar, E., Hein, C., & Stack, J. (2020). ZRBF Outcome Monitoring Survey: Round Two Program Learning Report.

UNDP. (2020b). ZRBF Resilience Knowledge Hub: Gash, M., Stack, J., & Ruparanganda, L. (2020). The Role of Savings Groups in Resilience Building: A Scoping Study.

van Aswegen, M., Retief, F. P., & Drewes, E. (2020). Regional resilience in peripheral South Africa: The northern cape case. *Town and Regional Planning, 77*, 1 17.

Zarepour Moshizi, M., Yousefi, A., Amini, A. M., & Shojaei, P. (2023). Rural vulnerability to water scarcity in Iran: An integrative methodology for evaluating exposure, sensitivity and adaptive capacity. *GeoJournal, 88*(2), 2121–2136.

Zhu, X., & Shang, X. (2023). Positioning as discursive struggle for equity: A critical discourse analysis of the nationally determined contributions (NDCs) of African countries. *Critical Discourse Studies*, 1–16.

8 Prospects and Challenges for Climate-Smart Communities in Rural Southern Africa

Munyaradzi Chitakira, Leonard Chitongo, and Kosamu Nyoni

8.1 Introduction

In recent decades, the phenomenon of climate change and climate variability and the related impacts on humans and the natural environment have been central to the research agenda as scholars seek to find solutions to the challenges posed by this phenomenon. Governments at all levels, including central and local government levels, industries, homes, organisations and individuals, have been making efforts to go green or become more resilient to the impacts of climate change (Revermann et al., 2018; Department of Environment, Forestry, and Fisheries RSA, 2019). The concept of climate smartness is understood in light of the Food and Agriculture Organization (FAO) of the United Nations (2013)'s definition of climate-smart agriculture, which is built on three pillars which are (i) sustainable increase in productivity and income, (ii) improving the resilience of livelihoods and ecosystems to climate change and (iii) reducing or removing greenhouse gas emissions from the atmosphere. Therefore, a climate-smart rural area refers to a community or region that employs strategies to address the impacts of climate change and variability and enhance the resilience of the community while promoting environmentally sustainable production and development to reduce the environmental footprint (Akamani, 2021; Chirisa & Matamanda, 2019; Defe & Matsa, 2012). The concept of 'climate smart' puts emphasis on having to adapt to the variable and changing climate, implement measures to lessen the impacts and ensure that development activities cause minimal adverse impact on the environment (Mpala & Simatele, 2024).

The concept of climate-smart villages developed by the Consultative Group on International Agricultural Research (CGIAR) Research Program on Climate Change, Agriculture and Food Security (CCAFS) depicts a model of climate-smart rural areas. Climate-smart villages 'are sites where researchers, local partners, and farmers collaborate to evaluate and maximize synergies in a portfolio of climate-smart agricultural interventions' (Aggarwal et al., 2013). This model is intended to ensure food security, encourage adaptation and mitigation and build resilience to climate-induced stresses. The

M. Chitakira (✉)
Department of Environmental Sciences, University of South Africa, Florida, South Africa
e-mail: chitam1@unisa.ac.za

L. Chitongo
School of Social Sciences, College of Humanities, University of KwaZulu-Natal, Durban, South Africa
e-mail: ChitongoL@ukzn.ac.za

K. Nyoni
Department of Environmental Sciences, University of South Africa, Florida, South Africa

Gary Magadzire School of Agriculture and Engineering, Great Zimbabwe University, Masvingo, Zimbabwe
e-mail: knyoni@gzu.ac.zw

M. Matsa et al. (eds.), *Climate Change Resilience in Rural Southern Africa*, Sustainable Development Goals Series, https://doi.org/10.1007/978-3-031-73600-1_8

components of the climate-smart village in this regard include climate information services, local knowledge and institutions, village development plans and climate-smart technologies.

The impacts of climate change on communities around the world, such as floods or droughts, water scarcity, biodiversity loss, impact on agriculture and on tourism and psychological impacts, have been extensively investigated and are well-documented (Tubiello & Van der Velde, 2007; Altieri & Koohafkan, 2008; Gwenzi et al., 2021; Zamasiya et al., 2017; Dube & Nhamo, 2018, Berrang-Ford et al., 2021; Abbass et al., 2022; Karimi et al., 2022; Sibiya et al., 2022; Shoko, 2023). However, more research is needed to establish context-specific solutions to the challenges caused by climate change in rural communities. This chapter, based on a literature review, seeks to contribute to the ongoing search for solutions by evaluating the prospects for rural communities in Southern Africa to become climate smart. The chapter also examines the challenges in this regard and what could be done to address such challenges. This chapter is structured as follows: it presents an introduction, the conceptual underpinnings and the resilience framework. It then outlines the methodology adopted in the study and presents the results, followed by a discussion of the results, and ends with conclusions and practical recommendations.

8.2 Conceptual Underpinnings

Climate-smart rural areas are distinguished by a range of characteristics, as detailed in Table 8.1. These characteristics include local adaptation strategies and mitigation practices, sustainable agricultural production and inclusive local community participation that ensures equity. They are also defined by sustainable natural resource management, infrastructure resilience and diversification of livelihoods.

Table 8.1 Features defining climate-smart rural communities

Feature	Description
Adaptation strategies	Implementing measures to address the impacts of climate change and variability, including changes in temperature, precipitation patterns and extreme weather events. Strategies may include designing climate-smart agricultural practices, water management systems and infrastructure that can withstand climate change impacts (Gwenzi et al., 2021)
Mitigation practices	Implementing measures to reduce greenhouse gas emissions and carbon footprint. This could involve the use of renewable energy sources, reforestation projects and sustainable land-use practices
Sustainable agriculture	Adoption of climate-resilient and sustainable agricultural practices that enhance food security; conserve water, soil and other natural resources; and reduce the environmental impact of farming activities (Imperiale & Vanclay, 2023)
Community engagement	Involvement of local communities in decision-making processes and the development and adoption of climate-smart practices. This involves raising awareness, building capacity and promoting community-based adaptation and mitigation efforts, including indigenous knowledge systems and practices (Chitakira, 2024a, b)
Natural resource management	Measures/efforts to promote sustainable management of natural resources, including water, soil and biodiversity, to ensure the long-term health and productivity of rural ecosystems (Kotze & Wood, 2021)
Infrastructure resilience	Planning and building infrastructure that can withstand the impacts of climate change and variability, such as resilient roads, bridges and buildings that are designed to cope with extreme weather events
Diversification of livelihoods	Encouraging the diversification of livelihoods, including on-farm and off-farm, to reduce reliance on climate-dependent or climate-sensitive livelihood activities. Promotion of multiple streams of income and enhancement of the ability of local communities to adapt to changing conditions (Berrang-Ford et al., 2021; Muzorewa & Chitakira, 2022)

Source: Author compilation from Tubiello and Van der Velde (2007), Aggarwal et al. (2013), Adesipo et al. (2020), Hussain et al. (2022) and Chitakira (2024a, b)

Climate change has proven to be the most aggressive threat in the fight against poverty, hunger and malnutrition in Africa due to its impact on agricultural production, access to food and financial success (Enete & Amusa, 2010). The most affected countries are those with an agro-based economy (Enete & Amusa, 2010). To this end, climate change has seriously threatened the food security and livelihoods of smallholder farmers in sub-Saharan Africa. The impact of climate change on food security has been exacerbated by existing structural and institutional challenges in sub-Saharan Africa (Clover, 2003). Agricultural development policies are currently faced with various challenges, such as a rapidly expanding population and declining soil fertility and agricultural frontiers. The main drivers of the expansion of agricultural frontiers in emerging countries are the current global demand for agricultural products, the flexibility of environmental regulations, foreign investments and the reduction of production costs, as well as poverty, energy and food dependency, which consequently aggravate environmental problems.

Against the backdrop of challenges facing agricultural development in sub-Saharan Africa, an integrated approach is needed to address these intertwined problems. Climate-smart agriculture (CSA) fits well as a holistic approach that can be used to achieve sustainable agricultural development and socioecological goals (Behnassi et al., 2021). CSA incorporates technologies and practices that are context-specific, such as sustainable intensification. The sustainable intensification approach uses innovations to intensify agricultural productivity on existing agricultural land without negative socio-economic and environmental impacts. Recent agricultural policies have focused on CSA to address socio-economic and environmental setbacks. Although CSA technologies and practices have shown a potential to minimise socio-economic and environmental impacts, their adoption has faced a myriad of challenges, such as a lack of proper policies and political will, limited awareness of CSA technologies among smallholder farmers and institutional and financial constraints (Dyantyi & Njenga, 2022).

It is worth mentioning at this point that previous agricultural development strategies in the developing world focused on irrigated agriculture to boost agricultural productivity, reduce food insecurity and poverty and promote economic growth. The rationale behind the adoption of this approach was that the resultant economic growth would subsequently deliver significant 'trickle-down' benefits for people living in rural areas, including marginalised drylands (Sibiya et al., 2022). The flaw behind this approach, however, was that it left rural areas with marginal economic growth, impoverished households that are prone to hunger and a degrading natural resource base (Dyantyi & Njenga, 2022).

As a result of the failure mentioned above, new emphasis is now placed on enhancing agricultural productivity and people's livelihoods while promoting environmental conservation in marginalised areas (Sibiya et al., 2022). Research has shown that investing in dryland agriculture could lead to higher agricultural productivity and socio-economic returns than if the same investment was made in high-potential areas (Kpadonou et al., 2017). To further support the preceding, Enu et al. (2023) highlighted that there were no agrohydrological restrictions to significantly increase yields of staple crops in drylands. However, to increase agricultural productivity, close attention should be paid to important limiting factors such as low soil fertility and water stress while at the same time taking care of food security, poverty alleviation and sustainable natural resource management (Ruben & Pender, 2004). This thrust is squarely in line with the main pillars of CSA, that is, sustainably increasing agricultural productivity and income, adapting and building resilience to climate change and reducing or removing greenhouse gas emissions (FAO, 2013). This has been taken into account to address food security and poverty eradication issues sustainably (Kpadonou et al., 2017). In light of the preceding observations, the present chapter seeks to contribute to knowledge development on how rural communities can become more climate smart and resilient to increasing climatic risks and shocks.

8.3 Resilience Framework

Resilience is the ability to regain the original shape or position after bending, stretching, compression or other deformation (Korber et al., 2017). Resilience analysis has been discussed in various contexts in relation to climate change and sustainability. Researchers have focused on the role of resilience in reducing the vulnerability of communities to natural disasters. Revell (2010) defines resilience as the ability of a system to overcome shocks and uncertainties in a sustainable manner. Therefore, in this chapter, the resilience to climate change is based on the development of rural communities. Meriläinen (2023) identifies three key conceptualisations of resilient communities, which are resilient communities of belonging, resilient communities of practice and resilient communities as an object of governance. These relate to the ability to withstand extreme climatic conditions and maintain socioecological and livelihood development. Smart communities are enhanced by enhanced networks and the formation of professional mentor networks, accepting that change is a part of life and avoiding seeing crises as insurmountable. Individuals can navigate their way to the psychological, social, cultural and physical resources that sustain their well-being and their ability, individually and collectively, to negotiate for resources to be provided and experience culturally meaningful ways.

8.4 Research Design

In this study, a systematic review of the literature was conducted to reduce bias in the literature review process. To identify and analyse the relevant literature on climate-smart rural communities, the Preferred Reporting Items for Systematic Reviews and Meta-Analyses (PRISMA) were adopted. Figure 8.1 illustrates the four-phase flow diagram that followed. A careful selection of keywords was attempted to identify the relevant literature. The keywords "resilient communit*" or "climate-smart communit*" AND "South* Africa" were entered into the Scopus search engine on 6 October 2023. This identified a total of 547 sources. One article was duplicated, so there were 586 potential sources. Of these, 381 were excluded by title and 137 were excluded after reading the abstract.

The selection of sources for study from these 586 was based on relevance to the research topic and the context of the rural area. Sources that were based on the urban or other nonrural area context were excluded. Sources that were not specific to Southern Africa, such as general discourses that could include cases or examples from rural Southern Africa or from which valuable insights could be derived, were included in the final selection. Review articles were excluded from the study. No article was excluded after a full-text review. Therefore, the final number of sources included in the study was 68 (Fig. 8.1).

8.5 Presentation of Findings

8.5.1 Prospects for Climate-Smart Communities in Rural Southern Africa

Several prospects for climate-smart communities in rural Southern Africa were identified. These include improved agriculture and agribusiness. Smart rural areas in Southern Africa can benefit from technology-driven agricultural practices. Examples include precision farming, IoT-based crop and livestock monitoring and the use of data analytics for better decision-making. Access to market information and e-commerce platforms can also enhance agribusiness opportunities for rural communities. A literature-based study by Castleden et al. (2011) has revealed that agricultural diversification in the Sahel region since the 1970s has helped to restore the resilience of the region to the effects of drought.

It also emerged that the Southern African region can take advantage of the prospects for renewable energy solutions. Many marginal rural communities in the Southern African region rely primarily on fuelwood for energy, such as the Habu community in Botswana, and such dependence on natural resources exposes communities to climatic risks

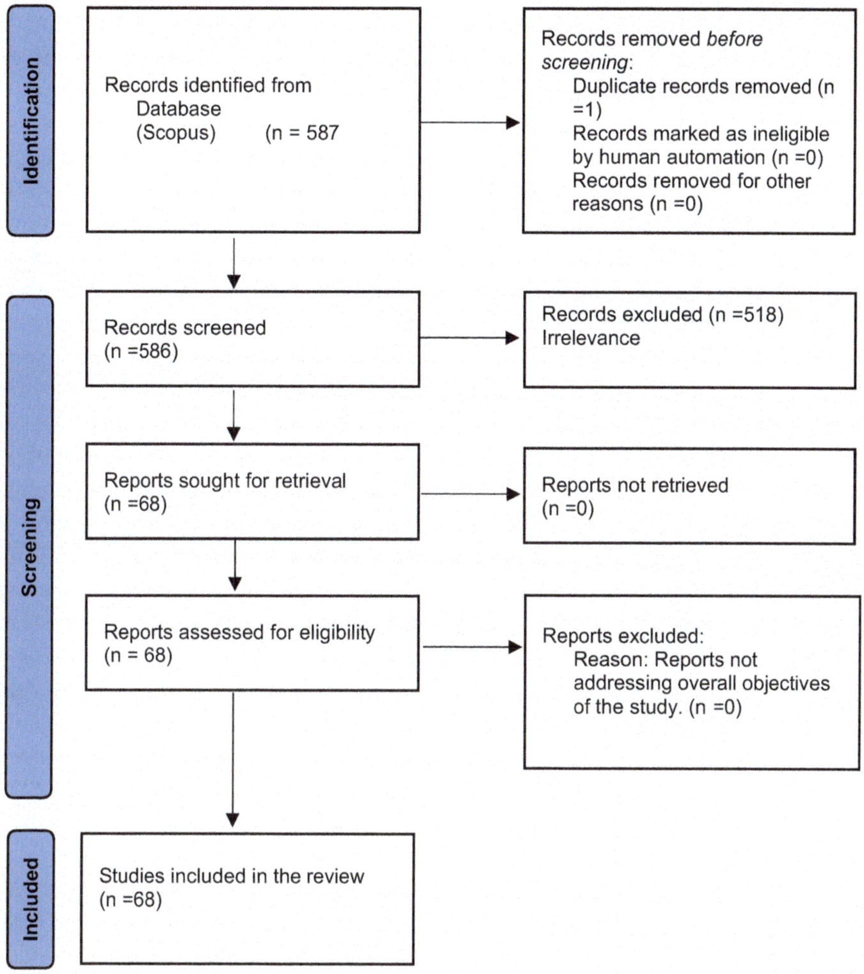

Fig. 8.1 PRISMA 2020 flow diagram. (Adapted and modified (Moher et al., 2007))

and shocks (Cassidy & Barnes, 2012). The region has abundant renewable energy resources, including solar and wind. It has abundant sunlight and can implement smart energy solutions such as off-grid solar power and microgrids, which can improve access to electricity in remote rural areas. Access to electrical energy can support community services and economic activities, enabling the diversification of livelihoods, which is important for improving the resilience of communities.

Efficient water management is one of the opportunities emerging in the Southern African region. Smart technologies can be employed for efficient water management, particularly in regions prone to droughts. Sensor-based monitoring systems, data analytics and water conservation practices can contribute to sustainable water use in agriculture and daily life. A study by Kruger (2016) focusing on agricultural activities by small-scale farmers in Malawi, Madagascar and Mozambique showed that small-scale farmers generally plant with the first rains. However, this alone would not ensure optimum production. Therefore, the study recommended early planting and the implementation of water management techniques or the use of residual moisture retention in the event of late planting to support crops during dry periods.

Smart communities require affordable and easily accessible health facilities. Many rural areas in Southern Africa are not readily accessible. This makes the provision of healthcare to these communities a challenge. Rural healthcare provision can benefit from improved healthcare and education through telehealth and e-learning initiatives. Telemedicine (the provision of medical care/advice remotely to patients) can provide access to medical expertise, while e-learning platforms can enhance health awareness and educational opportunities for rural populations. Buse et al. (2022) noted that synthesising the existing health frameworks with community-oriented approaches is helpful in rural and remote community settings. Such a holistic approach is required to deal with these complex environments.

Investment in smart infrastructure, including smart roads, connectivity and digital services, can improve the overall quality of life in rural areas. This includes the establishment of digital hubs and the provision of access to the Internet and digital technologies. Writing on solutions for the revitalisation of rural areas in South Africa, Meyer (2014) asserts that strong and healthy rural communities should be characterised by characteristics such as access to modern information communication technology, substantial levels of entrepreneurship, industrial specialisation, job opportunities and export sectors with a strong economic base. Therefore, there is a renewed focus on rural development supported by policy frameworks to enhance infrastructure development to increase rural communities' resilience.

8.5.1.1 Climate Resilience and Social Networks

Given the susceptibility of Southern Africa to the impacts of climate change, smart rural areas can focus on climate-resilient practices. This may involve early warning systems for extreme weather events, sustainable land management and climate-smart agriculture. To minimise the impacts of climate change on human well-being, governments, development agencies and civil society organisations have made substantial investments in improving people's capacity to adapt to change. To date, these investments have tended to focus on a narrow understanding of adaptive capacity. As such, Cinner et al. (2018) proposed an approach to build adaptive capacity across five domains. The domains include the assets that people can draw upon in times of need, the flexibility to change strategies, the ability to organise and collaborate, the learning to recognise and respond to change and the agency to determine whether to change or not.

The literature shows that investments in social networks and relationships and social assistance programmes can strengthen communities' resilience to a range of disasters (Agrawal et al., 2020). Households that have savings or surplus money after expenses and durable assets tend to have better absorptive resilience against shocks than poorer households. As such, social assistance through cash transfers (or grant programmes) becomes necessary, as this can strengthen the absorptive resilience of poor communities. For example, Lesotho's Child Grants Program has been shown to improve the resilience of low-income families (d'Errico et al., 2020). Despite their usefulness, cash transfers in sub-Saharan African countries are generally tiny amounts, less than USD25 per person per year. However, some countries such as Angola and South Africa are investing more significantly in social assistance, spending more than USD100 per person per year (Agrawal et al., 2020).

A study by Castleden et al. (2011) on resilience thinking in health protection revealed that resilient communities should be less dependent on external help in times of disaster. Rural households in Botswana have adopted a wide range of livelihood strategies to reduce their vulnerability to shocks and stresses. Cassidy and Barnes (2012) conducted a study of the Habu community in Botswana to analyse the influence of social connectivity on the resilience of households to shocks such as human illness and death, crop damage and livestock disease. The study concluded that households with more social networks are more resilient as they tend to have a more comprehensive range of livelihood strategies, higher levels of other forms of social capital and higher overall capital. This conclusion was confirmed by Zamasiya et al. (2017) in a study of 400 randomly selected farmers in the Hwedza district in Zimbabwe, which revealed that small farmers who were members of social groups were more likely to have a positive attitude towards climate change and variability and to adopt new adaptation strategies due to the influence of their peers. A related study in eastern Zimbabwe has shown that numerous activities in the informal sector have mushroomed in rural and urban communities in response to climate variability challenges, implying that households and individuals are adjusting livelihood strategies (Muzorewa & Chitakira, 2021). A lesson learned from these observations is that strengthening social networks and livelihood diversity are potential ways through which rural communities in the Southern African region can become more resilient and climate-savvy.

8.5.1.2 Community Engagement, Inclusivity and Incorporation of Indigenous Knowledge (IK)

Successful implementation of smart rural initiatives requires community engagement and participation. Inclusivity in decision-making processes ensures that the benefits of technology reach all segments of the population. A collaborative, participatory process involving local communities engenders the best-expected results on mitigation, preparedness and recovery (Haque & Burton, 2005). In Botswana and Zimbabwe, for example, there has been an increasing recognition of IK by governing authorities and development agencies as accurate and valuable (Cassidy et al., 2011; Zinhiva et al., 2021). This recognition presents an opportunity to enrich policies by incorporating IK. However, Cassidy et al. (2011: 77) observed that IK "has a long way to go to be fully recognised as a credible and respected knowledge system in its own right". Furthermore, a study by Zinhiva et al. (2021:85) in southern Zimbabwe revealed mixed responses by different development agencies pertaining to the application of indigenous ecological knowledge in the planning and management of biotic resources in communal areas.

8.5.2 Barriers to Having Climate-Intelligent Communities in Rural Southern Africa

Several barriers or challenges can impede the implementation of climate-smart initiatives in rural Southern Africa. Barriers vary between countries and within countries. The following passages discuss the challenges identified in the region.

8.5.2.1 Infrastructure and Technology Limitations

The Southern African region suffers from limited infrastructure, including reliable power and Internet connectivity, and this can hinder the access to and adoption of climate-smart technologies. Limited infrastructure, including electricity and Internet connectivity, can hinder the deployment of advanced technologies for sustainable agriculture, weather forecasting and other climate-resilient practices. In Malawi, irrigation schemes are an important part of the national agenda. However, the design and management of these schemes have not taken advantage of emerging approaches, such as multiple-use water services (MUS), that can improve their performance (Stedman et al., 2018). Stedman et al. (2018) focused on rural irrigation systems in the Nkhata Bay district in Malawi, evaluating the usage of the system, the profiles of the irrigators and barriers to irrigation to identify opportunities to start MUS using existing organisational structures. Infrastructure and technology adoption requires a holistic, integrated approach involving community-level irrigation and borehole committees, private sectors, local government ministries and development partners. However, such an integrated, holistic approach is still lacking.

8.5.2.2 Affordability

Given that poverty generally prevails in the region, many rural communities may face challenges in accessing and affording smart technologies. Financial constraints limit their ability to invest in climate-smart practices and technologies. The upfront costs of implementing sustainable agricultural methods, renewable energy solutions and resilient infrastructure can be prohibitive for resource-constrained communities. Studies in Zambia and South Africa have shown that the potential of Internet of Things (IoT) technologies to reduce poverty and enhance agricultural development in rural areas is limited due to affordability (Dlodlo & Kalezhi, 2015). Regarding renewable energy adoption, prohibitive prices and inability to pay have been found to be the main barriers to solar energy adoption in African countries such as South Africa and Kenya (Adenle, 2020). It is suggested that government initiatives to support accessing smart technologies must be inclusive and consider economic disparities between rural and urban areas and between the different rural regions.

8.5.2.3 Insufficient Knowledge and Awareness

The lack of awareness and understanding of climate change and its impacts can hinder the adoption of climate-smart practices. Educational programmes and awareness campaigns may be needed to inform rural communities about the benefits of climate-smart approaches and how to implement them. Chirisa and Matamanda (2019) note that Africa is one of the regions suffering from the effects of climate-induced disasters and risks and argue for scientific communication to advance environmental education on climate change. Attention should be paid to science communication, which must be packaged in ways that make it easier for the generality of citizens to interpret and understand.

8.5.2.4 Insufficient Capacity

Building local capacity to understand, operate and maintain smart technologies provides an important foundation for long-term success. In a study of the Mbire community in northern Zimbabwe, Mavhura et al. (2022) observed that riparian communities lack flood risk preparation. Capacity building is required in terms of resource mobilisation, evacuation plans, early warnings, flood education/training/awareness, floodproofing information, post-flood recovery and risk communication. Chikwiramakomo et al. (2021) underscore the importance of geospatial techniques in flood-hazard modelling, which is the key input in sustainable land-use planning. They argue that spatial analytical techniques play a key role in flood early warning systems aimed at supporting and building resilient communities in the face of floods induced by climate change.

8.5.2.5 Inconsistent Policy and Regulatory Frameworks

Clear policies and regulatory frameworks are essential to support the deployment of smart technologies in rural areas. Inconsistent or inadequate policies and regulations related to climate-smart practices can hinder progress. Clear and supportive policy frameworks are crucial to creating an enabling environment for the adoption of climate-smart initiatives. Sub-Saharan African countries are increasingly prioritising strengthening the adaptation capacity of smallholder farmers and integrating climate adaptation and mitigation strategies such as climate-smart agriculture into their national policy frameworks (Zseleczky & Yosef, 2014). Efforts have been made in countries such as Malawi, South Africa and Zimbabwe, which include the integration of vulnerability and adaptation to climate change into sustainable development policy planning and implementation (Mapfumo et al., 2014). However, a notable challenge is that policies do not address important issues such as land tenure and advancements in the labour force (Taylor, 2018).

8.5.2.6 Inadequate Institutional Support

Limited institutional support and capacity at the local and national levels can hinder the development and implementation of climate-smart policies. The literature shows that in Namibia, many communities experiencing the impacts of floods could not receive support and training from local extension officers due to a lack of human and financial capacity at the local level. In South Africa, local agricultural structures in Limpopo province and in the Western Cape provinces were not receiving sufficient resources, which limited the capacity of extension officers to offer services (Spear et al., 2018).

Although efforts have focused on developing resilient communities through the development of adaptation strategies, barriers and limits to managing climate change are usually not given enough space in development programmes. Gwenzi et al. (2021) examined barriers and limitations to managing climate change as experienced by smallholder farmers in UMP, Zimbabwe. Adaptation strategies to the major hazards that affect agricultural activities include conservation agriculture, mulching, crop diversity, staggered planting, response farming and supplementing crop water requirements. The results showed that adaptation strategies such as conservation agriculture failed during long and severe dry spells and droughts and poor water resources limited the capacity to supplement crop water requirements. Gwenzi et al. (2021) argued that the language in weather forecasts and advisories was too technical, and the forecasts given were too coarse to have significant meaning at the local level.

Institutional challenges are also experienced in the tourism industry. For instance, Dube et al. (2023) mapped and evaluated the impact of flood hazards on tourism in South African national parks. They argued that the central government remained unwilling to devolve power over wildlife management to subdistrict levels even though they earned huge revenue by leasing sport hunting and nonconsumptive tourism rights to operators from the private sector. Strengthening local institutions, providing technical assistance and promoting multistakeholder collaboration are essential for overcoming the barrier of inadequate institutional support for local communities.

8.5.2.7 Climate Change Vulnerability

Southern Africa is prone to climate variability and extreme weather events, such as floods and various forms of drought (meteorological, hydrological, agricultural and socio-economic drought). These events can disrupt agricultural activities and infrastructure, making it challenging for rural communities to plan and implement long-term climate-smart strategies. To address this situation, rural communities must have access to the necessary information to enhance their preparedness for disasters. However, access to information is a challenge. For example, a study by Ncoyini et al. (2022) on the access and use of climate information by small-scale sugar cane farmers in South Africa revealed that most small-scale sugarcane farmers did not access climate information and that the use of climate information was minimal. A study by Mabuku et al. (2018) in parts of Namibia and Zambia employed several additive scales to measure different social capitals and flood preparedness. The results of the independent t-test showed that Namibians had higher levels of flood preparedness than Zambians ($p < 0.05$). Flood preparedness differed from household to household and was influenced by various social and demographic factors such as sense of community, perception of risk, level of education, marital status, access and land size. Institutional and policy support should take note of these factors to help communities acquire and use climate information and adaptation strategies to improve disaster preparedness and resilience.

8.5.2.8 Challenges to Climate-Intelligent Agriculture

Agricultural production is a major source of income in rural Southern Africa, and the change to climate-smart farming will contribute immensely to climate-smart communities in the region. There are several barriers to the adoption of climate-smart agriculture, including the following:

(i) Land tenure challenges: In some cases, unclear or insecure land tenure can hinder investments in sustainable land management practices (Taylor, 2018). Farmers may be reluctant to adopt climate-smart agriculture if they lack secure land rights, as this could affect their long-term benefits.
(ii) Limited agricultural extension services. Regarding production, insufficient extension services and outreach to rural communities tend to hamper the dissemination of knowledge and best practices related to climate-smart agriculture. There are reports of a lack of capacity on the part of extension officers to disseminate climate and adaptation information to sugar cane farmers in South Africa such that capacity

building is required in this regard (Ncoyini et al., 2022). Strengthening extension services can play a key role in improving knowledge, building farmer capacity and promoting the adoption of sustainable practices.

(iii) Market access and related challenges. Limited access to markets and value chains can discourage farmers from adopting climate-smart practices, as they may be uncertain about the economic viability of such approaches. Lack of government support, lack of pro-poor financial schemes and poor road infrastructure limited the capacity to which smallholder farmers in Uzumba-Maramba-Pfungwe (UMP) in Zimbabwe could manage climate extremes (Gwenzi et al., 2021). Improving market access and creating incentives for sustainable agriculture can address this barrier.

8.5.2.9 Social and Cultural Barriers

Although weakening in many parts of the region, traditional practices and cultural norms can be a significant barrier, especially if they conflict with or hinder the adoption of new climate-smart technologies and practices. As such, it is important to promote sustainable development without interfering with or violating local norms, ways of life or value beliefs. As the pace of global climate change accelerates, residents of the arid and semi-arid lands of northern Kenya, for example, face significant obstacles to maintaining viable communities (Wright, 2019). Colonial and post-colonial political governance and economic models have broken down traditional mechanisms to cope with ecological stress. An archaeological example from the Chalbi Desert is presented as evidence of the prolonged persistence of human occupations through extreme climatic fluctuations. Occupations spanning from cal. 5000 to 600 years BP indicate the presence of robust regional exchange networks and either continuous or repeated occupation of this hot, dry, mosaic woodland. These systems are contrasted against the extrinsically imposed models that began in the late 1800s with the arrival of colonialism, in which political economies are more focused on ethnic-territorial and global interests rather than regional cooperation (Wright, 2019). A long-term ecological model of how traditional systems evolved to cope with the vulnerability associated with extreme climate conditions is needed to build resilient communities.

From the above discussion, the success of smart rural initiatives depends on a combination of factors, including technological innovation, community involvement and supportive policies. Addressing the myriad of challenges outlined above requires a comprehensive and collaborative approach involving all key stakeholders or interested parties (Mapfumo et al., 2014; Spear et al., 2018). These include governments, non-governmental organisations, the private sector and local communities. Tailoring solutions to the specific needs and context of each community is crucial to the successful implementation of climate-smart initiatives in rural Southern Africa.

8.5.3 Suggested Measures for Climate-Intelligent Communities

Taking from the preceding discussion, several potential prospects for the transition to climate-smart rural communities in Southern Africa can be compiled. These are presented in Table 8.2.

8.6 Discussion of the Findings

8.6.1 Need for Community Engagement

Community involvement is central to the successful implementation of the suggested measures towards climate-smart rural communities (Table 8.2). Efforts are necessary to foster community ownership of projects, which is important for ensuring community buy-in, sustainability and long-term success. Community resilience is likely affected by poor planning, which calls for institutional capacity building. Based on the demonstrated failure and costs of contingency disaster management over the past

Table 8.2 Suggested action to promote climate-intelligent rural communities

Suggested solution	Explanation
Enhancing access to technology	Investment in rural infrastructure, including reliable electricity and internet connectivity Facilitate technology transfer and knowledge sharing through partnerships with research institutions and technology providers
Financial support and incentives	Providing financial support or incentives for adopting climate-smart technologies Establishing financial mechanisms such as grants, subsidies and low-interest loans to support climate-smart initiatives Could encourage the participation and investment of the private sector in sustainable projects through incentives Develop innovative financing models, such as community-based revolving funds or public-private partnerships
Capacity building, education and awareness	Provide training and capacity-building programmes for community members, extension workers and local leaders on climate-smart agriculture and other climate-resilient practices Could implement educational programmes at the community level to raise awareness of climate change and the benefits of adopting climate-smart practices
Climate-resilient infrastructure	Designing and implementing infrastructure projects that incorporate climate resilience Investing in smart infrastructure that can withstand climate-related challenges, such as climate-resilient roads and bridges
Institutional capacity	Empowering and capacitating local institutions responsible for agricultural extension, natural resource management and climate adaptation Promoting collaboration between governmental agencies, non-governmental organisations and community-based organisations to enhance coordination and support
Appropriate policy development and implementation	Develop and enforce clear policies and regulations that support climate-smart initiatives and sustainable rural development Integrating climate considerations into broader development policies and plans Ensuring community participation in the development of policies and incorporation of local perspectives
Livelihoods and market diversification	Promoting livelihood diversification can reduce dependence on activities sensitive to climate change and variability Improving access to markets and value chains, facilitating the sale of climate-smart products Creating incentives for businesses to invest in rural areas, contributing to economic diversification
Effective and efficient extension services and farmer support	Boosting agricultural extension service provision to equip farmers with the knowledge and skills demanded by climate-smart practices Promoting farmer-to-farmer knowledge sharing and peer learning networks Establishing demonstration farms to showcase the benefits of climate-smart agriculture
Addressing land tenure and resource management issues	Implementing policy reforms as appropriate to address land tenure issues and ensure secure land rights for farmers Promoting sustainable land management practices to enhance soil health and reduce degradation
Role of research and innovation	Investing in research and innovation development is the foundation for locally appropriate climate-smart solutions Partnerships with research institutions and participation of the private sector could promote innovation in agriculture and rural development

Source: Compilation of authors by Cassidy and Barnes (2012), Mapfumo et al. (2014), Spear et al. (2018), Dintwa et al. (2022) and Hussain et al. (2022)

three decades, the successful mainstreaming of disaster management in local development planning requires several factors to be considered (Chirisa & Matamanda, 2019). These factors include institutional capacity building, building political will for long-term investment in disaster adaptiveness as opposed to relief and cultural transformation, where development planners have to see disaster management as a cornerstone to delivering their overall rural development targets rather than view it as separate from development. It is argued that incorporating indigenous knowledge and cultural practices into climate-smart initiatives is important in building resilience (Dintwa et al., 2022). Indigenous knowledge systems (IKS) are important in the maintenance of livelihoods of local communities. To reduce risks and future catastrophes after disasters, there is a need for a thorough implementation of risk reduction strategies.

8.6.2 Regional Collaboration and the Water-Energy-Food Nexus

In addition to involvement at the local/grassroots level, regional collaboration and knowledge sharing between governments of Southern African countries and other role players can create a platform to address common challenges collaboratively. Research on the water-energy-food (WEF) nexus in Southern Africa (Mpandeli et al., 2018) has emphasised the need for regional integration, considering that agro-ecological zones stretch across political boundaries and sovereignty. Additionally, it is argued that the integration of climate change adaptation strategies into the WEF nexus opens room for resource coordination, harmonisation of activities across all sectors, improved resilience and reduced vulnerability. Such efforts are believed to promote the achievement of regional development goals, as well as the Sustainable Development Goals of the United Nations, such as poverty alleviation, zero hunger, access to water and sanitation and access to affordable energy. Furthermore, Nhamo et al. (2020: 16) note: "The WEF application of the nexus analytical model has been successfully applied in Southern Africa to transform rural livelihoods and in South Africa to evaluate resource use and management to inform policy on priority areas for intervention". It is noted that nexus planning facilitates the development of sustainable livelihoods, which can improve the resilience of local communities.

8.6.3 Need for a Change in Approach and Perspective

The study findings point to the need to change from traditional and insincere community-based natural resource management (CBNRM) to a more serious smart community approach (CBA). The latter is a more robust approach that considers the interests of various population groups in society. Gebreyesus et al. (2022) assert that civil society is a crucial governance actor that carries a significant role to undertake in climate policy-making including local area adaptation planning. Consequently, this study revealed the importance of education, training and building resilience to climate change impacts. There is a need for a paradigm shift in changing perceptions that are more likely to impact preparedness behaviour. This, in turn, will improve community responses to climate-related disasters such as floods, planning for floods and preparedness (Mabuku et al., 2018).

Climate change manifestation, preparedness and response strategies are essential matters to consider prior to the implementation of a sustainable community-based climate change adaptation programme. Reflected knowledge gaps regarding climate change and wildlife depletion reveal the need for public awareness campaign programmes (Nkombi & Wentink, 2022). The South African National Environmental Management Act states that every individual "has a right to access environmental information and the right to participate in the promulgation and implementation of legislative, policy and other measures that prevent pollution and environmental degradation". For this reason, all members of the community are free to perceive environmental matters in their way.

Southern Africa is becoming more arid, and the growth of drought-resistant crops is a viable livelihood option. Due to the limited rainfall amount, there is restricted agricultural activity in the semi-arid parts of the Matabeleland South province in Zimbabwe. Therefore, wildlife conservation has been undertaken in these areas since the late 1970s (Chirisa & Matamanda, 2019). The move to focus on wildlife is also necessitated by the fact that wildlife remains res nullius (lacking a formal owner). Since the sector is state-owned, the government allows entities to register and manage wildlife in their localities under the community-based programme if all necessary conditions are met.

The widespread acceptance of prudent CBNRM as a sustainability measure has been opined in a myriad of community studies. What community members usually detest is a situation where the management typically develops top-down, with benefits from the sale of resources becoming fraudulent. Based on the Hwange CAMPFIRE experience, Shereni and Saarinen (2021) point to the declining acceptance of CBNRM due to a continued change in the political and economic environment in the country that has prompted locals to lose out on trophy benefits.

In order to achieve climate-smart communities in rural Southern Africa, there is a need to overcome natural and financial barriers that limit the level of adaptation strategies. Therefore, the development of adaptation strategies requires a holistic approach that is fully supported by all stakeholders (Gwenzi et al., 2021). The trend towards vivid dislike of traditional leadership, particularly by the younger generation, is a common scenario in Zimbabwe, given the politicoeconomic crisis that has had deleterious impacts on community stability. The findings resonate well with those of Syahrir et al. (2021) or Karimi et al. (2022), who argue that beyond the pressure from the central government, there is also hidden "politics of appropriation of conservation and development projects which is itself embedded in larger institutions of governance". Such differences call for more proactive policy-making, as Nkombi and Wentink (2022) advocate that the Katlehong township accentuates the significance of public participation and the engagement of all stakeholder groups in the formulation of any viable climate change legislation for the future. Therefore, a more viable approach to the development of smart communities for the benefit of the entire community would entail the interests of various members of the community.

8.7 Conclusions

The purpose of this chapter was to assess the opportunities for rural communities in Southern Africa to become more climate-smart and resilient to the impacts of climate change and variability, such as droughts and floods. The chapter also intended to assess the barriers to becoming climate-smart and how such barriers can be addressed in order to achieve climate-smart rural communities. What emerged from the study is that several conditions in Southern Africa promote the transition to climate-smart rural communities. The rural communities in the region can become climate-smart by adopting climate-smart agriculture and improved agribusiness, adopting renewable energy solutions and sustainable water management and making health facilities affordable and easily accessible in all areas through the use of modern technology. Opportunities for transition are also in the areas of infrastructure development, utilisation of social networks, incorporation of indigenous knowledge and inclusive community engagement to enhance resilience against droughts, floods and other disasters. Despite the many opportunities available in the region, the transition to climate-smart communities and the increase in resilience to extreme weather conditions are happening at a low pace. Several existing barriers in the Southern African region could impede the faster adoption of climate-smart initiatives and transition. The barriers revealed in the study include poorly developed infrastructure in rural areas, as well as limited access to modern technology. On the one hand, rural dwellers tend to have insufficient knowledge and awareness of climate change adaptation strategies.

On the other hand, there is limited capacity on the part of public extension service providers to support rural communities in adopting climate-smart strategies adequately. Inadequate institutional support has been revealed as a major challenge in the region, such that even where policies are in place, there is often a lack of implementation. Additionally, there are barriers embodied in the social and cultural values of some communities that hinder the adoption of adaptation strategies and transition to smart livelihoods based on climate. Opportunities and challenges tend to be shared between countries in the region, although there may be variations between countries due to differing contexts.

Based on the opportunities and challenges revealed in this study, this chapter would like to recommend potential measures to address the challenges and use the opportunities available to improve the climate smartness of rural communities in the region in focus. Respective national governments and their partners could invest resources to increase rural communities' access to appropriate technology, including Internet facilities. Additionally, governments, with the support of relevant stakeholders, must make capacity building, education and awareness of smart adaptation strategies a priority and ensure the provision of effective extension services. Investment in building climate-resilient infrastructure is necessary to ensure that rural communities remain accessible and can always access the necessary services in times of disaster. The development and implementation of appropriate policies are necessary for such efforts. The old question of land tenure, which remains unresolved in many cases, should be given the proper attention to ensure secure land rights for farmers and promote more meaningful investment in communal areas. The support of research and innovation remains vital, and funders are urged to provide the necessary financial resources to enable informed and evidence-based transition to climate-intelligent communities.

References

Abbass, K., Qasim, M. Z., Song, H., Murshed, M., Mahmood, H., & Younis, I. (2022). A review of the global climate change impacts, adaptation, and sustainable mitigation measures. *Environmental Science and Pollution Research, 29*(28), 42539–42559.

Adenle, A. A. (2020). Assessment of solar energy technologies in Africa-opportunities and challenges in meeting the 2030 agenda and sustainable development goals. *Energy Policy, 137*, 111180.

Adesipo, A., Fadeyi, O., Kuca, K., Krejcar, O., Maresova, P., Selamat, A., & Adenola, M. (2020). Smart and climate-smart agricultural trends as core aspects of smart village functions. *Sensors, 20*(21), 5977.

Aggarwal, P., Zougmoré, R., & Kinyangi, J. (2013). Climate-Smart Villages: A community approach to sustainable agricultural development. Copenhagen, Denmark: CGIAR. Research Program on Climate Change, Agriculture and Food Security (CCAFS). Available online at: www.ccafs.cgiar.org

Agrawal, A., Kaur, N., Shakya, C., & Norton, A. (2020). Social assistance programs and climate resilience: Reducing vulnerability through cash transfers. *Current Opinion in Environmental Sustainability, 44*, 113–123.

Akamani, K. (2021). An ecosystem-based approach to climate-smart agriculture with some considerations for social equity. *Agronomy, 11*(8), 1564. https://doi.org/10.3390/agronomy11081564

Altieri, M. A., & Koohafkan, P. (2008). *Enduring farms: Climate change, smallholders and traditional farming communities* (Vol. 6). Third World Network (TWN).

Behnassi, M., et al. (2021). Enhancing resilience for food and nutrition security within a changing climate. In M. Behnassi, M. Barjees Baig, M. El Haiba, & M. R. Reed (Eds.), *Emerging challenges to food production and security in Asia, Middle East, and Africa*. Springer. https://doi.org/10.1007/978-3-030-72987-5_1

Berrang-Ford, L., Siders, A. R., Lesnikowski, A., et al. (2021). A systematic global stocktake of evidence on human adaptation to climate change. *Nature Climate Change, 11*, 989–1000. https://doi.org/10.1038/s41558-021-01170-y

Buse, C. G., Allison, S., Cole, D. C., Fumerton, R., Parkes, M. W., & Woollard, R. F. (2022). Patient- and Community-Oriented primary care approaches for health in rural, Remote and resource-dependent places: Insights for eco-social praxis. *Front Public Health, 10*, 867397. https://doi.org/10.3389/fpubh.2022.867397. PMID: 35692331; PMCID: PMC9178183.

Cassidy, L., & Barnes, G. D. (2012). Understanding household connectivity and resilience in marginal rural communities through social network analysis in the village of Habu. *Botswana. Ecology and Society, 17*(4).

Cassidy, L., Wilk, J., Kgathi, D. L., Bendsen, H., Ngwenya, B. N., & Mosepele, K. (2011). Indigenous knowledge, livelihoods and government policy in the Okavango Delta, Botswana. In *Rural livelihoods, risk and political economy of access to natural resources in the Okavango Delta, Botswana* (pp. 75–97).

Castleden, M., McKee, M., Murray, V., & Leonardi, G. (2011). Resilience thinking in health protection. *Journal of Public Health, 33*(3), 369–377.

Chazovachii, B., Chitongo, L., Tagarirofa, J., & Rukomwe, C. (2019). Resilient entrepreneurial poor in urban housing by the homeless people in chimanimani post-cyclone idai disaster in Zimbabwe. *The Fountain: Journal of Interdisciplinary Studies, 3*(1), 45–56.

Chikwiramakomo, L., Gumindoga, W., Shekede, M. D., Gara, T. W., & Chuma, T. (2021). Modelling flood hazard in dry climates of Southern Africa: A case of Beitbridge, Limpopo Basin, Zimbabwe. Water SA, 47(4 October). https://doi.org/10.17159/wsa/2021.v47.i4.3787

Chirisa, I., & Matamanda, A. R. (2019). Science communication for climate change disaster risk management and environmental education in Africa. In L. Wilson & C. Stevenson (Eds.), *Building sustainability through environmental education* (pp. 190–212). IGI Global. https://doi.org/10.4018/978-1-5225-7727-0.ch009

Chitakira, M. (2024a). Community visioning. In F. Vanclay & A. M. Esteves (Eds.), *Handbook of social impact assessment and management* (pp. 547–561). Edward Elgar.

Chitakira, M. (2024b). Community visioning. In F. Vanclay & A. M. Esteves (Eds.), *Handbook of social impact assessment and management* (pp. 547–561). Edward Elgar Publishing.

Cinner, J. E., Adger, W. N., Allison, E. H., et al. (2018). Building adaptive capacity to climate change in tropical coastal communities. *Nature Climate Change, 8*, 117–123. https://doi.org/10.1038/s41558-017-0065-x

Clover, J. (2003). Food security in sub-Saharan Africa. *African Security Review, 12*(1), 5–15.

d'Errico, M., Garbero, A., Letta, M., & Winters, P. (2020). Evaluating program impact on resilience: Evidence from Lesotho's Child Grants Programme. *The Journal of Development Studies, 56*(12), 2212–2234.

Defe, R., & Matsa, M. (2012). The contribution of climate-smart interventions to enhance sustainable livelihoods in Chiredzi District. *Climate Risk Management, 33*(1), 100338. https://doi.org/10.1016/j.crm.2021.100338

Department of Environment, Forestry and Fisheries RSA. (2019). *National Climate Change Adaptation Strategy*. Republic of South Africa.

Dintwa, K. F., Letamo, G., & Navaneetham, K. (2022). Vulnerability perception, quality of life, and indigenous knowledge: A qualitative study of the population of Ngamiland West District, Botswana. *International Journal of Disaster Risk Reduction, 70*, 102788. https://doi.org/10.1016/j.ijdrr.2022.102788

Dlodlo, N., & Kalezhi, J. (2015). May. The internet of things in agriculture for sustainable rural development. In *2015 international conference on emerging trends in networks and computer communications (ETNCC)* (pp. 13–18). IEEE.

Dube, K., & Nhamo, G. (2018). Climate variability, change and potential impacts on tourism: Evidence from the Zambian side of the Victoria Falls. *Environmental Science & Policy, 84*, 113–123.

Dyantyi, O., & Njenga, J. (2022). Awareness and perceptions of smart irrigation technologies by small scale farmers in rural South Africa, 2022 IST-Africa Conference (IST-Africa). *Ireland*, 1–10. https://doi.org/10.23919/IST-Africa56635.2022.9845613

Enete, A. A., & Amusa, T. A. (2010). Challenges of agricultural adaptation to climate change in Nigeria: A synthesis from the literature. *Field Actions Science Reports. The Journal of Field Actions, 4*.

Enu, K. B., Zingraff-Hamed, A., Rahman, M. A., Stringer, L. C., & Pauleit, S. (2023). Potential of nature-based solutions to mitigate hydro-meteorological risks in sub-Saharan Africa. *Natural Hazards and Earth System Sciences, 23*(2), 481–505.

Food and Agriculture Organization of the United Nations (FAO). (2013). *Climate-smart agriculture sourcebook*. FAO.

Gebreyesus, A., Tesfay, G., Tesfay, M., & Meheretu, Y. (2022). Opportunities and challenges of community-based tourism development in Southern Tigray, Ethiopia. *African Journal of Hospitality, Tourism and Leisure, 11*(3), 938–954. https://doi.org/10.46222/ajhtl.19770720-267

Gwenzi, J., Mashonjowa, E., & Mafongoya, P. L. (2021). Barriers and limits to climate change management for locally developed adaptation strategies: Case study of uzumba maramba Pfungwe District, Zimbabwe. In W. Leal Filho, J. Luetz, & D. Ayal (Eds.), *Handbook of climate change management* (pp. 1–17). Springer.

Haque, C. E., & Burton, I. (2005). Adaptation options strategies for hazards and vulnerability mitigation: an international perspective. *Mitigation and Adaptation Strategies for Global Change, 10*, 335–353.

Hussain, S., Amin, A., Mubeen, M., Khaliq, T., Shahid, M., Hammad, H. M., Sultana, S. R., Awais, M., Murtaza, B., Amjad, M., & Fahad, S. (2022). Climate smart agriculture (CSA) technologies. In *Building climate resilience in agriculture: Theory, practice and future perspective* (pp. 319–338).

Imperiale, A. J., & Vanclay, F. (2023). Understanding the social dimensions of resilience: The role of the social sciences in disaster risk reduction, climate action and sustainable development. Sustainable Development. John Wiley and Sons Ltd. https://doi.org/10.1002/sd.265

Karimi, V., Valizadeh, N., Rahmani, S., Bijani, M., & Karimi, M. (2022). Beyond climate change: Impacts, adaptation strategies, and influencing factors. In S. A. Bandh (Ed.), *Climate Change*. Springer. https://doi.org/10.1007/978-3-030-86290-9_4

Korber, S., & McNaughton, R. B. (2017). Resilience and entrepreneurship: a systematic literature review. *International Journal of Entrepreneurial Behavior & Research, 41*(6), 519–539.

Kotze, D. C., & Wood, A. P. (2021). Assessing the long-term ecological sustainability of dambo cultivation in Southern Africa: Ten-year case studies from Zambia and Malawi. *Wetlands, 41*, 21. https://doi.org/10.1007/s13157-021-01399-5

Kpadonou, R. A. B., Owiyo, T., Barbier, B., Denton, F., Rutabingwa, F., & Kiema, A. (2017). Advancing climatesmart-agriculture in developing drylands: Joint analysis of the adoption of multiple on-farm soil and water conservation technologies in West African Sahel. *Land Use Policy, 61*, 196–207.

Kruger, L. (2016). The timing of agricultural production in hazard-prone areas to prevent losses at peak-risk periods: A case of Malawi, Madagascar and Mozambique. *Jàmbá: Journal of Disaster Risk Studies, 8*(2), 1–9.

Mabuku, M. P., Senzanje, A., Mudhara, M., Jewitt, G., & Mulwafu, W. (2018). Rural households' flood preparedness and social determinants in Mwandi district of Zambia and Eastern Zambezi Region of Namibia. *International Journal of Disaster Risk Reduction, 28*, 284–297. https://doi.org/10.1016/j.ijdrr.2018.03.014

Makoelle, T. M., & Malindi, M. (2015). Resilience: A framework for inclusive pedagogy in a South African context. *Electronic Journal for Inclusive Education, 3*(3), 2.

Mapfumo, P., Jalloh, A., & Hachigonta, S. (2014). *Review of research and policies for climate change adaptation in the agriculture sector in Southern Africa*. Agricultural Research and Development (CORAF/WECARD) and Future Agricultures Consortium.

Mavhura, E., Manyangadze, T., Mudavanhu, C., & Pedzisai, E. (2022). An assessment of riparian communities' preparedness to flood risk: the case of Mbire communities in Zimbabwe. *GeoJournal, 87*, 1671–1693. https://doi.org/10.1007/s10708-020-10329-7

Meriläinen, E. (2023). *Conceptualizations of resilient community for social futures, nourishing tomorrow: Clean engineering and nature-friendly living*. World Scientific Publishing Co., pp. 281–316. https://doi.org/10.1142/9789811264375_0009

Meyer, D. F. (2014). Exploration of solutions for revitalisation of rural areas in South Africa. *Mediterranean Journal of Social Sciences, 5*(4), 613–625.

Moher, D., Tetzlaff, J., Tricco, A. C., Sampson, M., & Altman, D. G. (2007). Epidemiology and reporting characteristics of systematic reviews. *PLoS Medicine, 4*(3), e78.

Mpala, T. A., & Simatele, M. D. (2024). Climate-smart agricultural practices among rural farmers in Masvingo district of Zimbabwe: Perspectives on the mitigation strategies to drought and water scarcity for improved crop production. *Frontiers in Sustainable Food Systems, 7*, 1298908. https://doi.org/10.3389/fsufs.2023.1298908

Mpandeli, S., Naidoo, D., Mabhaudhi, T., Nhemachena, C., Nhamo, L., Liphadzi, S., Hlahla, S., & Modi, A. T. (2018). Climate change adaptation through the water-energy-food nexus in southern Africa. *International Journal of Environmental Research and Public Health, 15*(10), 2306.

Muzorewa, W., & Chitakira, M. (2021). Climate-smart livelihood strategies in rural and urban communities in eastern Zimbabwe: An in-depth literature study. *South African Geographical Journal, 103*(3), 395–410.

Muzorewa, W., & Chitakira, M. (2022). Evaluating the role of climate-smart agriculture towards sustainable livelihoods in Mutare district, Zimbabwe. *Future of Food: Journal on Food, Agriculture and Society, 10*(1). https://doi.org/10.17170/kobra-202110144900

Ncoyini, Z., Savage, M. J., & Strydom, S. (2022). Limited access and use of climate information by small-scale sugarcane farmers in South Africa: A case study. *Climate Services, 26*, 100285.

Nhamo, L., Ndlela, B., Mpandeli, S., & Mabhaudhi, T. (2020). The water–energy–food nexus as an adaptation strategy for achieving sustainable livelihoods at a local level. *Sustainability, 12*, 8582.

Nkombi, Z., & Wentink, G. J. (2022). The role of public participation in disaster risk reduction initiatives: The case of Katlehong township. *Jàmbá: Journal of Disaster Risk Studies, 14*(1), a1203. https://doi.org/10.4102/jamba.v14i1.1203

Revell, K. (2010, September). Working with informality: increasing resilience in cities of the Global South. In *46th ISOCARP Congress* (pp. 1–13).

Revermann, R., Krewenka, K. M., Schrniedel, U., Olwoch, J. M., Helmschrot, J., & Jurgens, N. (Eds.). (2018). *Climate change and adaptive land management in southern Africa - Assessments, changes, challenges, and solutions* (*Biodiversity & Ecology*) (Vol. 6). Klaus Hess Publishers.

Ruben, R., & Pender, J. (2004). Rural diversity and heterogeneity in less-favoured areas: the quest for policy targeting. *Food Policy, 29*(4), 303–320.

Shereni, N. C., & Saarinen, J. (2021). Community perceptions on the benefits and challenges of community-based natural resources management in Zimbabwe. *Development Southern Africa, 38*(6), 879–895.

Shoko, K. D. (2023). The psychosocial impact of climate change among smallholder farmers: A potential threat to sustainable development. *Frontiers in Psychology*, 1–12.

Sibiya, N., Sithole, M., Mudau, L., & Simatele, M. D. (2022). Empowering the voiceless: securing the participation of marginalised groups in climate change governance in South Africa. *Sustainability, 14*(12), 7111. https://doi.org/10.3390/su14127111

Spear, D. D., Haimbili, E., Baudoin, M. A. D., Hegga, S. D., Zaroug, M. D., Okeyo, A., & Angula, M. (2018). Vulnerability and adaptation to climate change in semi-arid areas in Southern Africa.

Stedman, S., Mnyimbiri, A. M., Kawonga, Z., Malota, M., Njera, D., Hall, R. P., & Holm, R. H. (2018). Using irrigation to kick-start multiple-use water services for small-scale farmers in Malawi: A case study of the Nkhata Bay District. *Irrigation and Drainage, 67*(5), 645–653.

Syahrir, R., Wall, F., & Diallo, P. (2021). Coping with sudden mine closure: The importance of resilient communities and good governance. *The Extractive Industries and Society, 8*(4), 101009. https://doi.org/10.1016/j.exis.2021.101009

Taylor, M. (2018). Climate-smart agriculture: What is it good for? *The Journal of Peasant Studies, 45*(1), 89–107.

Tubiello, F. N., & Van der Velde, M. (2007). Land and water use options for climate change adaptation and mitigation in agriculture. *SOLAW Background Thematic Report—TR04A*.

Wright, D. K. (2019). Long-term dynamics of pastoral ecology in northern Kenya: An old model for new resilience. *Journal of Anthropological Archaeology, 55*(101068), 1–15.

Zamasiya, B., Nyikahadzoi, K., & Mukamuri, B. B. (2017). Factors influencing smallholder farmers' behavioural intention towards adaptation to climate change in transitional climatic zones: A case study of Hwedza District in Zimbabwe. *Journal of Environmental Management, 198*, 233–239.

Zinhiva, H., Chitakira, M., & Mukwada, G. (2021). Factors associated with the utilisation of indigenous ecological knowledge in the planning and management of biotic resources in Masvingo Province, Zimbabwe. *Indilinga African Journal of Indigenous Knowledge Systems, 20*(1), 81–101.

Zseleczky, L., & Yosef, S. (2014). Are shocks becoming more frequent or intense? Resilience for food and nutrition security. In S. Fan, R. Pandya-Lorch, & S. Yosef (Eds.), *Resilience for food and nutrition security* (pp. 9–17). International Food Policy Research Institute (IFPRI).

Climate Resilience and Global Pandemics: The Southern African Perspective

9

Evans Chazireni, Gladman Jekese, and Isaac Nyambiya

9.1 Introduction

Many pandemics have impacted the world for millennia. These include the flu (1889–1890), Spanish flu (1918–1920), AIDS (1981–1988) and H1N1 swine flu pandemic (2009–2010) (Hilfiker & Ganguly, 2020). Such disease outbreaks sweep through many societies, causing widespread negative impacts and leaving significant marks on various aspects, such as economic, financial, social and environmental, inducing irreversible changes (Ganguly et al., 2020). Successful management of the pandemic depends on the measures taken to improve response and recovery procedures (Xu et al., 2021). Different aspects of human existence are affected by pandemics. This chapter places more emphasis on the social well-being of communities because, if they can survive the pandemic, they will be better able to recover and make up for the adverse effects of population growth and the economy (Alonge et al., 2019).

E. Chazireni · I. Nyambiya
Department of Physics, Geography and Environment Science, Great Zimbabwe University, Masvingo, Zimbabwe
e-mail: echazireni@gzu.ac.zw; inyambiya@gzu.ac.zw

G. Jekese (✉)
Department of Information and Marketing Sciences, Midlands State University, Gweru, Zimbabwe
e-mail: jekeseg@staff.msu.ac.zw

Cheshmezangi (2020) maintains that climate-resilient communities can deal with the impacts of pandemics more effectively and recover sustainably. Literally, resilience means 'return to the past'. It can be defined as the measure of how well a system can withstand shocks while maintaining its prior resistance. Climate resilience to pandemics indicates people's ability to continue operating at their best during a pandemic crisis, and it includes institutional, social, demographic and economic dimensions (Cheshmezangi, 2020; Jovanovich et al., 2020). Extensive literature on the ability of communities to withstand pandemics in the face of climate change and its determinants has been published. Current research has focused on the resilience to the threat of pandemics in an era of climate. This implies that there is a dearth of research that offers a suitable framework for assessment in order to promote pandemic resistance to climate change (Afrin et al., 2021; Büyükozkan et al., 2022). Wang et al. (2021) remark that a better understanding of factors that contribute to resilience to pandemics in a changing climate provides opportunities to better prepare for future pandemics.

There are similarities between pandemics and climate change in that both provide shocks and stresses that amplify already existing socio-economic problems. They both reflect the tragedy of the commons, are retrogressive, are risk multipliers, generate shocks and require the same fundamental shift in approach to be addressed.

M. Matsa et al. (eds.), *Climate Change Resilience in Rural Southern Africa*, Sustainable Development Goals Series, https://doi.org/10.1007/978-3-031-73600-1_9

The consequences are felt worldwide. While climate change is a factor in the emergence of pandemics, the root causes of both pandemics and climate change are unquestionably related to unsustainable development (van der Voorn et al., 2022). This broader relationship shows that climate change can immensely contribute to pandemics and vice versa.

Rural Southern Africa is more vulnerable to climate change and pandemics due to low levels of development (Abdool Karim, 2020; Makate et al., 2019). It is characterised by poor communities with women, children, older populations and those with pre-existing health conditions, which makes them more vulnerable to pandemics and climate change (Saidi et al., 2023). The population in most parts of rural Southern Africa depends on indigenous knowledge, which can make it challenging to communicate resilience to climate change (Nyahunda & Tirivangasi, 2019). In this review, we focus on resilience to pandemics in the context of climate change in rural Southern Africa. This chapter is organised as follows. In Sect. 9.2, we give an overview of the systematic review of the methodology used in the literature. We also highlight the results of our literature search. In Sect. 9.3, we discuss our findings. Finally, in Sect. 9.4, we offer a conclusion.

9.2 Materials and Methods

9.2.1 Selection of Literature

The literature for this chapter came from various scientific resources such as IEEE (www.ieeexplore.ieee.org), Web of Science (www.webofscience.com), Scopus database (www.scopus.com), Google Scholar (www.scholar.google.com) and PubMed (www.pubmed.ncbi.nlm.nih.gov), all accessed on June 10, 2023. These multidisciplinary databases were used since they have international scope and are reliable (Pranckutė, 2021). They also have extensive citation indexing coverage, allowing for the extraction of the best data from scientific papers (Zhu & Liu, 2020). The keywords used as search criteria were 'climate', 'resilience', 'global', 'worldwide', 'universal' and 'pandemics', with added context from rural and Southern Africa to find articles relevant to the topic, and the search string is based on research results published after 2001. Zoonotic exposures have been occurring with increasing rapidity in recent years (Milbank & Vira, 2022). Between the years 2000 and 2022, a total of nine zoonotic outbreaks including COVID-19 have been declared public health emergencies of international concern (PHEIC), these being SARS (2001), H1N1 (2009), MERS-CoV (2012), EVD (2014 and 2019), polio (2014), Zika (2016), COVID-19 (2020) and monkeypox (2022). Reports indicate that between 2001 and 2022, sub-Saharan Africa alone experienced at least 800 public health emergencies (Moyo et al., 2023), which in itself shows the vulnerability of the region against its capacity to cope compared to the rest of the world.

The research outputs contain publicly available research publications, review papers, book chapters and case studies. We focus more on pandemic evidence and climate resilience in our chapter. On June 10, 2023, a preliminary search of the databases above produced 1169 articles. However, an expeditious cooperative perusal of the titles of the extracted articles revealed that 87% of the publications focused on individual resilience. Consequently, we eliminated studies that had nothing to do with pandemics and climate change resilience by using the database filtering capability. By using a semisystematic literature procedure, documents from sources other than database searches were made possible. The collected data were analysed using the following procedures: finding relevant literature and keywords, choosing relevant literature, critically evaluating research, distilling the synthesis of previously acquired research and documenting key findings.

9.2.2 Literature Analysis Procedures

The chosen research outputs were comprehensively examined to extract the information required for this study. The authors created a spreadsheet with the publications chosen on the rows and columns to gather data on a variety of topics before carefully examining the contents of the publications. These topics included the proper definition and factors that influence resistance to pandemics in the wake of climate change. Three phases of the analysis were established.

The initial author collected and collated the data and produced the first draft of the paper, which was then analysed and proofread by the second. The third author responded to the reviewer's comments and ensured that the quality of the paper met the required standard for semisystematic review by attending to all the reviewer's comments.

9.3 Results and Discussions

Pandemics are one of the most dangerous and catastrophic events. They are detrimental to societies (Gong et al., 2020). Pandemics are known to have no spatial boundaries, and this implies that they can diffuse on a greater scale than any other disaster and wreak huge damage to societies. Examples and effects of pandemics show how important it is to create more resilient societies to climate change. According to a number of studies, policymakers have learned a lot from pandemic outbreaks and how to develop communities that are more climate resilient (Elcheroth & Drury, 2020).

Moraci (2020) maintains the resilience of societies to pandemics in the face of climate change, including mitigation, absorption and adaptation. The first capacities of resilience, mitigation and absorption take place either before or right after the pandemic breaks out. Mitigation involves the minimisation of damages, and absorption refers to the systems' capacity to withstand unexpected events. Adaptation is when a system adjusts to changes caused by disasters or a pandemic. The study's findings demonstrate that pandemic resilience has multiple dimensions, including social, institutional, demographic and economic dimensions. This implies that developing pandemic-resilient communities in a changing climate demands a comprehensive approach among policymakers (Austin & Gregory, 2021).

9.3.1 Institutional Resilience to Pandemics

When pandemics occur, communities may not be able to function properly due to the emergency nature of such pandemics. Consequently, due to the decline in services, lives are disrupted, and communities often fail to return to previous stable situations (Büyükozkan et al., 2022). In this sense, the actions of local governments become more vital. Massaro et al. (2018) and Loxton (2020) concur that governments observe common interests, minimise possible causalities and restore normal development processes. During a pandemic outbreak, many essential commodities became scarce due to greater demand, and people generally found it difficult to survive (Loxton et al., 2020). Governments adopt strategies to manage outbreaks in societies by distributing essential goods, particularly medicines (Duffey & Zio, 2020; Chu et al., 2021).

Government institutions play a significant role in the management of the pandemic crisis. The ability of the government to prevent the outbreak of a pandemic, limit the spread, deliver essential services and help people recover from damages caused by the pandemic is called institutional or organisational resilience. The main facets of the institutional resilience of climate to pandemics are effective leadership and local government support. Proper local government decision-making and coordination are essential institutional factors. Managing pandemics through climate resilience requires adequate cooperation between government and communities. Massaro et al. (2018) and Duffey and Zio (2020) argue that in societies with lower individual participation, the effectiveness of government actions to manage the pandemic crisis becomes less.

9.3.2 Social Resilience to Pandemics

One of the most important dimensions of resilience to any pandemic or crisis is social resilience (Saja et al., 2019), which is the ability of societal groups to manage pandemic damages and shocks

(Jewett et al., 2021). Social groups are highly vulnerable during pandemics. Research shows that pandemics have a number of detrimental effects on societal groups (Alizadeh et al., 2023). Social networks are also highly vulnerable during pandemics through measures such as social distancing and lockdowns in which members of the same families and communities are separated or unnecessarily forced to endure long periods of unnatural contact. Societal links are crucial in mitigating shocks and minimising damages (Bukuluki et al., 2020). Social resilience does not impact communities in isolation, but it complements other resilience during pandemic crises (Rockenbauch & Sakdapolrak, 2017). Kim and Bostwick (2020) and Kimhi (2020) put it differently, asserting that institutional, demographic and economic systems are rendered ineffective in curbing pandemics if sufficient social resilience is absent in communities.

Components of social resilience include communication, reciprocal support, community safety and protection, public education and awareness (Chiang & Huang, 2016; Saja et al., 2019). Partnerships and public willingness are critical factors of social resilience, which take the shape of supportive structures (Rockenbauch & Sakdapolrak, 2017). These factors are highly beneficial in improving communities' ability to adapt to pandemics, with education and public awareness being further ways to reinforce this.

In Southern Africa, pandemics are a significant threat to human life, have a social impact on communities and significantly impact public behaviour (Maison et al., 2021). Social networks are critical to improving community awareness, and, against this background, these networks are essential factors in a community that is socially resilient to pandemics. It has been observed that ensuring social cohesion in sub-Saharan Africa during and after COVID-19 is important to impart resilience to affected communities (Semo & Frissa, 2020). It is interesting that in a study of emerging black adults in sub-Saharan Africa, social resilience was a perception factor, and youth in these communities felt safer from the pandemic if they assumed that their neighbourhoods were of a better standard of living (Cockcroft et al., 2023).

Research in sub-Saharan Africa has also demonstrated that communities can be efficiently mobilised within themselves through participation, education and information, thus increasing their awareness of prevailing pandemics and the quick intervention methods they can employ to avert morbidity and mortality (Owoyemi et al., 2021). Community-level participation and connectedness in these strategies are invariably pivotal for successful outcomes (Frimpong & Paintsil, 2023; Giwa-Daramola & James, 2023). These findings manifest the importance of social climate resilience before and after pandemic crises.

9.3.3 Economic Factors Affecting Resilience to Pandemics

In Southern Africa, like other parts of the world, economic structures are disrupted during pandemics as a result of market dynamics related to supply and demand (Bassett et al., 2021). Challenges such as inflation, unemployment and weakening of the gross national product are common. In modern times, Sub-Saharan Africa has had disruptions caused by Ebola virus disease (EVD) since the first outbreak in 1976. Sierra Leone, Liberia and Guinea have had their economies eroded by up to \$1.6 billion, representing 12% of the gross domestic product because of the 2013 EVD outbreak (Evans, 2015). It has been estimated that COVID-19, for instance, would add about 27–40 million people reduced to poverty in SSA, with 9.1–9.4% of the world population affected by extreme poverty (Saidi et al., 2023). At least about 25% of sub-Saharan Africans were adversely affected by the COVID-19 pandemic (Mashige et al., 2021), and lockdowns have been observed to essentially erode the savings of at least 30% of the working population in sub-Saharan Africa due to long periods of lay-off and no productivity (Teachout & Zipfel, 2020). Studies also revealed that remittances from the diaspora community to Africa fell, and by 2020, foreign direct investment was expected to fall by about 42% (Anyanwu & Salami, 2021).

In addition, high recovery costs and hunger are some of the inevitable negative impacts of pandemics on the economy (Duguma et al., 2021; Fan et al., 2021). Although individual businesses or industries in Southern Africa may grow significantly during a pandemic, actual data indicates that pandemics usually cause societies' overall economic progress to fall (Qian & Fan, 2020; Dai et al., 2021). Rahmadana and Sagala (2020) define economic climate resilience as the community's ability to meet basic necessities, produce goods and means of subsistence, limit damage and continue with development during a pandemic.

The ability of an economy and financial institutions to provide necessities, healthcare and livelihoods during pandemic outbreaks is another definition of the resilience of the economic climate to pandemics. Leach et al. (2021) and Beninger and Francis (2021) interpret economic climate resilience as economic flexibility in recovering from damages, with food security and financial resources being key factors. Access to sufficient food resources is a critical concern during pandemics. Integrating technologies into economic activities can modify the impacts of pandemics and business development during pandemic periods (Kumar et al., 2020).

9.3.4 Demographic Factors Affecting Resilience to Pandemics

The ability of people to improve a region's standard of living conditions and demographic dynamics is referred to as demographic climate resilience. A healthy demography is important to increasing climate resilience. COVID-19 has allowed researchers to analyse how demographic factors play a role in disease correlations and severity. For instance, a younger population, a lower population density and a lower rate of urbanisation meant that the numbers of mortalities and morbidities in sub-Saharan Africa were less than in more developed countries (Nguimkeu & Tadadjeu, 2021).

Evidence also suggests that the general mitigating factor against adverse events during COVID-19 was the lower age range of the African population despite the much poorer health facilities within these countries (Adams et al., 2021). As such, COVID-19 has been termed a disease of affluence on account of the more robust correlation standard of living between societies with a variety of underlying conditions such as obesity, diabetes, cancer, cardiovascular conditions and morbidity and mortality, when the reverse has been experienced in Africa (Zinsou et al., 2023). Africa has also been identified as the most physically active region in the world with a less sedentary lifestyle, and as such, in the case of COVID-19, this has disfavoured susceptibility (Wachira et al., 2020)

In Southern Africa, having people with good mental, physical and social health is essential to living in a community that is resistant to pandemics (Saghapour et al., 2021). Numerous researchers confirm that in societies with healthier citizens, the mortality rate as a result of pandemic infections is lower in Southern Africa (Aghapour et al., 2019; Austin & Gregory, 2021). Numerous researchers concur that societies with poorer public hygiene and fewer access to health facilities are less climate-resistant to pandemics (Tavares & Betti, 2020; Algahtani et al., 2021). The demographic resilience of climate change to pandemics is the ability to prevent illness and lower the risk of infection through personal hygiene, mental and physical well-being and quality of life (Ferreira et al., 2020).

9.3.5 Climate Change-Health Systems Nexus: The Southern African Narrative

The effects of climate change are rapidly increasing globally. Southern Africa, with a population of about 69,000,000, is not spared from such climate change impacts. Southern Africa represents one of the global climate change hot spots due to its inconsistent climatic regimes and high climatic disasters. Climate change has led to significant changes in temperatures. In the period 1961–2015 alone, mean annual temperatures in Southern Africa increased between 1.04 and 1.44 °C annually; the frequency of hot days has increased, while that of cold ones has decreased (Trisos et al., 2022). Precipitation is projected to decrease by 20%. The impacts are exacerbated by the already high burden of diseases, fragmented healthcare system and food and water insecurity (Myers et al., 2017). Climate change can have an impact on human health. In Southern Africa, the risks of malnutrition, malaria and diarrhoea are likely to increase as temperatures increase and rainfall becomes more variable (USAID, 2017).

Climate change-induced environmental degradation is not a new phenomenon in Southern Africa. Southern Africa has remarkable biodiversity values, which are progressively threatened by climate change (Vogel et al., 2019). Climate change is a major threat to Southern Africa's biodiversity on all scales, from genes to large biomes. The increasing loss of biodiversity and concomitant ecosystem degradation are alarming. The concern comes from the fact that the vast population of Southern Africa depends on ecosystem services. It is estimated that more than 10,000 people in Southern Africa live in hazard-prone areas, and their livelihoods depend vitally on agricultural practices exposed to hazards (Global Drought Observatory, 2019). People in Southern Africa, particularly in rural areas, depend on the ecosystem for essential services, including water, food, medicine, recreational, cultural and spiritual services. In Southern Africa, along the coasts, climate change-induced rise in sea level affects coastal areas, causing increased erosion, degradation of coral reefs and salination of water in river systems and groundwater. The highly valued mangroves in Southern Africa are experiencing more significant destruction due to extreme river flooding and large-scale cyclonal rains (Nikolau et al., 2017). High temperatures, erratic rainfall and extreme evapotranspiration demand, as well as the high intensity of human activities, exceed the resilience limits of ecosystems and trigger irreversible environmental degradation (IPCC, 2019). Thus, persistent droughts and other harsh climate variables experienced in many parts of the region negatively impact ecosystems.

Climate change has also had numerous impacts on human health and healthcare systems in Southern Africa. The negative impact of climate change on health is fundamental and uncontested. Numerous studies have recognised the impacts of climate change on health (Chersich et al., 2018; Milfont et al., 2021). The challenges of food, water, vector-borne diseases, undernutrition and mental health consequences can be attributed mainly to high temperatures, erratic precipitation and other extreme weather events that cause health threats to the Southern African region. As climate hazards occur, human pathogenic diseases increase. In Southern Africa, climate-induced human migration due to droughts and extreme weather events has been associated with an increased prevalence of HIV. South Africa is still plagued by high rates of HIV (human immunodeficiency virus) and tuberculosis (Lieber et al., 2021).

Furthermore, poorer management of the HIV pandemic related to overcrowding, malnutrition and poverty as a result of climate change is likely to increase the susceptibility of individuals to tuberculosis and other HIV-related morbidities (Lieber et al., 2021). Healthcare personnel and health facilities have a mammoth task of dealing with these climate-induced health impacts. Climate-induced health impacts are quite enormous, particularly for vulnerable groups such as children, women and people with low incomes.

Climate change increases people's exposure to pollutants related to water, leading to numerous types of illnesses. Some illnesses are caused by pathogens in water, such as bacteria, viruses, protozoa and blue-green algae. The outbreak of cholera has often been associated with heavy rains and floods, particularly in Southern African countries. In South Africa, precipitation and temperature increase the number of diarrhoeal cases, resulting in many deaths (Abu & Codjoe, 2018). In addition, there is an increase in tick distribution and tick-borne diseases in Southern Africa. The impact of floods from severe precipitation can cause death, and victims are compelled to leave their homes in Southern Africa. The mental health consequences of severe climate events are not spared due to related anxiety, depression, social dysfunction and loss of confidence. Heatwaves and the health effects related to high temperatures have recently gained attention in Southern Africa.

9.4 Conclusions

In the present research, it emerged that although many studies deal with different forms of disaster resilience, such as floods and earthquakes, a few studies address the resilience of climate change to pandemics. The literature review revealed that resilience to pandemics in the era of climate change has different paradigms, particularly those that are institutional, social, economic and demographic, that work together to minimise the impacts of the pandemics. Communities in Southern Africa must cooperate with experts to reduce the likelihood of infections. An integrated action plan is necessary to build pandemic resilience. It will drive and oversee teamwork across Southern Africa's many systems to manage potential damages, absorb unexpected events and facilitate rebuilding processes.

Building resilient societies should be an ongoing endeavour rather than something that happens before or during pandemics. To determine the degree of resistance of various populations to pandemics in light of climate change, researchers should employ the framework utilised in this study in conjunction with traditional assessments and classification methodologies in the future. By examining professional viewpoints, the suggested framework can be adjusted to the unique characteristics of various Southern African populations, thereby increasing the assessment of climate resilience.

References

Abdool Karim, S. S. (2020). The South African response to the pandemic. *New England Journal of Medicine, 382*(24), e95. https://doi.org/10.1056/NEJMc2014960

Abu, M., & Codjoe, S. N. A. (2018). Experience and future perceived risk of floods and diarrheal disease in urban poor communities in Accra, Ghana. *International Journal of Environmental Research and Public Health, 15*, 2830.

Adams, J., MacKenzie, M. J., Amegah, A. K., Ezeh, A., Gadanya, M. A., Omigbodun, A., Sarki, A. M., Thistle, P., Ziraba, A. K., Stranges, S., & Silverman, M. (2021). The conundrum of low COVID-19 mortality burden in Sub-Saharan Africa: Myth or reality? *Global Health Science and Practice, 9*(3), 433–443. https://doi.org/10.9745/GHSP-D-21-00172

Afrin, S., Chowdhury, F. J., & Rahman, M. (2021). COVID-19 pandemic: Rethinking strategies for resilient urban design, perceptions and planning. Available from 32, https://doi.org/10.3389/frsc.2021.668263. Accessed on 10 /10/2024.

Aghapour, A. H., Yazdani, M., Jolai, F., & Mojtahedi, M. (2019). Capacity planning and reconfiguration for disaster-resilient health infrastructure. *Journal of Building Engineering, 26*, 100853. https://doi.org/10.1016/j.jobe.2019.100853

Algahtani, F. D., Hassan, S. N., Alsaif, B., & Zrieq, R. (2021). Assessment of the quality of life during COVID-19 pandemic: A cross-sectional survey from the Kingdom of Saudi Arabia. *International Journal of Environmental Resilience and Public Health, 18*(3). https://doi.org/10.3390/ijerph18030847

Alizadeh, H., Sharifi, A., Damanbagh, S., Nazarnia, H., & Nazarnia, M. (2023). Impacts of the COVID-19 pandemic on the social sphere and lessons for crisis management: A literature review. *Natural Hazards, 117*(3), 2139–2164. https://doi.org/10.1007/s11069-023-05959-2

Alonge, O., Sonkarlay, S., Gwaikolo, W., Fahim, C., Cooper, J. L., & Peters, D. H. (2019). Understanding the role of community resilience in addressing the Ebola virus disease epidemic in Liberia: A qualitative study (community resilience in Liberia). *Global Health Action, 12*(1). https://doi.org/10.1080/16549716.2019.1662682

Anyanwu, J. C., & Salami, A. O. (2021). The impact of COVID-19 on African economies: An introduction. *African Development Review*, 1–16. https://doi.org/10.1111/1467-8268.12531

Austin, Z., & Gregory, P. (2021). Resilience in the time of pandemic: The experience of community pharmacists during COVID-19. *Research in Social and Administrative Pharmacy Journal, 17*(1), 1867–1875. https://doi.org/10.1016/j.sapharm.2020.05.027

Bassett, H. R., Lau, J. G., Suri, S. K., Advani, S., & Sharan, S. (2021). Preliminary lessons from COVID-19 disruptions of small-scale fishery supply chains. *World Development, 143*(2021), 105473. https://doi.org/10.1016/j.worlddev.105473

Beninger, S., & Francis, J. N. P. (2021). Resources for business resilience in a COVID-19 world: A community-centric approach. *Business Horizons*. https://doi.org/10.1016/j.bushor.02.048

Bukuluki, P., Mwenyango, H., Katongole, S. P., Sidhva, D., & Palattiyil, G. (2020). The socioeconomic and psychosocial impact of COVID-19 pandemic on urban refugees in Uganda. *Journal of Humanities and Social Sciences, 1*, 100045. https://doi.org/10.1016/j.ssaho.100045

Büyükozkan, G., Ilıcak, M., & Feyzio, O. (2022). A review of urban resilience literature. *Sustainable Cities and Society, 77*, 103579. https://doi.org/10.1016/j.scs.2021.103579

Chersich, M. F., Wright, C. Y., Venter, F., Rees, H., Scorgie, F., & Erasmus, B. (2018). Impacts of climate change on health and well-being in South Africa. *International Journal of Environmental Research and Public Health, 15*(9), 1884.

Cheshmezangi, A. (2020). The city in need: Urban resilience and city management in disruptive disease outbreak events. In *The City in Need, 2020.* [8] Mishra, S. V., Gayen, A., & Haque, S. M. COVID-19 and urban vulnerability in India. *Habitat International Journal, 103*, 102230. https://doi.org/10.1016/j.habitatint

Chiang, Y., & Huang, C. (2016). Exploring social resilience: Insights into climate change adaptation gaps, Taiwan. *Journal of Marine Science and Technology, 24*(6), 1081–1092. https://doi.org/10.6119/JMST-016-0928-2

Cockcroft, K., Greyling, M., Fouché, A., Ungar, M., & Theron, L. (2023). South African emerging adults' capacity for resilience in the face of COVID-19 stressors. *Journal of Health Psychology*. https://doi.org/10.1177/13591053231208620

Dai, R., Mookherjee, D., Quan, Y., & Zhang, X. (2021). Industrial clusters, networks and resilience to the COVID-19 shock in China. *Journal of Economic Behavior and Organization, 183*, 433–455. https://doi.org/10.1016/j.jebo.2021.01.017

Duffey, R. B., & Zio, E. (2020). COVID-19 pandemic trend modeling and analysis to support resilience decision-making. *Journal of Biology, 9*(7), 1–13. https://doi.org/10.3390/biology9070156

Duguma, L. A., van Noordwijk, M., Minang, P. A., & Muthee, K. (2021). COVID-19 pandemic and agroecosystem resilience: Early insights for building better futures. *Journal of Sustainability, 13*(3), 1–22. https://doi.org/10.3390/su13031278

Elcheroth, G., & Drury, J. (2020). Collective resilience in times of crisis: Lessons from the literature for socially effective responses to the pandemic. *British Journal of Social Psychology, 59*(3), 703–713. https://doi.org/10.1111/bjso.12403

Evans, D. (2015). *The economic impact of Ebola on Sub-Saharan Africa*. World Bank Group.

Fan, S., Teng, P., Chew, G., & Smith, L. (2021). Food system resilience and COVID-19—Lessons from the Asian experience. *Global Food Security, 28*, 100501. https://doi.org/10.1016/j.gfs.2021.100501

Ferreira, R. J., Buttell, F., & Cannon, C. (2020). COVID-19: Immediate predictors of individual resilience. *Sustainability, 12*(16). Available from https://doi.org/10.3390/su12166495. Accessed 10/10/2023.

Frimpong, S. O., & Paintsil, E. (2023). Community engagement in Ebola outbreaks in sub-Saharan Africa and implications for COVID-19 control: A scoping review. *International Journal of Infectious Diseases, 126*, 182–192. https://doi.org/10.1016/j.ijid.2022.11.032

Ganguly, S., Gupta, A., Arora, S., & Arora, N. (2020). Impact of pandemics. In *Integrated risk of pandemic: Covid-19 impacts, resilience and recommendations. Disaster resilience and green growth*. Springer. https://doi.org/10.1007/978-981-15-7679-9_5

Giwa-Daramola, D., & James, H. S. (2023). COVID-19 and microeconomic resilience in Sub Saharan Africa: A study on Ethiopian and Nigerian households. *Journal of Sustainability (Switzerland), 15*(9). https://doi.org/10.3390/su15097519

Global Drought Observatory. (2019). GDO Analytical Report: Drought in Southern Africa–January 2019.

Gong, H., Hassink, R., Tan, J., & Huang, D. (2020). Regional resilience in times of a pandemic crisis: The case of COVID-19 in China, Tijdschr. *Journal of Economic and Social Geography, 111*(3), 497–512. https://doi.org/10.1111/tesg.12447

Hilfiker, L., & Ganguly, S. (2020). A review of pandemics. In *Integrated risk of pandemic: Covid-19 impacts, resilience and recommendations. Disaster resilience and green growth*. Springer. https://doi.org/10.1007/978-981-15-7679-9_2

IPCC. (2019). Summary for policymakers. In P. R. Shukla, J. Skea, E. Calvo Buendia, V. Masson-Delmotte, H. O. Portner, D. C. Roberts, P. Zhai, R. Slade, S. Connors, R. van Diemen, M. Ferrat, E. Haughey, S. Luz, S. Neogi, M. Pathak, J. Petzold, J. Portugal Pereira, P. Vyas, E. Huntley, K. Kissick, & I. Belkacemi (Eds.), *And Malley, Journal of Climate Change and Land: An IPCC special report on climate change, desertification, land degradation, sustainable land management, food security, and greenhouse gas fluxes in terrestrial ecosystems, in press*.

Jewett, R. L., Mah, S. M., Howell, N., & Larsen, M. M. (2021). Social cohesion and community resilience during COVID-19 and pandemics: A rapid scoping review to inform the United Nations research roadmap for COVID-19 recovery. *International Journal of Health Services, 51*(3), 325–336. https://doi.org/10.1177/0020731421997092

Jovanovich, P., Klimek, O., Renn, R., Schneider, K., Øien, J., Brown, M., DiGennaro, Y., Liu, V., Pfau, M., Jeli'c, T., Rosen, B., Caillard, S., Chakravarty, P., & Chhantyal, A. (2020). Assessing resilience of healthcare infrastructure exposed to COVID-19: Emerging risks, resilience indicators, interdependencies and international standards. *Environment Systems and Decisions, 40*(Issue 2), 252–286. https://doi.org/10.1007/s10669-020-09779-8

Kim, S. J., & Bostwick, W. (2020). Social vulnerability and racial inequality in COVID-19 deaths in Chicago. *Health Education & Behavior, 47*(4), 509–513. https://doi.org/10.1177/1090198120929677

Kimhi, S., Marciano, H., Eshel, Y., & Adini, B. (2020). Resilience and demographic characteristics predicting distress during the COVID-19 crisis. *Social Science*

& Medicine, 265, 113389. https://doi.org/10.1016/j.socscimed.2020.113389

Kumar, S., Luthra, S. K., Mangla, Y., & Kazançoglu, A. (2020). COVID-19 impact on sustainable production and operations management. *Journal of Sustainable Operations and Computers, 1*, 1–7. https://doi.org/10.1016/j.susoc.2020.06.001

Leach, M., MacGregor, H., Scoones, I., & Wilkinson, A. (2021). Post-pandemic transformations: How and why COVID-19 requires us to rethink development. *World Development, 138*, 105233. https://doi.org/10.1016/j.worlddev.2020.105233

Lieber, M., Chin-Hong, P., Whittle, H. J., Hogg, R., & Weiser, S. D. (2021). The synergistic relationship between climate change and the HIV/AIDS epidemic: A conceptual framework. *AIDS and Behavior, 25*(7), 2266–2277.

Loxton, M., Truskett, R., Scarf, B., Sindone, L., Baldry, G., & Zhao, Y. (2020). Consumer behaviour during crises: Preliminary research on how coronavirus has manifested consumer panic buying, herd mentality, changing discretionary spending and the role of the media in influencing behavior. *Journal of Risk Finance and Management, 13*(8), 166. https://doi.org/10.3390/jrfm13080166

Maison, D., Jaworska, D., Adamczyk, D., & Affeltowicz, D. (2021). The challenges arising from the COVID-19 pandemic and the way people deal with them. A qualitative longitudinal study. *PLoS One, 16*(10), e0258133. Available from https://doi.org/10.1371/journal.pone.0258133. Accessed 09/09 2023.

Makate, C., Makate, M., Mango, N., & Siziba, S. (2019). Increasing resilience of smallholder farmers to climate change through multiple adoption of proven climate-smart agriculture innovations. Lessons from Southern Africa. *Journal of Environmental Management, 231*, 858–868. https://doi.org/10.1016/j.jenvman.2018.10.069

Mashige, K. P., Osuagwu, U. L., Ulagnathan, S., Ekpenyong, B. N., Abu, E. K., Goson, P. C., Langsi, R., Nwaeze, O., Timothy, C. G., Charwe, D. D., Oloruntoba, R., Miner, C. A., Ishaya, T., Ovenseri-Ogbomo, G. O., & Agho, K. E. (2021). Economic, health and physical impacts of covid-19 pandemic in sub-Saharan African regions: A cross sectional survey. *Risk Management and Healthcare Policy, 14*(June), 4799–4807. https://doi.org/10.2147/RMHP.S324554

Massaro, E., Ganin, A., Perra, N., Linkov, I., & Vespignani, A. (2018). Resilience management during large-scale epidemic outbreaks. *Scientific Reports, 8*(1), 1–9. https://doi.org/10.1038/s41598-018-19706-2

Milbank, C., & Vira, B. (2022). Wildmeat consumption and zoonotic spillover: Contextualising disease emergence and policy responses. *The Lancet Planetary Health, 6*(5), e439–e448. https://doi.org/10.1016/S2542-5196(22)00064-X

Milfont, T. L., Zubielevitch, E., Milojev, P., & Sibley, C. G. (2021). Ten-year panel data confirm generation gap but climate beliefs increase at similar rates across ages. *Nature Communications, 12*(1), 1–8.

Moraci, F., Errigo, M. F., Fazia, C., Campisi, T., & Castelli, F. (2020). Cities under pressure: Strategies and tools to face climate change and pandemic. *Journal of Sustainability, 12*(18), 1–31. https://doi.org/10.3390/su12187743

Moyo, E., Mhango, M., Moyo, P., Dzinamarira, T., Chitungo, I., & Murewanhema, G. (2023). Emerging infectious disease outbreaks in sub-Saharan Africa: Learning from the past and present to be better prepared for future outbreaks. *Frontiers in Public Health, 11*. https://doi.org/10.3389/fpubh.2023.1049986

Myers, J., Young, T., Galloway, M., Manyike, P., & Tucker, T. (2017). Responding to climate change in southern Africa–the role of research. *South Africa Medicine Journal, 11*, 820–822. https://doi.org/10.10520/EJC67456. PMID: 22272964.

Nguimkeu, P., & Tadadjeu, S. (2021). Why is the number of COVID-19 cases lower than expected in sub-Saharan Africa? A cross-sectional analysis of the role of demographic and geographic factors. *World Development, 138*, 105251. https://doi.org/10.1016/j.worlddev.2020.105251

Nikolau, D. K., Macamo, C. D., Bandeira, S. O., Taju, A., & Mabilana, H. A. (2017). Mangrove change detection, structure and condition in a protected area of eastern Africa: The case of Quirimbas National Park, Mozambique. *WIO Journal of Marine Science, 16*, 47–60.

Nyahunda, L., & Tirivangasi, H. M. (2019). Challenges faced by rural people in mitigating the effects of climate change in the Mazungunye communal lands, Zimbabwe. *Journal of Disaster Risk Studies, 11*(1), 1–9. https://doi.org/10.4102/JAMBA.V11I1.596

Owoyemi, A., Okolie, E. A., Omitiran, K., Amaechi, U. A., Sodipo, B. O., Ajumobi, O., Nnaji, C. E., & Okedo-Alex, I. N. (2021). Importance of community-level interventions during the COVID-19 pandemic: Lessons from sub-Saharan Africa. *American Journal of Tropical Medicine and Hygiene, 105*(4), 879–883. https://doi.org/10.4269/ajtmh.20-1533

Pranckutė, R. (2021). Web of Science (WoS) and Scopus: The titans of bibliographic information in today's academic world. *Publications, 9*(1). https://doi.org/10.3390/publications9010012

Qian, Y., & Fan, W. (2020). Who loses income during the COVID-19 outbreak? Evidence from China. *Research in Social Stratification and Mobility, 68*, 100522. https://doi.org/10.1016/j.rssm.2020.100522

Rahmadana, M. F., & Sagala, G. H. (2020). Economic resilience dataset in facing physical distancing during COVID-19 global pandemic. *Data in Brief, 32*, 106069. Available from https://doi.org/10.1016/j.dib.2020.106069. Accessed on o9/10/2023.

Rockenbauch, T., & Sakdapolrak, P. (2017). Social networks and the resilience of rural communities in the global south: A critical review and conceptual reflections. *Ecology and Society, 22*(1). https://doi.org/10.5751/ES-09009-220110

Saghapour, T., Giles-Corti, B., Jafari, A., & Qaisrani, G. (2021). Supporting pandemic disease preparedness:

Development of a composite index of area vulnerability. *Health & Place*. Available from https://doi.org/10.1016/j.healthplace.102629. Accessed on 09/08/2023.

Saidi, Y., Labidi, M. A., & Ochi, A. (2023). Economic growth and extreme poverty in sub-Saharan African countries: Non-linearity and governance threshold effect. In *Journal of the Knowledge Economy (issue 0123456789)*. Springer US. https://doi.org/10.1007/s13132-023-01421-7

Saja, A. M. A., Goonetilleke, A., Teo, M., & Ziyath, A. M. (2019). A critical review of social resilience assessment frameworks in disaster management. *International Journal of Disaster Risk Reduction, 35*, 101096. https://doi.org/10.1016/j.ijdrr.2019.101096

Semo, B. W., & Frissa, S. M. (2020). The mental health impact of the covid-19 pandemic: Implications for sub-Saharan Africa. *Psychology Research and Behavior Management, 13*, 713–720. https://doi.org/10.2147/PRBM.S264286

Tavares, F. F., & Betti, G. (2020). The pandemic of poverty, vulnerability, and COVID-19: Evidence from a fuzzy multidimensional analysis of deprivations in Brazil. *World Development, 139*(2021). https://doi.org/10.1016/j.worlddev.105307

Teachout, M., & Zipfel, C. (2020, May 17). *The economic impact of COVID-19 lockdowns in sub-Saharan Africa*. International Growth Centre.

Trisos, C. H., Adelekan, I. O., Totin, E., Ayanlade, A., & Efitre, J. (2022). Africa. In H.-O. Portner, D. C. Roberts, M. Tignor, E. S. Poloczanska, K. Mintenbeck, A. Allegri, et al. (Eds.), *Climate change: Impacts, adaptation and vulnerability. Contribution of Working Group II to the sixth assessment report of the intergovernmental panel on climate change* (pp. 1285–1455). Cambridge University Press. https://doi.org/10.1017/9781009325844.011

USAID. (2017). *Risk expands, but opportunity awaits: Emerging evidence on climate change and health in Africa*. Available at: https://www.climatelinks.org/sites/default/files/asset/document/ 170414_ATLAS%20project_SSA%20Health

van der Voorn, T., van den Berg, C., Quist, J., & Kok, K. (2022). Making waves in resilience: Drawing lessons from the COVID-19 pandemic for advancing sustainable development. *Current Research in Environmental Sustainability, 4*, 100171. https://doi.org/10.1016/j.crsust.2022.100171. Epub 2022 Jun 13. PMID: 35720270; PMCID: PMC9189097.

Vogel, C., Steynor, A., & Manyuchi, A. (2019). Climate services in Africa: Re-imagining an inclusive, robust and sustainable service. *Climate Services, 15*, 100107.

Wachira, L., Arena, R., Sallis, J. F., Lambert, E. V., Martin, O., Laddu, D. R., Onywera, V., & Oyeyemi, A. L. (2020). Why are COVID-19 effects less severe in sub-Saharan Africa? Moving more and sitting less may be a primary reason. *Progress in Cardiovascular Diseases, 71*(January), 103–105. https://doi.org/10.1016/j.pcad.2022.04.012

Wang, C., Zhang, T., Xu, W., Ruan, H., & Tang, J. (2021). Social capital, technological empowerment, and resilience in rural China. *International Journal of Environmental Research and Public Health, 18*(22), 11883. https://doi.org/10.3390/ijerph182211883

Xu, W., Xiang, L., Proverbs, D., & Xiong, S. (2021). The influence of COVID-19 on community disaster resilience. *International Journal of Environmental Research and Public Health, 18*(1), 1–18. https://doi.org/10.3390/ijerph18010088

Chu, Z., Cheng, M., & Song, M. (2021). What determines urban resilience against COVID-19: City size or governance capacity? *Sustainable Cities and Society, 75*, 103304. https://doi.org/10.1016/j.scs.2021.103304

Zhu, J., & Liu, W. (2020). A tale of two databases: The use of web of science and scopus in academic papers. *Scientometrics, 123*(1), 321–335. https://doi.org/10.1007/s11192-020-03387-8

Zinsou, B. E., Letourneur, D., Siko, J., de Souza, R. M., Adjagba, F., & Pineau, P. (2023). Main modulators of COVID-19 epidemic in sub-Saharan Africa. *Heliyon, 9*(1). https://doi.org/10.1016/j.heliyon.2022.e12727

10 Climate Resilience Strategies and Policies in Southern Africa: A Review of Literature

Rameck Defe and Gideon Mutanda

10.1 Introduction

Resilience is increasingly seen as a unifying concept and policy instrument that uses development approaches to address the chronic vulnerability of populations exposed to recurrent shocks and stressors (Zseleczky & Yosef, 2014). Global economic losses from disasters are increasing largely because of the increasing vulnerability of people and their assets, and there is need for resilience planning (Oxley, 2014). Resilience planning is a process that communities can undertake to identify potential hazards and threats, and then establish adaptation, mitigation, and recovery plans (Sharifi, 2016). Resilience building has emerged as a way to capture how communities can prepare for disasters, which helps reduce the vulnerability and risk associated with natural and man-made hazards. Planning for resilience can reduce future disaster-related response and recovery costs and improve recovery time following natural or human-caused hazard events. The goal of the resilience planning programme is to plan, design, and execute solutions that improve the security and resilience of critical infrastructure against various threats in collaboration with officials from the federal, state, local, tribal, and territory governments as well as infrastructure owners and operators (Lewis, 2019; Wiseman et al., 2014).

Resilience planning requires a community to identify one or more hazards that could disrupt life and create hardship. Hazards of greatest concern will be unique to individual, communities/utilities. Many factors influence the types of hazards considered during the resilience planning process (Hung et al., 2016). The Sixth Assessment Report by the Intergovernmental Panel on Climate Change (IPCC) predicts that the average temperature in Southern Africa will rise by close to 4 °C. Thus, increased surface temperature is one of the major hazards facing Africa in general, and Southern Africa in particular (IPCC, 2021a, b). Human-induced climate change will cause African countries to experience more frequent extreme weather events like floods and droughts (Mashura, 2018). The vulnerability of the African continent will be worsened by chronic poverty, limited infrastructure, and a limited capacity to cope with climate-related emergencies (Mabvurira & Chikwaiwa, 2018). Such issues identified in Southern Africa justify adoption of resilience planning initiatives (Mitchell, 2013; Nhamo et al., 2019; Ngwenya, 2021).

R. Defe (✉)
Department of Geography, Environmental Sustainability and Resilience Building, Midlands State University, Gweru, Zimbabwe
e-mail: defer@staff.msu.ac.zw

G. Mutanda
Institute for Corporate Citizenship, University of South Africa, Pretoria, South Africa
e-mail: mutangw@unisa.ac.za

M. Matsa et al. (eds.), *Climate Change Resilience in Rural Southern Africa*, Sustainable Development Goals Series, https://doi.org/10.1007/978-3-031-73600-1_10

Africa is the most vulnerable continents and as such climate change has increased widespread adverse impacts on nature and people increasing the vulnerability of the sub-region as it is predominantly dependent on climate-sensitive sectors such as agriculture, tourism, biodiversity, and ecosystems (SADC, 2010a, b; IPCC, 2021a, b). The Southern Africa which is already bearing the brunt of human-induced climate change and is already burdened with developmental challenges is vulnerable to climate change. The vulnerability of Africa is further exacerbated by its considerable development constraints (IPCC, 2021a, b).

The extreme climatic events being experienced in Southern Africa, especially the El Nino-related droughts, are negatively impacting the inhabitants and economies of the region (SADC, 2010a, b). UNEP and ICRAF (2006) argue that, more than anywhere else, understanding the link between climate change and development in Southern Africa is crucial as the region's national economies are anchored on agriculture and other climate-sensitive sectors. The warming of Southern Africa's climate more than the global average negatively affects the sub-region's chances of meeting Agenda 2030 and realising the Africa Agenda 2063 of "an integrated, prosperous, and peaceful Africa driven by its citizens" (African Union, 2022). Climate change, which is already wreaking havoc on vulnerable communities, is increasingly impeding Africa's ability to achieve its Agenda 2063 and SDG goals. In response to these threats, the African Union developed the African Union Climate Change and Resilient Development Strategy and Action Plan (2022–2032) to unlock the continent's potential in building climate-resilient communities and economies (African Union, 2022).

However, realization of Africa's vision requires proactive collective sub-regional efforts, policies, and strategies to make climate-resilient communities and economies (Kenny, 2017; Addaney et al., 2023). Resilience is recognized as central to achieving the sustainable development goals (SDGs), that is, it is reflected in a range of SDG targets (Croese et al., 2020). Development initiatives being implemented targets to achieve SDG1-no poverty, SDG2-zero hunger, SDG3-good health and well-being, and SDG13-climate action. Building resilience requires an integrated approach and a long-term commitment to improving three livelihood critical capacities that is absorptive, adaptive, and transformative capacity (Bene et al., 2012; Frankenberger & Nelson, 2014).

Southern Africa is urged to think differently, think far ahead, and respond to the situation of tomorrow by investing in long-term programming to address this long-term climatic challenge (WFP, 2021). Yet, according to Naidoo et al. (2013), one of the biggest gaps in responding to climate change in some countries in Southern Africa is the lack of national policies, related strategies, and action plans to guide efforts at adaptation and mitigation. Thus, the region's response strategy to climate change should be informed by enabling policies and strategies for building the resilience of communities to the vagaries of climate change. Yet, currently, literature focusing on climate change policy landscape and resilience building is limited in Southern Africa. This paper therefore seeks to analyze climate resilience strategies and policies adopted to reduce communities' vulnerability to climate change adversities.

10.2 Methodology

Research articles and publications considered for this review were obtained from peer-reviewed journals published in high-impact journals on disaster risk reduction, environmental planning, climate change management, and rural community development and policies. To search for relevant articles, keywords and phrases were used to ensure that the adopted articles fall within the confines of the study objectives. This study sought to address the following research questions: (1) Is there climate resilience planning in the context of Southern Africa's national development frameworks? (2) What climate change-related and sector-specific policies are being implemented in Southern African countries to promote resilient societies, ecosystems, and economies? (3) What are the significance and limitations facing resilience policies in Southern Africa? The study is based on a desk review of literature on climate resilience planning and regional development policies that integrate climate change issues. These were reviewed, taking note of how they acted as guidelines in the formulation of climate change resilience policies. It also reviewed literature on climate change-related policies, sector-specific policies, and frameworks that were partly formulated with climate action in mind. Documents on regional policies related to climate change and sector-specific policies that mainstreamed issues on climate change adaptation, mitigation, and resilience were obtained from the Internet. The study targeted websites for the African Union, SADC, Southern African countries' policy frameworks (national development plans and country's sector-specific policies) and development partners such as UNEP, UNDP, and USAID, as they have interests in development issues and threats to development such as climate change.

10.3 Summary of Documents Reviewed

Research articles and publications considered for this review were obtained from peer-reviewed journals, organization websites, and progress reports. The review focused on Southern Africa, since the topic and study objectives focused on climate resilience policies and strategies for the region. After screening and evaluating all sources consulted, it was established that, 69% were research papers, 12.3% were review articles, 9.2% were book chapters, 5% were company/resilience-based organizations, and 4.6% were conference papers.

Developing countries are facing developmental challenges that have led them to formulate developmental plans to holistically deal with these challenges. Climate change as a crisis multiplier is exacerbating the magnitude and impact of these development challenges. People and places with significant development limits are particularly vulnerable to climate dangers (IPCC, 2021a, b). Climate change, if left unaddressed, is likely to intensify poverty, inequality, food insecurity, and health issues, amongst other difficulties, which could undo years of development progress in certain nations (World Economic and Social Survey, 2016).

This then stresses the importance of embedding climate change issues into the developmental plans and objectives of developing countries. Treating climate change issues in isolation from developmental plans will not only derail the resolution of development challenges but also make it difficult to achieve the SDGs as they are indivisible and interdependent. There is a need for developing countries to ensure that their development plans at all levels of government integrate climate change issues if economies and communities are to be climate resilient. Developing nations can increase their economic growth and achieve SDGS while strengthening their ability to withstand climate change by addressing climate challenges in the context of their development objectives (USAID, 2013).

Integration of climate change policies is an appreciation that development planned outside the context of climate change is doomed to fail as climate change affects every sector of the economy and the livelihoods of people. As climate change threatens development objectives and is influenced by development decisions, planning for climate resilience is a step away from "development as usual," which unintentionally increases vulnerability (Organisation for Economic Co-Operation and Development, 2009). The United Nations Economic Commission (2020) reports that the 2030 Agenda's SDGs 11.a and 11.b call for governments to adopt and implement integrated climate change measures into national policies and plans. This strategy aids developing nations in maintaining development benefits despite a changing climate. Developing countries cannot truly develop themselves when climate change affects and jeopardizes their national development efforts; hence, climate change issues should be included in development programming and plans (USAID, 2013). Since climate change impacts cut across multiple scales and sectors, planning for climate change resilience requires input from various levels of government, including the federal government, state government, and local government (Berke et al., 2021). The central government should set the tone and direction by mainstreaming climate change issues into the country's national blueprints and sector-specific policies as part of climate resilience planning. The national level is critical for mainstreaming climate change issues, as strategic decisions that affect all sectors and scales are made at this level. It is at this level that medium- to long-term development and poverty reduction strategies and targets are defined through national visions, national development plans, and national development strategies. When national policies and planning frameworks are designed through the prism of climate change, this can result in the identification of particularly climate change-vulnerable communities, geographic areas, or sectors. It is an emerging trend today to integrate climate dimensions into policy-making to ensure that development enhances climate resilience so as to minimise the risk of inadvertent increases in exposure or sensitivity and, in turn, support the achievement of development objectives (OECD, 2014). Table 10.1 shows findings on the inclusion of climate change issues in SADC's national development policies. All the countries have embedded climate change issues in their medium- to long-term development plans, an indication of the realization of climate change as a serious threat to development in Southern Africa. Countries like South Africa, Namibia, Botswana, and Zambia have embedded environmental sustainability and climate change issues as a complete chapter, while Zimbabwe and Lesotho have integrated them as cross-cutting issues affecting all economic sectors. Mauritius's national development plan integrated climate change and disaster risk management issues though it did not dedicate a chapter on that. However, it is indisputable to argue that countries in Southern Africa are realizing that climate change and associated disasters are real, and there is a need for appropriate responses from the national to the local level.

10.4 Climate Resilience Policies in Southern Africa

The Pan-African vision of an integrated, prosperous, and peaceful Africa cannot be achieved in the face of global environmental challenges like climate change, which reverse developmental gains. Africa is made up of sub-regions such as Southern Africa; hence, an integrated Africa is achievable if each region and individual countries domesticate the Pan-African vision in their developmental and climate change policies, strategies, and action plans. Under this aspiration, the agenda seeks to have climate-resilient economies and communities by 2063, a realization of the importance of mainstreaming climate change in development issues and different sectors of the economy (African Union, 2022). The African Union further developed the Climate Change and Resilient Development Strategy and Action Plan, which provides an outline for a harmonized and coordinated approach to responding to climate change, setting out priorities and action areas for enhanced climate cooperation and climate resilient development pathways for the continent in pursuit of its objective of building resilient African communities, ecosystems, and economies (African Union, 2022). As a sub-region of the African continent, SADC had to tailor all its regional policies and programmes in order to feed into the 2063 Pan-African vision and related climate change action plan.

10.5 SADC's Climate Change and Resilience Frameworks

To achieve the aspiration of Agenda 2063 and the SADC vision, SADC developed the Climate Change Strategy and Action Plan (2015). The strategy lists the following priority interventions: development of appropriate meteorological technologies, promotion of climate-smart agricultural techniques, adoption of strategies for wide water conservation, development of climate-proofing strategies for infrastructure, mainstream climate change strategies in all sectors, promotion of research that builds climate resilience in

Table 10.1 Integration of climate change issues in national development frameworks

Author (s) and year	Findings
Government of Zimbabwe (2020)	The government has incorporated climate change considerations into development planning, including the NDS1 as one of its 12 objectives, which seeks to promote climate resilience. NDS1 mentions climate change and extreme climatic events as some environmental challenges facing the country. To promote SDG13 and some of its development objectives, NDS1 calls for climate action climate change adaptation, and climate change mitigation
Government of South Africa (2010)	Chapter 1 of NDP mentions climate change as one of the environmental challenges intertwined and closely linked to poverty, water scarcity, and food shortages in the country. The "primary approach to adapting to climate change is to strengthen the nation's resilience," which will be achieved once socio-economic objectives are met. NDP calls for climate change adaptation, mitigation, and a transition to a low-carbon economy, while making sure that the poor and vulnerable are protected from transitional costs, and considering the overall economic implications
Government of Botswana (2017)	Chapter 7 of the development plan recognizes climate change and extreme climate events as threats to the success of NDP 10 and future government programmes and different sectors. So NDP 11 through a climate change policy to be formulated will continue focusing on building sustainable and resilient economic sectors, societies, and institutions
Government of Lesotho (2018)	Lesotho's National Strategic Development Plan identifies climate change and extreme climate events as domestic threats affecting the country's development objectives and negatively impacting on agriculture, food security, poverty, and health. NSDP II calls for mainstreaming of climate Change across all sectors, adopt climate change adaptation and mitigation to promote climate resilient development
Government of Eswatini (2023)	National Development Plan for Eswatini identifies climate change and various extreme climate events as threats to the water, agriculture, industry, health, infrastructure, and energy sectors, which have led the country to adopt various policies. Climate change and disaster risk management are considered cross-cutting issues, while Chapter 8 of the plan identifies community resilience to climate change and natural disasters as a sectoral outcome. However, many emergency preparedness and climate response mechanisms remain underfunded, fragmented, and not well coordinated
Government of Zambia (2022)	Strategic Development Area 3 of the National Development Plan (8NDP) of Zambia places a strong emphasis on addressing climate change through mitigation and adaptation strategies, enhancing resilience to the impacts of climate change, improve adaptive capacity, reducing vulnerability, and enhancing resilience for human and biophysical systems
Government of Malawi (2017)	The policy aims to enhance resilience to climate change impacts and disasters, improve disaster preparedness, response, and recovery, and integrate climate change adaptation and mitigation strategies across various sectors. The policy reflects a commitment to proactive and coordinated efforts to manage climate change and disaster risks, with the goal of reducing vulnerability and enhancing the resilience of the population
Government of Mauritius (2021)	Given the country's vulnerability to climate change threatening sectors like agriculture, fisheries, forestry, and tourism, the National Development Plan (NDP5) is focusing on enhancing resilience and sustainable management of the environment. The country is transitioning towards a low-carbon and climate-resilient economy, focusing on climate sensitive sectors

Source: Authors

all sectors, etc. (SADC, 2020b). SADC adopted the SADC Resilience Framework 2020–2030 in 2020, whose aim is "to provide a broad strategic framework to act as a guide as member states and partner organisations develop and/or align their resilience strategies and prepare a number of mechanisms to guide DRR" (GIZ, 2020). The framework stresses that building more resilient societies and economics is crucial for attaining the Pan-African and SADC vision and objectives of the RISDP in the face of increasing disaster risks, pandemics, and conflicts in the region. It therefore calls for the adoption of an integrated approach to sustainable development, disaster risk and pandemic reduction, and promoting resilience to climate change if the SADC vision is to be actualized. Table 10.2 show findings on the adoption of climate change and resilience policies by some SADC countries. The few selected countries had comprehensive legal and policy frameworks covering climate change and disaster risk management issues that guided different sectors and levels of government in the domestication of climate change issues. However, implementation of the policies was a challenge due to poor coordination between different sectors and lack of financial support for effective implementation of the policies.

10.6 SADC Regional Agricultural Policy (RAP)

SADC's agricultural sector is highly vulnerable to the impacts of climate change and variability, as most smallholder farmers practice rain-fed agriculture (CCARDESA). In response, SADC developed the RAP in 2014, a sector-specific framework that was partly created with climate action and adaptation in mind to help SADC member states' agricultural sectors integrate climate issues (AFSA, 2020). One of RAP's four specific objectives is to reduce the vulnerability of the region's food and nutrition security status in a changing economic and climatic environment (CCARDESA). The provisions of the RAP on food security are further amplified by the SADC Food and Nutrition Security Strategy 2015–2025 (AFSA, 2020). The strategy outlines one of the priority intervention areas as dissemination of climate information, sharing of best practices on adaptation and mitigation of climate change (AFSA, 2020). Table 10.3 shows that all Southern African countries' agricultural policies advocated for various responses to ensure a climate resilient agricultural sector. Some of the stated measures to ensure climate-resilient agriculture include improving access to information, adopting drought-tolerant crop varieties, promoting research, and extending technologies and practices that mitigate risks, as well as developing early-warning and insurance systems. Many countries' agricultural policies are being revised to be in line with the RAP and also to mainstream climate change and disaster risk management if the agricultural sector is to remain vibrant and productive.

10.7 SADC Health Policy Framework

The SADC Health Policy Framework is a sector-specific integrated framework that mainstreams disaster risk reduction management as the region has not created a protocol on DRR or management (UN-Habitat, 2022). The SADC Health Policy Framework complements the RISDP's Social and Human Capital Development pillar, which calls for the development of modern and well-resourced health systems that are accessible and responsive in addressing the burden of disease and emergencies emanating from climatic extreme hazards and disasters. Objective 5 of the RISDP's Social and Human Capital Development pillar on cross-cutting issues calls for improved disaster risk management in support of regional resilience (SADC, 2020b). In alignment with these pillars, the SADC Health Policy Framework's Article 25 on emergency health services and disaster management states that parties shall (i) cooperate and assist each other in the coordination and management of disaster and emergency situations; (ii) develop awareness, risk reduction, preparedness, and management plans for natural and human-induced disasters;

Table 10.2 Findings on adoption of climate change and resilience policies by SADC countries

Author (s) and year	Findings
Government of Zimbabwe (2020)	Zimbabwe has adopted climate change and disaster policies geared towards adaptation, resilience, and sustainable development, with a focus on improving agricultural practices and water management though their effectiveness is challenged by governance issues and the need for better implementation and enforcement mechanisms
Nyahwo (2020)	While Zimbabwe has established policies and strategies to manage climate change and disasters, there are significant challenges in awareness, implementation, and support for vulnerable groups
Khavhagali et al. (2023)	The findings revealed South Africa adopted National Climate Change Adaptation Strategy of 2019, and the National Disaster Management Framework of 2005 though the policy is misaligned in setting and prioritizing objectives, lack of coordination during implementation, and inadequate monitoring mechanisms. The two policies do not fully align with the Sendai Framework for Disaster Risk Reduction
Beaumont (2017)	The country has several policies addressing climate change issues, including the National Climate Change Response Strategy (NCCRS) of 2004 and the Climate Change Bill, with a particular focus on adaptation strategies. SA's policies span across various levels of government, including national, sectoral, provincial, and metropolitan areas. The focus has shifted to sector-specific policies targeting the agriculture, health, and biodiversity sectors to ensure they mainstream climate change

Source: Authors

and (iii) develop mechanisms for cooperation and assistance with emergency services (UN-Habitat, 2022; SADC, 1999). Table 10.4 shows findings on the mainstreaming of climate change issues and disaster risk management in Southern Africa country-specific health policies. Table 10.4 shows that half of the countries (Malawi, Angola, and Namibia) had implicitly embedded climate change and disaster risk management in their national health policies, despite the IPCC (2021b) reporting with high confidence that climate change, extreme events, and climate-related food-borne and water-borne diseases will greatly impact human health, human mortality, and morbidity.

10.8 Climate Change Adaptation in SADC: A Strategy for the Water Sector

It is projected that for Southern Africa, climate change impacts would mostly be felt through reductions in the quality and quantity of water resources, thereby negatively impacting food production and energy generation, and indirectly impacting nutrition, health, sanitation, and water conflicts, amongst other social challenges (Lesolle, 2012). This is a regional strategy developed in 2010 that seeks to mainstream climate change in the SADC water sector. The main goal of the climate change adaptation (CCA) strategy is to improve climate resilience and the adoption of a multidimensional approach to climate change adaptation in Southern Africa through integrated and adapted water resources management at regional, river basin, and local levels (SADC, 2010a). Mainstreaming climate change issues in the water policy is in line with the priorities outlined in the Regional Indicative Strategic Development Plan. Water is regarded as being at the nexus of Sothern Africa's development challenges since societies and main economic sectors rely on water-intensive sectors such as agriculture and fishing. The SADC water strategy considers water and climate change concerns as a complicated array of layers, such as the three-dimensional cube (Lesolle, 2012). The strategy calls for the implementation of adaptation measures at different levels (regional, national, and local), at different stages of the adaptation process (before, during, and after a water crisis or extreme event), and in different areas of intervention (SADC water adaptation cube) (SADC, 2010a). To improve climate resilience and reduce climate vulnerability in the SADC region, the following priority or adaptation measures were identified: water gov-

Table 10.3 Integration of climate resilience in national agricultural policies

Author (s) and year	Findings
World Bank (2021)	The National Agricultural Policy of Malawi recognizes the significant vulnerability of the country to the effects of climate change and extreme events particularly due to its reliance on rain-fed agriculture. The policy sets forth a clear objective to plan and implement urgent interventions to minimize the future adverse effects of climate change on agriculture, given Malawi's low economic capacity to cope with these changes
Government of Swaziland (2005)	The Comprehensive Agriculture Sector Policy of 2005 captures several issues related to climate change, weather extremes, and associated natural disasters affecting Swaziland's agricultural sector. The CASP outlines several strategies to promote resilience such as adaptation strategies in the crop and livestock sector, implementing drought mitigation and prevention techniques, adoption of drought-tolerant crop varieties
Alliance for Green Revolution in Africa (2022)	The Strategic Plan for the Development of the Agricultural Sector in Mozambique has been revised to address contemporary issues, including climate change. The document aligns agricultural policy with global emerging issues such as climate change, sustainable development, and agriculture climate-smart varieties as part of the response to climate change challenges. However, it is not explicit on disaster management in the agricultural sector
Government of Zambia (2016)	The Second National Agricultural Policy of Zambia places a strong emphasis on addressing climate change and disaster management through a variety of strategies aimed at increasing resilience, promoting sustainable practices, and enhancing the capacity to manage and mitigate climate-related risks
Government of Botswana (2015)	Botswana's agricultural policies, as outlined in the policy brief, do indeed consider climate change issues
Government of Zimbabwe (2019)	The policy framework outlines strategic objectives and initiatives aimed at building resilience to climate change and mitigating its effects on agriculture. It presents a comprehensive approach to addressing climate change and disaster risk management by promoting resilience, sustainable practices, and the adoption of climate-smart technologies, dissemination of climate information and enhancing the capacity of farmers and extension workers to adopt resilient, sustainable agriculture practices
Government of South Africa (n.d)	The policy position on climate change and disaster risk management within the South African agricultural sector emphasizes the need to shift from government-led relief programmes to a system that encourages farmers to adopt practices that reduce their vulnerability to climate-related risks and manage disasters more sustainably. The government's role is redefined to support farmers in building resilience against climate change and managing disaster risks through information access, promoting research and extension of technologies and practices that mitigate risks, developing a national early-warning system for disaster management

Source: Authors

Table 10.4 Integration of climate resilience in national health policies

Author (s) and year	Findings
Government of Lesotho (2017)	The National Health Policy 2017 of Lesotho places a significant emphasis on addressing climate change and disaster risk issues within the context of public health. The policy recognizes the importance of strengthening health systems to enhance emergency preparedness, be more resilient and capable of responding to public health emergencies that may arise from climate-related events and other disasters
Government of Malawi (2018)	The National Health Policy of Malawi does not explicitly focus on climate change though it acknowledges the impact of climate change on health by exacerbating challenges in providing social services such as healthcare
Government of Angola (2012)	National Health Development Plan of Angola does not explicitly include climate change issues although it mentions the importance of dealing with emergencies and health-related challenges, of which some may be climate change-induced
Government of Namibia (2010)	The National Health Policy Framework does not explicitly mention climate change but focuses primarily on improving service delivery and environmental health issues, such as the need for sanitation facilities, water quality, and the management of environmental and occupational hazards
Government of Botswana (2011)	The National Health Policy of Botswana acknowledges the impact of climate change on health outcomes, as the climate crisis will influence the prevalence of diseases such as malaria and cholera. Rising temperatures and conducive breeding conditions for mosquitoes are likely to increase malaria cases, while reduced rainfall could lead to food production challenges, resulting in malnutrition over the long term
Government of Zambia (2011)	Zambia's National Health Policy's environmental health section includes measures to strengthen the capacity to respond to the effects of climate change by reducing environmentally related diseases

Source: Authors

ernance, water management, and infrastructure development (SADC, 2010a). At each level, all relevant actors and their roles and responsibilities to implement the climate change strategy should be identified to reduce the vulnerability of communities, ecosystems, and economies. Table 10.5 shows findings on the mainstreaming of climate resilience and disaster risk management in some Southern African countries. Out of the seven listed countries only Zimbabwe, South Africa, Swaziland, Zambia, and Malawi had national water policies that were explicit on the relationship between the water sector, climate change, and water-related disasters. Policies from these countries call for research and collaboration on climate change effects, importance of managing water resources to mitigate water-related extreme events, the need for climate-resilient water management strategies, protection of the population and the environment from the adverse effects of water-related disasters, and alignment of water sector legislation to reflect these priorities.

10.9 Significance of Resilience Planning and Policies in Southern Africa

As risks intensify for the world's poorest people, building resilience to shocks and stressors has been prioritized as an important cross-cutting outcome area across development programming. The concept is influential in informing development initiatives, climate change adaptation (CCA), and disaster risk reduction (DRR). This is achieved through the major resilience outcome of assessing the three critical livelihood capacities, which are absorptive, adaptive, and transformative.

The programmes on resilience building are associated with social change processes. Resilience empowers segregated community groups and classes of people. There are processes that promote and enhance gender justice, participation of women and youth, and conflict resolution. This is mainly achieved through some initiatives that are securing, enhancing human,

Table 10.5 Mainstreaming of climate resilience in national water policies

Author (s) and year	Findings
Government of Zimbabwe (2012)	The Zimbabwe National Water Policy (2012) recognizes the significant impact of climate change and extreme events on water resources including rainfall. To address this, the policy mandates responsible institutions to integrate climate change and disaster risk management considerations into all water resource planning and design activities, support research and collaboration on climate change effects, and update legislation to reflect these priorities
Government of South Africa (1997)	The policy outlined in the provided document recognizes the unpredictable nature of the South African climate, which can lead to natural disasters such as floods and droughts. It calls for the responsible institution to develop new strategies focusing on water-related disaster prevention. The policy acknowledges the importance of managing water resources to mitigate water-related extreme events, integrating climate change and disaster risk management into the broader framework of water resource management and that the nation's water resources are managed in a way that protects human life and property from the dangers of water-related disasters
Government of Lesotho (2007)	The policy takes an implicit position on climate change or disaster risk management through its focus on related issues such as sustainable water resources management, environmental protection, and the implementation of drought-relief strategies and flood management measures. These elements are indicative of an implicit recognition of the need to adapt to and mitigate the impacts of climate variability, which is often exacerbated by climate change
Government of Swaziland (2018)	The policy acknowledges the impact of climate change on water resources and the need for integrated strategies to mitigate and adapt to these changes. It stresses the importance of managing water-related disasters, the need for climate-resilient water management strategy, ensure sustainable water provision, and protect the population and the environment from the adverse effects of climate variability
Government of Zambia (2010)	The policy identifies the increasing incidence of water-related disasters as consequences of a changing climate. To manage the disasters, the policy outlines several measures that included promotion of public awareness on climate change mitigation and adaptation strategies, strengthening community resilience to water-related disasters. It also advocates for a state of disaster preparedness to mitigate the impacts of extreme water-related events and the development of an effective management of water resources, which includes disaster risk management
Government of Namibia (2000)	The National Water Policy White Paper implicitly mentions a climate change or disaster risk management where it seeks to address the management of extreme hydrological risk and vulnerability, which can be closely related to the impacts of climate change and disaster risk management
Government of Malawi (2005)	Malawi's water policy calls for integration of disaster risk management into the broader framework of water resources management, with a focus on preparedness, early warning, mitigation, and zoning of disaster-prone area. It calls for measures that ensure that the country's water resources are managed in a way that minimizes the adverse effects of climate change and water-related disasters on the population and the economy

Source: Authors

social, natural, physical, financial, and household asserts based on the sustainable livelihood framework.

The concept of resilience planning is a longer term and more cost-effective strategy for substantially improving local capacities (absorptive, transformative, and adaptive capacity) to withstand shocks and stressors. Resilience capacities need to be enhanced to achieve resilient development outcomes, that is, the realization of rights and well-being in spite of shocks, stressors, and uncertainty. The three critical capacities are interconnected and exist at multiple levels, that is, individual, household, community, national, and ecosystem levels. The capacities strengthened by resilience policies and strategies are as follows: Resilience building approaches that strengthen DRR approaches depends on a system composed of different capitals. As a result, to build resilience capacities, there is a need to implement strategies that consider the different capitals. The capitals are financial, human, natural, physical, and social capital (Table 10.1).

10.10 Challenges of Resilience Policies and Strategies in Africa

Efforts to produce resilient societies have proved one of the most complex challenges of sustainable development endeavours, because much of the traditional policies and models aimed at building resilience are based on a linear and oversimplified understanding of resilience. For instance, the Malawi National Resilience Strategy 2018–2030 and the draft Implementation Plan (2018–2023) emphasize breaking the cycle of poverty and food insecurity. The Eswatini Resilience Strategy has a specific focus on addressing hazards and disaster risk drivers, the Lesotho National Strategic Resilience Framework emphasizes resilient livelihoods, while the Zimbabwe Resilience Strategic Framework (Government of Zimbabwe, 2015) focuses on household well-being and community development. On the other hand, the FAO Resilience Strategy 2018–2021 emphasizes the full value chain of food security.

Africa, particularly Southern Africa, is exposed to a wide range of hazards which can trigger disasters, and it is projected that climate-related hazards will increase in severity and frequency.

The main hazards experienced in the Southern Africa region are climate-related (drought, floods, and tropical cyclones), although diseases (pandemics and epidemics), pest infestations, fires, transport, and industrial accidents and, more rarely, conflict and earthquakes also occur. The limited ability to adapt to these hazards is exacerbated by poverty. Poverty increases vulnerability, which is socially differentiated amongst the population, and currently exhibits significant gender differences between men and women and boys and girls.

There are also political, institutional, and technical capacity challenges that hinder the SADC region from effectively addressing these risks (Nemakonde & Van Niekerk, 2023). Disaster risk reduction is currently limited in its ability to reduce current risk and reduce the creation of new risks. In addition, there is disproportionate emphasis on disaster risk management. There has been a focus on preparation for emergency relief, response, and recovery, rather than proactive investment in resilience-building to equitably reduce disaster risk.

10.11 Conclusion

Overall, this study sought to answer the following research questions: (1) What is resilience planning in the context of developing countries? (2) What climate change-related and sector-specific policies are being implemented in Southern Africa to promote resilient societies, ecosystems, and economies? (3) What are the significance and limitations facing resilience policies in Southern Africa? Our study revealed that the desired outcome for all development actors and country governments in building resilience is to make the target groups to better resist and recover from the cyclical shocks and stressors. Resilience building initiatives target strengthening the critical livelihood capacities (absorptive, adaptive, transformative) in a manner that the targeted beneficiaries will be able to withstand future shocks and reverse developmental gains undermined by the shocks and stressors. The manifestation of shocks and stressors has an impact on the GDP of a country or region. Shocks and stressors undermine or reverse developmental gains to be accrued by the vulnerable community. This is because the occurrence of a climate-induced shock results in the outbreak of other forms of shocks. Shocks and stressors occur simultaneously, that is, one shock can contribute to occurrence of another form of a shock. Resilience building initiatives generally aim at alleviating poverty; achieve food security amongst the vulnerable communities. Resilience building is in line with the global sustainable development goals, especially 1 and 2 on poverty and food security.

References

Addaney, M., Jarbandhan, D. B., & Dumenu, W. K. (Eds.). (2023). *Climate change in Africa: Adaptation, resilience, and policy innovations*. Springer Nature.

African Union. (2022). *Africa's climate change and resilient development strategy and alignment toolkit for the SADC regional resilience framework 2020–2030*.

Alliance for Food Sovereignty in Africa (AFSA). (2020). *The Regional Policy Environment for And EU-Africa climate cooperation*.

Béné, C., Godfrey-Wood, R., Newsham, A., & Davies, M. (2012). *Resilience: New utopia or new tyranny? Reflection about the potentials and limits of the concept of resilience in relation to vulnerability reduction programmes* (IDS working paper 405). Institute of Development Studies.

Berke, P., Kates, J., Malecha, M., Masterson, J., Shea, P., & Yu, S. (2021). Using a resilience scorecard to improve local planning for vulnerability to hazards and climate change: An application in two cities. *Cities, 119*, 103408.

Croese, S., Green, C., & Morgan, G. (2020). Localizing the sustainable development goals through the lens of urban resilience: Lessons and learnings from 100 resilient cities and cape town. *Sustainability, 12*(2), 550.

Deutsche Gesellschaft für Internationale Zusammenarbeit (GIZ). (2020). *Regional – National department of environmental affairs (South Africa) climate change adaptation perspectives for the SADC*.

Frankenberger, T., & Nelson, S. (2014). *Design, monitoring and evaluation of resilience interventions: Conceptual and empirical considerations. IDS working paper* (Vol. 2015, p. 495). Institute of Development Studies.

Hung, H. C., Yang, C. Y., Chien, C. Y., & Liu, Y. C. (2016). Building resilience: Mainstreaming community participation into integrated assessment of resilience to climatic hazards in metropolitan land use management. *Land Use Policy, 50*, 48–58.

IPCC. (2021a). *Sixth assessment report: Synthesis report of the IPCC sixth assessment*.

IPCC. (2021b). *The IPCC's sixth assessment report: Impacts, adaptation options and investment areas for a climate-resilient southern Africa: Factsheet Southern Africa*.

Kenny, L. R. (2017). *A systematic review of South Africa's policy response to climate change*.

Lesolle, D. (2012). *SADC issue paper on climate change: Assessing the policy options for SADC member states: SADC research and policy paper series 01/2012*.

Lewis, T. G. (2019). *Critical infrastructure protection in homeland security: Defending a networked nation*. John Wiley & Sons.

Mabvurira, V., & Chikwaiwa, B. K. (2018). Impact of climate-change on the livelihoods activities of former commercial farm workers in Mazowe South Constituency, Zimbabwe. In K. Nyikahadzoi & B. B. Mukamuri (Eds.), *Building climate change resilient communities in Africa insights from Zimbabwe's urban and rural areas*. Konrad-Adenauer-Stiftung.

Mashura, S. (2018). Climate change and crop production in Zimbabwe: An error correction modelling approach. In K. Nyikahadzoi & B. B. Mukamuri (Eds.), *Building climate change resilient communities in Africa insights from Zimbabwe's urban and rural areas*. Konrad-Adenauer-Stiftung.

Mitchell, A. (2013). *Risk and resilience: From good idea to good practice*.

Naidoo, S., Davis, C., & Archer van Garderen, E. (2013). *Forests, rangelands and climate change in Southern Africa* (Forests and climate change working paper no. 12). Food and Agriculture Organization of the United Nations.

Nemakonde, L. D., & Van Niekerk, D. (2023). Enabling conditions for integrating government institutions for disaster risk reduction and climate change adaptation in the SADC region and beyond. *Risk, Hazards & Crisis in Public Policy, 14*(1), 6–26.

Ngwenya, S. (2021). *The influence of livelihood projects on food security resilience levels in Zimbabwe* (Doctoral dissertation), North-West University (South Africa).

Nhamo, L., Mabhaudhi, T., & Modi, A. T. (2019). Preparedness or repeated short-term relief aid? Building drought resilience through early warning in southern Africa. *Water SA, 45*(1), 75–85.

OECD. (2014). *Climate resilience in development planning: Experiences in Colombia and Ethiopia*. OECD Publishing. https://doi.org/10.1787/9789264209503-en

Organisation For Economic Co-Operation and Development. (2009). *Policy guidance on integrating climate change adaptation into development co-operation report (AR6) summary for policymakers*.

Oxley, M. (2014). A "people-centered principles-based" post-hyogo framework to strengthen the resilience of nations and communities. *International Journal of Disaster Risk Reduction, 4*, 1–9.

SADC. (1999). *Protocol on health in SADC*. SADC Secretariat.

SADC. (2010a). *Climate change adaptation in SADC: A strategy for the water sector SADC*. SADC Secretariat.

SADC. (2010b). *Southern Africa sub-regional framework of climate change programme: Adaptation and mitigation actions, supported by enabling measures of implementation*.

SADC. (2020a). *SADC RISDP 2020–2030*. SADC Secretariat.

SADC. (2020b). *SADC RISDP 2020–2030 Blueprints*. SADC Secretariat.

Sharifi, A. (2016). A critical review of selected tools for assessing community resilience. *Ecological Indicators, 69*, 629–647.

UNEP & ICRAF. (2006). *Climate change and variability in the Southern Africa: Impacts and adaptation strategies in the agricultural sector.*

United Nations Economic Commission for Latin America and the Caribbean. (2020). Planning for resilience: An integrated approach to tackle climate change in the Caribbean, Policy Brief LC/CAR/2020/1.

United Nations Human Settlements Programme (UN-Habitat). (2022). Regional assessment on urban vulnerability and resilience in Southern African development community member states: Strengthening capacities for reducing urban vulnerability and building resilience in Southern Africa.

USAID. (2013). *National Adaptation Planning Fact Sheet.*

Wiseman, E., McLaughlin, T., & Mrad, N. (2014). Critical infrastructure protection and resilience literature survey: State of the art. *DRDC Project# CSSP-2013-TI-1039, Ottawa.*

World Economic and Social Survey. (2016). *Climate change resilience: An opportunity for reducing inequalities.* Department of Economic and Social Affairs, United Nations.

World Food Programme. (2021). *Climate change in Southern Africa: A position paper for the world food programme in the region.*

Zseleczky, L., & Yosef, S. (2014, May). *Are Shocks Really Increasing? A selective review of the global frequency, severity, scope and impact of five types of shocks.* 2020 Conference Paper 5. Washington, D.C. IFPRI.

11 Climate Resilient Development in Rural Southern Africa: Gaps and Prospects

Hardlife Zinhiva, Susan Dube, and Clarice Mudzengi

11.1 Introduction

Climate change has the potential to have the most devastating effects on world society. Developing climate resilience is critical to mitigating the negative effects of climate change. Unfortunately, climate change impacts frequently disproportionately affect marginalized rural areas, emphasizing the importance of equitable and proactive resilience planning and resource allocation. Climate resilience is defined as the ability to predict, plan for, and respond to climate-related hazards, trends, or disturbances. Improving climate resilience in rural regions entails understanding how climate change will produce new or alter existing climate-related risks, as well as implementing efforts to better manage economic and social risks. Local communities, development partners, stakeholders, and governments have an important role to play in mobilizing local resources and technologies, securing finances and livelihoods, updating infrastructure, and climate-smart planning to mitigate the adverse climate change impacts. There are varied strategies for climate action meant for creating climate resilient rural communities. This chapter assesses the approach and progress of climate resilient development in rural Southern Africa, noting the gaps and prospects associated with the noble climate action.

11.2 Methodology

To gather information for the current study, a survey of the body of literature on climate resilient development in rural Southern Africa was done. Most of the desk evaluation of scholarly publications served as the foundation for the content analysis of the literature. The keywords "Climate resilient development*" or "CRD in Southern Africa*" or "CRD gaps and prospects*" were explicitly targeted in the search strategy. Following that, the relevance of the studies returned by the search strategy was assessed. Firstly, English-language publications were reviewed. Secondly, the publications that focused on climate resilient development as their main topic were chosen. Because the analyzed papers had undergone some peer review, the findings were deemed credible. We thematically restricted

H. Zinhiva (✉)
Department of Physics, Geography & Environmental Science, Great Zimbabwe University, Masvingo, Zimbabwe
e-mail: hzinhiva@gzu.ac.zw

S. Dube
Department of Environmental Studies, Zimbabwe Open University, Masvingo, Zimbabwe
e-mail: dubes@zou.ac.zw

C. Mudzengi
Gary Magadzire School of Agriculture and Engineering, Great Zimbabwe University, Masvingo, Zimbabwe
e-mail: cmudzengi@gzu.ac.zw

M. Matsa et al. (eds.), *Climate Change Resilience in Rural Southern Africa*, Sustainable Development Goals Series, https://doi.org/10.1007/978-3-031-73600-1_11

the review to the areas of climate, resilience, development, gaps, and prospects based on the results of the preliminary searches. Although there is a wealth of literature in each of these areas, for the sake of this study, we only included articles that specifically addressed the term "climate resilient development." The justification for this is that we believed this corpus of work to be the best representative of what is currently known about the idea of climate resilient development.

11.3 Concept of Climate Resilient Development

Climate resilient development (CRD) is described as the process of implementing climate action, such as greenhouse gas mitigation and risk reduction adaption strategies, to promote sustainable development for all (IPCC, 2022; Trisos et al., 2022). It is motivated by the need to address impacts of climate change, either due to natural or anthropogenic long-term shifts in temperature and weather patterns. Human actions that contribute to climate change include the use of fossil fuels, deforestation, and agriculture. They are exacerbated by other worldwide issues such as overconsumption, population expansion, fast urbanization, land degradation, biodiversity loss, poverty, and inequality. Increases in temperature and variations in precipitation due to climate change have already been shown to cause geographic shifts and increase the incidence of vector-borne diseases that affect both humans and animals. Another direct adverse effect is on food and nutrition security, through reductions in rain-fed crop yields and quality. Climate-induced floods and sea level rise to destroy infrastructure and property and damage important coastal ecosystems. These are some of the effects of climate change, which might jeopardize development goals such as enhancing economic growth, reducing poverty, providing access to education and health for all, and protecting the environment (USAID, 2014). Climate action to address these impacts includes various adaptation and mitigation strategies. Comprehension of the multifaceted nature of the relationships between development and climate action is thus pertinent for progress towards CRD, particularly in rural areas in developing regions such as Southern Africa.

To address the effects of climate change, CRD recognizes the importance of combining adaptation with mitigation and development in collaborative, inclusive, just, and equitable ways across different arenas of interaction (Stringer et al., 2022). The relationship between development and climate action is such that, while development processes can increase vulnerabilities to climate impacts by weakening the ability to adapt and mitigate, poorly designed climate change action can derail sustainable development efforts. To quicken and intensify the enforcement of sustainable development initiatives and procedures, it is suggested that CRD should consider a wider array of sustainable development priorities. This would make climate action and sustainable development mutually dependent (IPCC, 2022). According to the IPCC (2022), four dimensions promote progress towards CRD: equity and justice, inclusivity, knowledge diversity, and ecosystem management. Consideration of these in CRD can increase the probability of attaining development goals such as the provision of primary healthcare, adequate clean water and sanitation, and food security and nutrition, yet achieve objectives of climate action. However, CRD requires a holistic approach involving all relevant stakeholders.

11.4 Climate-Resilient Development in Rural Areas: A Global Perspective

Climate variability has an impact on crop production and rural populations, posing hazards to food security and escalating rural poverty. As a result, boosting rural families' ability to adapt to climate unpredictability has emerged as one of the most pressing concerns for international and national institutions (Nunez et al., 2023). Nunez et al. (2023) investigated the effects of rural households' livelihood practices on their sensitivity and adaptation to climate variability. The study found that farmers with the most capital and diversified livelihood choices were the least sensitive to climate change's negative effects. Liu et al. (2023) investigated whether adopting soil nutrient management strategies through soil testing and formulated fertilizer would improve the capacity of farmers to absorb and recuperate from climate-related traumas and stressors. The study found that implementing soil testing and formulated fertilization considerably improved the climate resilience of rice farmers in Central China. The use of soil nutrient management technology enhanced net agricultural income, rice output, and overall productivity. Ren et al. (2023) agree that boosting green growth in agriculture is critical for tackling environmental issues, achieving sustainable development, and establishing a sustainable future. This study investigated the influence of innovation, natural resource availability, and climate change in encouraging green growth in the agriculture sector of China.

Siyao and Sanga (2023) investigated factors influencing access to and adoption of climate change adaptation knowledge to raise awareness among smallholder tomato growers in Tanzania's rural areas. According to the study's conclusions, the most important information for climate change adaptation is drought- and disease-tolerant tomato seedlings, as well as the use of pesticides and fertilizers. Smallholder tomato farmers prefer to learn about climate change adaptation via families, friends, neighbours, and other sources such as the radio and cell phones. Marzieh and Hojjatallah (2023) conducted research on scaling up climate-smart regenerative farming for the recovery of degraded agroecosystems in impoverished countries. This study helps to establish resilient farming systems by finding the most effective strategies for restoring soil and water resources and researching the key drivers of resilient farming. The study concludes that climate-smart regenerative agriculture (CSRA) is a viable and prominent strategy for recovering degraded agroecosystems; hence, attempts to scale up CSRA in soil and water-vulnerable areas are essential.

Quarshie et al. (2023) evaluated the potential of climate-smart agriculture (CSA) techniques to improve the results of sustainable food systems (SFS) in Ghana's smallholding agriculture environment. The study shows that rural farmers regard CSA as helping to "food and nutrition security," "economic performance," "social equality," "environmental stewardship," 'avoid pest and disease infestations," and "raise human capital knowledge" in that descending sequence. Climate-smart agriculture strengthens small-scale farmers' resilience by producing crops that are more adaptable to climate change and less harmful to the environment. Their finding supports the implementation of climate-relevant elements including climate information, climate policy, and combining agricultural research and development with indigenous knowledge. Indeed, countries in the Global South should get access to improved information management systems to promote broader adoption of climate change adaptation and climate-smart agriculture practices. As a result, the key finding of Quarshie et al.'s study (Quarshie et al., 2023) is that CSA improves farming system resilience in the face of climate-related shocks, resulting in a win-win situation for food security and the preservation of the environment.

11.5 Climate Resilient Development (CRD) in Rural Southern Africa: Progress to Date

Southern Africa is one of the world's most vulnerable regions to climate change. Climate-related phenomena such as floods, droughts, and tropical cyclones affect most of the local population, limiting development and the availability of natural resources. Climate change impacts such as increasing temperatures, uncertain rainfall patterns, and more intense and frequent extreme weather events, particularly in semi-arid regions, put rural livelihoods at risk (Salik et al., 2015). Over the past 10 years, the region has experienced many episodes of heat waves, droughts, and floods, which have had a significant impact on agricultural production due to decreased crop productivity and crop failure (Hussain, 2010). A thorough evaluation of the literature reveals that developing adaptive capability and integrating climate resilience into development initiatives are crucial for ensuring long-term sustainability in rural Southern Africa.

The work of numerous academics and professionals has helped to establish the idea of climate resilience development (CRD). To achieve climate-resilient development in rural Southern Africa, several efforts have been made by governments in conjunction with NGOs to lessen exposure to and susceptibility to climate risks, reduce greenhouse gas emissions, and preserve biodiversity. Efforts made involve skilfully traversing the numerous interactions between the various systems such as industry, health, nutrition, water, urban growth, housing, and transport so that decisions made in one area do not have negative consequences in other areas and that chances are taken advantage of to speed up the transition to a safer, more equitable world.

Given that droughts influence all aspects of human life, building resilience requires increasing efforts to reconstruct rural communities so that they are versatile, adaptable, and responsive to new ideas (CARE International 2016). Building climate resilience in the most vulnerable areas is crucial for the survival of humanity. Climate resilience development is primarily concerned with a system, society, or civilization's ability to accomplish its human, social, ecological, and economic goals while addressing the long-term consequences of climate change in a mutually reinforcing manner. Resilience is required when bracing for the changes and adjusting to them. This can be accomplished through increasing the ability of people, households, and communities to absorb information, adapt to changing conditions, and translate that information into solutions for dealing with those problems.

Numerous success stories from rural Southern African communities attest to the importance of enhancing community resilience to climate change. Many nations have made investments in a variety of initiatives or programmes as a means of enhancing adaptive capacities and increasing resilience to climate change. A good example was the three-district Climate Resilient Building Project Consortium in Zimbabwe, which was being led by ActionAid in the Zambezi Valley. The districts are Kariba, Mbire, and Binga. It was a national initiative backed by the Department for International Development (DFID) of the United Kingdom (UK), the European Union (EU), the Swedish Embassy, the United Nations Development Program (UNDP), and the Ministry of Lands, Agriculture, Water, Climate, and Rural Resettlement. The Zimbabwe Environmental Law Association (ZELA), Afrosoft Holdings (Private) Limited, and African Breeders Services Total Cattle Management (Private) Limited were additional ZRBF-ZVA members involved in the Zambezi Valley consortium.

The programme's principal objective was to increase the receptive, flexible, and transformative capacities of Zambezi Valley farming communities by fostering individual, household, and community resilience, as well as enhanced well-being in the face of shocks and stress caused by climate change in a sustainable manner (Moyo, 2021). Communities in the Zambezi Valley saw some of the most severe shocks brought on by droughts and floods due to climate change. The ZRBF Zambezi Valley Alliance (ZVA) consortium organized community-based training of three sub-aqua/water diver units at the district level and provided them with the appropriate diving gear to save lives to solve the flooding crisis. The major goal was to get ready for, respond to, and recover from floods that have afflicted the region throughout the years. Other climate-resilient projects started in the Zambezi Valley by ZRBF included greenhouses offering conditions for the nurturing and development of seedbanks supporting horticultural projects, innovation hubs, artificial insemination services to improve livestock, and solar-powered piped water schemes at neighbourhood schools and villages which resulted in vibrant horticultural projects in the three districts. After 6 months of cooperative work, the combination of interventions was given to communities, leaving them in possession of nutrition centres, solar-powered water schemes, and innovation centres that help the villagers to improve the breeds of livestock (Action Aid Newsletter, 2022).

Furthermore, climate-resilient development attempts to continuously improve outcomes for development in a way that is resistant to both long- and short-term challenges brought out by climate change. According to a thorough evaluation of the literature, the Zambezi Valley has shown climate-resilient development, with people in the three districts increasing their ability to protect developmental gains and improve well-being outcomes in the face of climate shocks and strains. The initiatives have also helped to lessen the requirement for ongoing inflows of humanitarian aid. According to the ActionAid Zimbabwe Newsletter (2022), food insecurity dropped from 65.4% in 2019 to 50.7% in 2020 in the three districts. In addition, compared to provincial averages, smallholder farmers' small grain yields increased by 83% when they adopted agro-ecological practices. Around the Zambezi Valley, significant progress has been made in the construction of climate-resilient infrastructure. Some of the project's initiatives included assisting smallholder farmers, particularly women and young people, in improving their abilities, thereby allowing them to absorb, recover, and adapt to the effects of shocks and stresses in a way that protects rural livelihoods, enhancing recovery gains, and supporting resilient sustainable agriculture. ZRBF's efforts are commendable because the communities' absorptive ability has risen, allowing people to aid themselves after a disaster rather than relying on humanitarian agencies. This is true notwithstanding the challenges the three districts face in making these projects viable.

In Zimbabwe, agro-ecology has been successfully used to manage natural resources and increase resilience to climate change using catchment-based strategies and integrated water resource management. There is evidence that following Cyclone Idai in 2019, the areas that were using agro-ecological land management practices recovered well from the cyclone. Due to its numerous possibilities for managing soil and land, agro-ecology promotes sustainable food production. According to a recent comprehensive review of agro-ecology by Amoak et al. (2022), agro-ecological interventions such as diversification of crops, soil management, livestock integration, and pest management are positively related to food security and nutrition on a global scale. Crop diversification, for example, ensures that farmers may get food from their fields for a longer period of time since the crops mature at different rates. As a result, such rural areas are more resilient to climate change, boosting household resistance to weather-related climate change. Benefits from such activities have the overall impact of making households and communities more resilient to climate change and its consequences.

Another project to take note of that was carried out to promote climate-resistant development in rural communities is the Shashe initiative in Matabeleland South, Zimbabwe. An irrigation system was established for the Shashe farming community in the Maramani Communal Area (4000 people in 20 villages in Beitbridge District in Zone V) (Cesvi, 2019). On more than 90 hectares of common property, residents actively participated in the cultivation of more than 22,000 oranges. Young farmers were connected to professional fruit juice manufacturers, national markets, and marketing channels through an innovative private public community partnership (PPCP). This project's success was partly attributed to its inclusion philosophy and use of modern technologies. Although small-scale citrus farming has not been effective, a commercial citrus business has been established. This shows how important community involvement in CRD is to create long-lasting interventions.

The Shashe programme marked a substantial departure from conventional agricultural interventions in favour of a more creative system built on cutting-edge technology, market connections, and local ownership. The project through its approach of incorporating cutting-edge technology, revitalized an ageing irrigation system that had declined over the previous 30 years, primarily because of low technical and financial capacity within the community after central government support ended after independence. It also introduced a new commercial crop (citrus) into the local community-managed agriculture. Contemporary and more sophisticated water-efficient irrigation equipment (submersible pumps in the Shashe River; overhead irrigation from three centre pivots, which saves 50% more water than traditional methods) was utilized to replace obsolete irrigation techniques and a collapsed irrigation system (Moyo, 2021). Following the community's request, intercropping between the young orange trees produced vegetables (seed beans, squash, sweet potatoes, turnips, cabbage, and maize) for sale and consumption. Later in the project, contract farming was implemented as a financial innovation. Farmers could borrow inputs such as seeds and fertilizer as part of their arrangement with the project's agribusiness companies and repay their loans with the proceeds from the crop. Because of the cheaper initial capital investment and guaranteed crop sales, even the poorest households were able to buy fertilizer and grow high-value crops (i.e. seed multiplication rather than food). The project's cooperative development with the community and partners offered an innovative approach to sustainability based on financial autonomy and localized democratization of authority. Through connections to the larger national and regional economies, these collaborations aimed to build resilience, while local administration and accountability guaranteed that the initiative remained responsive to community needs. The Shashe initiative points to evidence that progress has been made towards building climate resilience in Southern Africa. Increased community resilience and adaptation capacity to climate shocks resulted in better and more sustainable living circumstances for residents of Beitbridge and Mwenezi districts.

Climate-smart agriculture (CSA) is one of many sustainable agricultural practices that are helping rural communities in Southern Africa develop climate resilience in order to withstand the negative effects of climate change and unpredictability on farming systems (Makate et al., 2018). The Ministry of Agriculture and Climate Change in Zimbabwe has embraced the Intwasa/Pfumvudza agricultural model which demands the use of tiny pieces of land and the right use of agronomical techniques for higher earnings, this project significantly contributes to CRD. Additionally, it is founded on conservation agriculture concepts, which attempt to protect agricultural production from climate change while also tackling concerns with low productivity and poor profitability of farming among smallholder farmers who confront several difficulties. The breeding of small animals and the cultivation of small grain crops are essential methods for managing climate change for smallholder farmers in Zimbabwe and the rest of Southern African countries. Their additional and complementary efforts to adapt to climate change are vitally important for farm output and profitability. Significant progress has been made in encouraging and implementing the cultivation of small grain crops to mitigate the effects of climate change. Sorghum, millet, and rapoko are the little grains that are chosen above maize because they are more suited for semi-arid climates. Small grains have a lengthy shelf life, can withstand dry conditions, and are rarely affected by insects or other pests. This has helped most rural populations in Southern African nations build climate resilience in certain respects (Phiri et al., 2021). According to the IPCC (2014) report, conservation agriculture offers a practical way to increase the adaptability of agroecosystems and livelihoods. In most Southern African nations, climate-smart agricultural practices have been shown to have advantages for smallholder farmers. For example, zero tillage in Tanzania, particularly in rural regions, has dramatically enhanced food security by keeping out-of-season rainfall in the soil while also lowering greenhouse gas emissions by eliminating soil disturbance. Farms utilizing terracing, which conserves soil moisture, have seen financial benefits; these benefits were increased when mingled with numerous other climate-smart farming techniques, such as minimal tillage and cover crops. Because of the increased food security and livelihoods, this in a way enables communities to adjust to and survive despite undergoing shocks and pressures caused by climate change.

Through the capture, storage, and rerouting of rainfall flow, rainwater harvesting (RWH) has frequently been designated as a key step for addressing climate change and boosting the productivity of smallholder agriculture in the majority of rural regions. In rain-fed settings, RWH has enhanced yields and reduced crop failure during dry spells. Communities are able to maintain sustainable livelihoods even during droughts once they invest towards RWH techniques.

Progress has also been made by some Southern African countries on building climate resilient infrastructure. A notable example of a country in the area where roads and bridges were damaged by unforeseen floods brought on by heavy rain is Zambia. To protect infrastructure from the severe effects of climate change, Zambia has invested in the construction of climate-resilient roads, such as the Kalomo-Dundumwezi-Itezhi-tezhi-Namwala road project in the country's rural areas (Arslan, 2019). The cost of the 237-km road network, which was supported by a loan from the African Development Bank (AfDB), was US$21 million. Elevating embankments and other elements on the climate-resilient road serve to protect it from flooding. The drainage systems' climatically resistant criteria aid in reducing soil erosion.

To strengthen the resilience of communities and vulnerable ecosystems, the Malawian government has prioritized adaptation (Kamdonyo, 2019). The "Green Belt Initiative" (GBI), a project to use water from water bodies in the country to improve agricultural production, food, nutrition, and income security, the creation of a national climate change management structure, and assisting people in rural areas increase their ability to adapt through empowerment with methods and techniques under the Climate Adaptation for Rural Livelihoods and Agriculture (CARLA) programme are some of the key initiatives. Furthermore, the Kenyan government has established policies designed to mitigate the effects of climate change, promote adaptation strategies, and ensure the best farming methods in the nation through the dissemination of information, extension services, and financial assistance to economically disadvantaged farmers all over the country (Hesse & Pattison, 2013).

11.6 Prospects for Climate Resilient Development in Rural Southern Africa

There is tremendous potential for developing resilient rural communities in Southern Africa through a variety of innovative solutions, one of which is supplying solar electricity to the grid. As humanity fights climate change and tries to transition to a just, sustainable, and carbon-free future, solar power is an important component of the puzzle. From 15 GW in 2008 to 505 GW in 2018, the total installed capacity of solar photovoltaic systems generated 2.4% of the world's electricity yearly (Ren, 2019). Solar and other climate-friendly renewables presently account for 11% of the global primary energy supply, but there is nevertheless plenty of work to be done before they can replace 85% of primary energy currently derived from carbon-intensive fossil fuel sources (BP, 2019). People will need to make daring promises to hasten the adoption of renewable energy in a range of demographic, socioeconomic, geopolitical, and environmental contexts.

Developing nations in Southern Africa have high prospects of expanding solar-powered electricity for home consumption using rooftop solar systems as well as for industrial and commercial usage due to their advantage of location. This demonstrates that there are numerous cutting-edge potentials to use solar solutions to promote climate mitigation, support adaptation to climate change, foster community resilience, and aid in achieving the sustainable development goals of the United Nations (Shammin & Enamul Haque, 2019). Solar home systems (SHSs), which are mounted on the rooftops of individual homes, provide a practical way to electrify rural areas while also reducing greenhouse gas emissions. SHS offers a variety of advantages in addition to serving as a backup power source. A long-term climate-resilient development is formed, as is access to new income-generating enterprises, educational benefits for children, enhanced security, reduced indoor air pollution, television access, and mobile phone access (Komatsu et al., 2011). Rural communities will be able to endure in the case of climate change shocks of any magnitude due to several income-generating activities carried out by the villagers to maintain sustained food security. These initiatives show that SHS could improve community climate resilience and address many SDGs.

Integrating climate resilience into development planning is another approach to guarantee that Southern African nations invest in CRD in their rural areas. By mainstreaming, the policy and planning environment would be altered so that climate-resilient development happens naturally rather than requiring extra effort. High-level plans are influenced by institutional inertia and several political forces that these adjustments must overcome. A crucial tool for interacting with high-level officials and showcasing the advantages that climate resilience can bring to economies, livelihoods, and access to finance is using the framework to construct no- or low-regrets investment programmes.

11.7 Gaps Identified and Defined

Various gaps need addressing to achieve CRD in Southern Africa. These include limited knowledge of climate change science, and a lack of advanced and appropriate methodologies and tools to support research on CRD. The IPCC (2022) notes that Africa confronts serious data restrictions as a result of inadequate investment in weather observation stations, research, and data sharing. This is likely to affect the knowledge required for policy formulation and development, adoption, and implementation of adaptation and mitigation strategies to climate-related challenges. The lack of advanced and appropriate methodologies and tools to support research on CRD is often attributed to shortages of resources such as expertise or equipment. As the climate continues to change, so should adaptation and mitigation approaches. However, this should be guided by current research findings and recommendations in methodologies and structures for monitoring, recording, assessing, and learning from experience, as well as identifying "best practices" and their features (NRC, 2010). Where there is certainty or more knowledge, the correct action is more likely to be employed (Coast Adapt, 2017). It is also critical to address research problems regarding ways for implementing adaptive management and risk reduction for development. For example, how might autonomous response behaviour and policy actions be integrated to achieve the optimum outcomes? Furthermore, there are constraints in considering equity and justice, inclusion, knowledge diversity, and ecological stewardship in CRD. According to the IPCC (2012), concerns of social values/climate justice/equity/participation, as well as how they intersect with the deployment of mitigation, adaptation strategies, and sustainable development policy in various regions and sociopolitical circumstances, should be investigated further. Ziervogel (2020) further note that while Southern African nations may possess well-crafted national adaptation plans, such as climate change legislation, if resilience building has not progressed past the policy-making stage, this would result in a shortfall in CRD achievement.

It has been demonstrated that youth participation in initiatives to slow down climate change is low for a variety of reasons, including self-exclusion brought on by a lack of awareness or interest. One smart agricultural initiative, the Pfumbvudza/Intwasa project, for example, mostly involved elderly women, illustrating the substantial labour required to engage young people in such activities. The programme's goal was to secure increased yields despite climate change.

Lack of financial help and support from the government, corporations, and, in particular, banks have played a big role in young people's lack of interest in developing sustainable communities. For example, the Youth Empower Bank has attempted to provide loans to young people who wish to start their businesses, but the bulk of these banks are in major cities, and the collateral requirements are stringent, leaving young people in rural areas at a disadvantage.

For farmers to construct an innovative resilience mechanism against climate change shocks like droughts and floods, scientific information is necessary. The capacity of rural communities in

Southern African nations to respond to food-related climatic disasters is weakened by the slow reception of scientific knowledge. According to a thorough evaluation of the literature, communities have little ability to embrace and maintain climate-resilient technologies and approaches. Due to rising climatic variability and catastrophic occurrences, conventional adaption mechanisms and techniques are becoming ineffective. Small-scale farmers and livestock owners lack sufficient access to the information, resources, and links needed to adopt climate-resilient practices and technologies sustainably. Although small-holder farmers rely heavily on agrarian extension services; they currently do not have the personnel or financial resources necessary to help rural communities adapt to climate change in a way that is effective. Due to limited access to other energy sources, rural people are heavily dependent on biomass energy (mostly firewood and charcoal), which puts additional strain on the environment and perpetuates a pervasive sense of energy insecurity.

Furthermore, cooperation between farmer organizations and business sectors including input suppliers, dealers, and processors is lacking, as is access to post-harvest technology. In contrast to the dispersed physical infrastructure (such as roads) and market connections, smallholder farmers, particularly women, have restricted access to post-harvest innovations, insurance, knowledge, extension services, and inputs. Furthermore, a key impediment to both post-harvest loss reduction and value chain growth in most rural communities is a lack of appropriate access to dependable energy infrastructure for value-added operations such as processing and cooling. Rural residents frequently struggle to keep their produce intact until the best market prices and profits are reached because they have limited access to post-harvest technologies. Additionally, increased effects of climate change can lead to increased mycotoxin contamination, pest and disease outbreaks, and spoilage in perishable products, especially in the horticulture industry without adequate storage facilities. Mumtaz et al. (2023) noted that there is limited literature on climate adaptation policy implementation at the subnational level in the agricultural sector when assessing the effectiveness of its implementation.

11.8 Factors Associated with Progress in Climate-Resilient Development in Rural Southern Africa

The whole purpose of CRD is to successfully cope with and manage the impacts of climate change while averting the worsening of those impacts. In Southern Africa, there are factors associated with progress in CRD such as ineffective government policies, poor institutional frameworks, inadequate financial, human, and technical resources, increasing desertification, high dependency on agriculture, and political instability. For instance, mitigation strategies aim to address climate change from its root cause. As an example, addressing the contribution of the energy sector towards carbon emissions requires effective policies that support shifts towards alternative renewable energy sources, as well as regulate industrial emissions. Financial support therefore becomes important. If there are no facilities to provide the capital needed, then such mitigation strategies will not work. Similarly, with adaptation strategies, adoption needs support from all stakeholders concerned, as well as financial and other resources.

11.9 Conclusions and Recommendations

Climate resilient development should be a personal, household, local community, national, and global priority response to the vagaries being brought about by climate change and its induced environmental change. Human development is anchored in stable and rich ecosystems, which largely inspire and encourage innovations and enthusiasm for prosperity and limitless civilizations. However, as change is experienced and inevitable, every human community should be supported materially, technologically, socio-psychologically, and intellectually to empower and promote local development, a pillar for climate resilient development. Farmers, natural resource entrepreneurs, indigenous people, and local communities in Southern Africa have largely benefitted from research, innovations, and development outputs from renowned scholars, the private sector, NGOs, and quasi-government entities whose support goods and services have built local adaptive capacity. It is notable that climate resilient development is marred by variability in both spatial and temporal aspects of equitable and just distribution of resources and scientific support services. The need for supporting multi-stakeholder involvement, adequate provision of technical and financial resources, and cooperation between private and public sectors cannot be overemphasised. It should therefore be the goal of every human community to fight the climate crisis and bring a permanent and sustainable solution to its vagaries through earnest climate resilience development.

References

Action Aid Zimbabwe. (2022). Newsletter, Promoting Resilience in Communities to cope with shocks and stresses. *Newsletter, September–December, 12 December 2022.*

Amoak, D., Luginaah, I., & McBean, G. (2022). Climate change, food security, and health: Harnessing Agroecology to build climate resilient communities. *Sustainability 2022, 14*, 13954. https://doi.org/10.3390/su142113954. Academic Editor: Sean Clark Received: 22 August 2022 Accepted: 24 October 2022 Published: 27 October 2022.

Arslan, A. (2019). *Climate resilience in rural Zambia: Evaluating farmers response to El Nino induced drought*. International Fund for Agricultural Development (IFAD).

BP Statistical Review of World Energy. (2019). https://www.bp.com/content/dam/bp/business-sites/en/global/corporate/pdfs/energy-economics/statistical-review/bp-stats-review-2019-full-report.pdf

Cesvi (2019) Annual Report 2019. Financial Report. Cesvi Fondazione Onlus. Bergamo. www.cesvi.org

Coast Adapt. (2017). *Barriers to adapting to climate change*. https://coastadapt.com.au/barriers-toadaptingclimatechange#:~:text=There%20is%20a%20very%20wide,resources%20and%20lack%20of%20leadership

Hesse, C., & Pattison, J. (2013). *Ensuring devolution supports adaptation and climate resilient growth in Kenya*. IIED Briefing Paper-International Institute for Environment and Development, 17161, 4 p. https://pubs.iied.org/17161iied

Hussain, S. (2010). *Food security and climate change assessment in Pakistan*. Special Program on Food Security and Climate Change, Oxfam Novib.

Intergovernmental Panel on Climate Change. (2014). *Climate Change 2014: Impacts, adaptation and vulnerability*. IPCCWGIIAR5 Technical Summary. Available at http://ipccwg2.gov/AR5/images/uploads/WGIIAR5-TS_FGDall.pdf

IPCC. (2012). Special report on managing the risks of extreme events and disasters to advance climate change adaptation. In C. B. Field, V. Barros, T. F. Stocker, D. Qin, D. J. Dokken, K. L. Ebi, M. D. Mastrandrea, K. J. Mach, G.-K. Plattner, S. K. Allen, M. Tignor, & P. M. Midgley (Eds.), *A Special Report of Working Groups I and II of the Intergovernmental Panel on Climate Change*. Cambridge University Press, 582 pp.

IPCC. (2022). Impacts, adaptation, and vulnerability. In H.-O. Pörtner et al. (Eds.), *Contribution of Working Group II to the Sixth Assessment Report of the Intergovernmental Panel on Climate Change* (pp. 3–33). Cambridge University Press.

Kamdonyo, D. R. (2019). *Malawi gears up to adapt to climate change*. http://napglobalnetwork.org/2019/09/malawi-gears-up-for-its-climate-change-adaptation/.

Komatsu, E., et al. (2011). Seven-year Wilkinson Microwave Anisotropy Probe (WMAP) observations: Cosmological interpretation. *The Astrophysical Journal Supplement Series, 192*, 18. https://doi.org/10.1088/0067-0049/192/2/18

Liu, Y., Ruiz-Menjivar, J., & Zhang, J. (2023). Do soil nutrient management practices improve climate resilience? Empirical evidence from rice farmers in Central China. *Environment, Development and Sustainability, 25*(9), 10029–10054.

Makate, C., Makate, M., & Mango, N. (2018). Farm household typology and adoption of climate-smart agriculture practices in smallholder farming systems of southern Africa. *African Journal of Science, Technology, Innovation and Development, 10*(4), 421–439. https://doi.org/10.1080/20421338.2018.1471027

Marzieh, K., & Hojjatallah, S. (2023). Scaling up climate-smart regenerative agriculture for the restoration of degraded agro ecosystems in developing countries. *Sustainable Production and Consumption, 38*, 159–173.

Moyo, S. (2021, October 28). *Towards sustainable and climate resilient rural communities in Zimbabwe anchored on youth inclusion*. Zimbabwe Environmental Law Association (ZELA).

Mumtaz, M., Ruppim de Oliveira, J., & A. (2023). Framework for analyzing the implementation of climate adaptation policies in the agriculture sector at the subnational level. *Environmental Science and Policy, 147*, 126–137.

National Research Council. (2010). Adapting to impacts of climate change. In *Report of the panel on adapting to the impacts of climate change*. NAS/NRC Committee on America's Climate Choices, National Research Council, 292 pp.

Nunez, A. P. B., Gutierrez-Montes, I., Hernandez-Nunez, H. E., Suarez, D. R. G., Garcio, G. A. G., Casanoves, F., Flora, C., & Sibelet, N. (2023). Diverse farmer livelihoods increase resilience to climate variability in southern Colombia. *Land Use Policy, 1*, 106731.

Phiri, K, Nhliziyo, M, Madzivire, S.I., Sithole, M & Nyathi, D (2021) Understanding climate smart agriculture and the resilience of smallholder farmers in Umguza district, .

Quarshie, P. T., Abdulai, S., & Fraser, E. D. G. (2023). (Re)assessing climate-smart agriculture practices for sustainable food systems outcomes in sub-Saharan Africa: The case of Bono East Region, Ghana. *Geography and Sustainability Open Access, 4*(2), 112–126.

Ren, X. (2019). *Global status report, renewable energy policy network for the 21st century*. https://www.ren21.net/wp-content/uploads/2019/05/gsr_2019_full_report_en.pdf

Ren, X., He, J., & Huang, Z. (2023). Innovation, natural resources abundance, climate change and green growth in agriculture. *Resources Policy, 85*, 103970.

Salik, K. M., Ishfaq, S., Saeed, F., Noel, E., & Syed, Q. (2015). *Pakistan country situation assessment*. Pathways to Resilience in Semi-Arid Economies.

Shammin, M. R., & Enamul Haque, A. K. (2019). *Small-scale solar solutions for energy resilience in Bangladesh* in the book Climate change and community resilience insights from South Asia, Chapter 14.

Siyao, P. O., & Sanga, E. E. (2023). Factors influencing access to and uptake of climate change adaptation information among smallholder tomato growers in Iringa and Morogoro regions—Tanzania. *Sustainability and Climate Change., 16*(3), 228–248.

Stringer, L. C., Simpson, N. P., Schipper, E. L. F., & Eriksen, S. H. (2022). Climate resilient developmmt pathways in global drylands. *Anthropocene Science, 1*, 311–319.

Trisos, C. H., et al. (2022). Impacts, adaptation, and vulnerability. In H.-O. Pörtner et al. (Eds.), *Contribution of Working Group II to the Sixth Assessment Report of the Intergovernmental Panel on Climate Change* (pp. 1171–1274). Cambridge University Press.

USAID. (2014). *Climate-resilient development: A framework for understanding and addressing climate change*. Report compiled for the United States Agency for International Development Global Climate Change Office, Climate Change Resilient Development Project. Washington, DC.

Ziervogel, G. (2020). Intergovernmental Panel on Climate Change (IPCC) Sixth Assessment Working Group II Report. *South African Journal of Science*.

Climate Finance for Building Resilience in Rural Southern Africa: A Systematic Review of the Status Quo, Gaps, and Prospects

12

Joseph Tinarwo and Beauty Dzawanda

12.1 Introduction

Ambitious climate change targets, as spelt in the United Nations Framework Convention on Climate Change (UNFCCC) conferences, require significant finance for adaptation and mitigation (Skovgaard et al., 2023). Taking urgent action to combat climate change demand stakeholders to mobilize large-scale investment flows from both domestic, international, and private financiers (Grigoriev et al., 2020; Hofisi et al., 2013). Globally, countries need approximately US$2 trillion in investments per year to limit global warming, boost resilience, and support a global transition to net zero emissions (World Bank, 2022). Notwithstanding causing less than 4% of the global greenhouse gas emissions, Africa is among the worst-affected continents by climate change globally (Zheng et al., 2023). The deleterious effects of climate change in Africa have coincided with other shocks such as the long-term impact of the COVID-19 pandemic, economic slowdowns and downturns, and the spillover effects of Russia–Ukraine war (Gray et al., 2023). Combined, these shocks have negatively impacted on Africa's growth trajectory and minimized its efforts to fight climate change.

Africa needs approximately US$2.8 trillion for addressing climate change between 2021 and 2030, without including the cost for loss and damage (AEO, 2023). However, Africa only receives US$30 billion annually in climate finance flows to its region (AEO, 2023). According to UNFCCC (2021), the climate finance gap in African countries is expected to be even broader as nations usually underestimate their financial needs, particularly in the context of adaptation, due to problems in costing their national adaptation plans (NAPs). Recently, the African Heads of State and Government adopted the Nairobi Declaration on Climate Change between the 4th and 6th of September 2023 in Nairobi, Kenya, with the objective to launch bold and innovative policy strategies to fund climate action. Key to the Nairobi Climate Pact was the adoption of a novel climate financing framework that is designed to effectively respond to the needs of the African region, including debt restructuring and relief, reforms to international financial institutions, and new global taxes to fund climate action. While Africa receives considerably less climate finance than what individual countries need, the Southern African region shoulders largest climate financing gap in real terms (Ganti et al., 2023). This financing gap is mainly attributed to the lowest levels of climate investment and other complex downside risks

J. Tinarwo (✉)
University of South Africa, Department of Financial Governance, Pretoria, South Africa

B. Dzawanda
Department of Geography, Environmental Sustainability and Resilience Building, Midlands State University, Gweru, Zimbabwe

M. Matsa et al. (eds.), *Climate Change Resilience in Rural Southern Africa*, Sustainable Development Goals Series, https://doi.org/10.1007/978-3-031-73600-1_12

like persistent global inflation and the geopolitical tensions that is spurring food and energy prices (Funder & Dupuy, 2022).

Climate finance is currently regarded as a "buzz" term in development literature. UNFCCC argues that climate finance entails the broad range of domestic and external financing from state and non-state actors with the objective to support climate change mitigation and adaptation (UNFCCC, 2019). Article 9 of the Paris Climate Agreement defines climate finance as all local, national, or external financial flows from any source directed to advance efforts on mitigation and adaptation against climate change (UNFCCC, 2015). Although the above definitions appear to be impartial on the meaning of climate finance, they are circuitously formed on normative notions that refer to all forms of finance from diverse sources intended to take concerted efforts to address the negative impacts of climate change. The definitions of climate finance obviously have a normative content and are guided by what the organization or author contributing them deems acceptable. Against this background, this research adopts the definition offered by the UNFCCC because of its comprehensive nature. Therefore, climate finance is a very complex term that incorporates all investments and interventions that promote environmental sustainability and development such as the attainment of the worldwide sustainable development goals (SDGs). Conversely, resilience building in the context of climate change is defined as the ability and capacity to predict climate change threats and risks, cope with shocks and stresses, as well as restructure and transform long-term development pathway (UNFCCC, 2019). Understanding the climate finance landscape is crucial to help governments and other stakeholders to build climate resilience by assessing the financial needs required for adaptation and mitigation.

This chapter examines the status quo, gaps, and prospects of climate finance for building resilience in rural Southern Africa and seeks to answer the following research objectives:

(a) To examine the nature and context of climate finance in rural Southern Africa.
(b) To analyze the gaps and challenges of climate finance in rural Southern Africa.
(c) To develop a framework that can be used to tackle the challenges of climate finance in rural Southern Africa.

12.2 Methodology

This review adopted a systematic approach, focusing exclusively on climate finance documents in rural areas of Southern Africa, as presented in Fig. 12.1. Such documents were mostly obtained from Scopus database, Google Scholar, and government websites using Boolean combinations of the following keywords: climate finance, rural areas, Southern Africa, Zambia, South Africa, Malawi, and Zimbabwe, for the period from 2015 to the present. The reason for targeting sources from 2015 was because this was when the Paris Agreement, a legally binding international treaty on climate change was adopted. It was a landmark in the multilateral climate change process because it brought all nations together to combat climate change and adapt to its effects.

A total of 125 documents related to this study were identified, out of which 59 articles were finally selected for analysis after the inclusion and exclusion selection criteria. On Scopus database, the keywords that were used to initiate the search process resulted in 32 documents as shown in Table 12.1. After that the csv file was extracted from Scopus database, with a list of 32 documents for further screening process. The csv file was used to screen the relevant documents based on title selection and this gave a total of 27 documents. Then screening based on abstracts was done and a total of 24 documents were selected. From the 24 documents, the researchers had no access to 3 documents and 21 were successfully retrieved. After full text screening, 3 documents were excluded based on results that were not relevant; therefore, only 17 documents, from Scopus database, were finally synthesized and reviewed to produce findings in this research.

Most of the assembled documents from Scopus database did not address climate finance

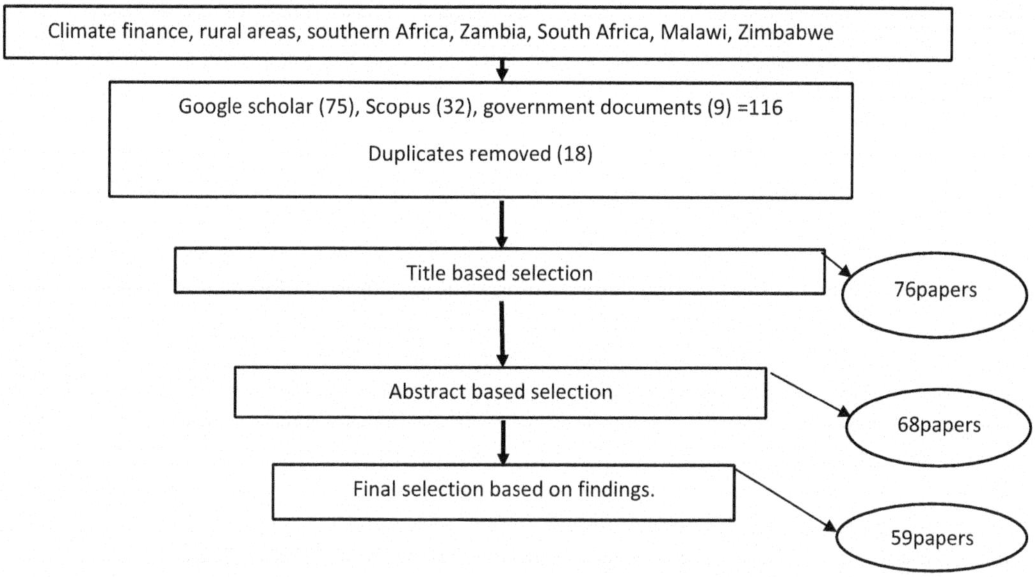

Fig. 12.1 Schematic flowchart of the review process. (Source: Authors)

Table 12.1 Scopus database selection criteria that resulted in 32 documents

TITLE-ABS-KEY ("climate finance" AND "southern Africa" OR "Zimbabwe" OR "South Africa" OR "Zambia" OR "Malawi")

in rural areas of Southern Africa but focused on climate finance in general. Hence, it was not possible to depend only on electronic databases, as there was a need to consult government-published documents and other published works on Google Scholar on the subject matter. This limitation was also solved by critically analyzing the findings of the selected documents and identifying areas where climate finance in rural areas have been addressed.

The same inclusion and exclusion criteria were adopted for 75 documents that were accessed from Google Scholar after literature search was done using the same keywords. Firstly, duplicate documents (17) to those selected from Scopus database were removed and this gave a total of 57 documents. These were further subjected to title screening and 12 documents were excluded from the list. Abstract screening was done, and 7 documents were found irrelevant. After full text screening, 5 documents were excluded and 33 documents were finally synthesized and reviewed, from Google Scholar, to produce findings in this research. Nine documents were accessed from government and organization websites, and a total of 59 documents were used to produce findings, as shown in Fig. 12.1.

12.3 Results and Discussion

12.3.1 Flow Chart of Results Obtained from the Search and Screening Process

Figure 12.1 shows the findings of the literature search and screening procedure that was embarked upon in coming up with documents that were reviewed and consulted in this research.

12.3.2 Nature and Context of Climate Finance in Rural Southern Africa

Southern Africa is among the regions that are severely affected by climate change globally, particularly in the rural areas (Steffen et al., 2007; McDowell et al., 2019; Ngwenya & Simatele,

2020a). Rural areas in Southern Africa are adversely affected due to their dependency on climate sensitive sectors, high exposure to climate stresses, limited adaptive capacity, and high dependence on climate finance for climate mitigation and adaptation projects from developed countries (Filho et al., 2018; Dionne & Horowitz, 2016; Bowman & Minas, 2019; Ngwenya & Simatele, 2020b). There is no prioritization to adaptation measures by the governments of countries in the Southern African region (Hainaut & Cochran, 2018; Ngwenya & Simatele, 2020b). Yet the costs of the adaptation measures are rising far faster compared to the funding that is available. Thus, Southern Africa's hurdles are escalated by institutions that are too weak to efficiently deal with climate change due to decreased capacity and low resilience of inhabitants to manage the impacts of climate change (Bisaro et al., 2018; IPCC, 2018). Many communities in these countries are poor, particularly in rural areas, with little capacity to adapt. Their exposure to extreme and recurrent hazards, along with their reliance on natural resources, makes them extremely susceptible (Bowman & Minas, 2019). Henceforth, with worsening climate change impacts, these exposed countries will be affected severely. Therefore, there is urgent need for adaptation measures in this region, and this also means allocating and accessing a lot of funding to the rural areas in this region.

There is unequal distribution of climate funds worldwide. Between 2016 and 2019, Africa accessed only 26% of climate funding (Fonta et al., 2018; Ngwenya & Simatele, 2020b). In the African continent, South Africa is the one that received the most multilateral climate financing; however, this is mostly for mitigation measures with only 2% directed to adaptation (Fonta et al., 2018; Ngwenya & Simatele, 2020b). The countries that accessed moderately high amounts for adaptation measures are Zambia and Comoros, but the amounts are still insufficient to cater for their needs. Botswana did not receive any financing for adaptation. The 2020–2023 COVID-19 pandemic further compromised climate financing for Africa in general, and Southern Africa in particular, exacerbating the region's extent of vulnerability (Nhamo & Chapungu, 2024). This clearly shows that vulnerability status is not considered when allocating climate funds although climate financiers frequently speak about prioritizing the most vulnerable countries. Germanwatch's Global Climate Risk Index 2021 ranked Mozambique, Zimbabwe, and Malawi the 1st, 2nd and 5th most affected countries in the world in 2019 (Watson & Schalatek, 2019; Mungai et al., 2022). However, they were placed 32nd, 108th, and 75th in climate financing received. This means that the rural areas which mostly need climate financing also suffer as a result of this ranking.

Most nations that have high risks are left behind, especially those in Southern Africa. The general trend is that countries that have donors and strong institutional capability to implement projects are the ones that receives climate finance. There is reluctance among investors when it comes to investing in nations or regions that have poor institutional capacity and records of corruption. They always want to work with predictable environments so as to generate returns. As a result, the little that reaches Southern African countries tends to benefit urban areas a lot as compared to rural areas, which justifies why there is scanty literature on empirical studies on climate financing in rural areas in this sub-region. Watson and Schalatek (2019) argued that when vulnerable countries receive adaptation finance, the communities at risk do not access it. Adaptation funding is usually determined at an international level and then directed nationally with slight contribution at local level. For example, more than 70% of respondents in a research of people displaced by Cyclone Idai in Zimbabwe indicated that they were not aware of climate change or adaptation (Chirisa et al., 2021; Dzingirai et al., 2023). And in many cases, before funds reach the beneficiaries they are lost to corruption. However, despite these hurdles in Southern Africa, there is some progress being noted in climate adaptation. As of 2021, 72% of Southern African countries had at least one instrument on national adaptation planning (Watson & Schalatek, 2019; Mungai et al., 2022). More so, almost all countries in Southern Africa, excluding Botswana and Angola, adopted national plans and Botswana launched an adaptation framework.

Rural communities in Zimbabwe experience severe climate-induced disasters just like many countries in Southern Africa. This intensely affects the country due to its dependence on climate-sensitive sectors like rain-fed agriculture. About 70% of rural communities in Zimbabwe are reliant on rain-fed agriculture; hence, decoupling rain-fed agriculture from climate change is one of the country's adaptation main concerns to develop resilient communities (Zvirevo, 2019; Chirisa et al., 2021). With rising back-to-back droughts, these rural communities are in continual poverty since their yields and earnings are impacted by the recurrent reduction in rains. This aggravates the communities' adaptive capabilities and makes their livelihoods to be affected continuously over time.

Efforts to reduce susceptibility to climate change and develop lasting coping capabilities of rural populations in dealing with climatic impact have been made by the Zimbabwean government (Zvirevo, 2019; Chirisa et al., 2021; Dzingirai et al., 2023). Initiatives like the accreditation to the adaptation fund came as vital enablers to deliver the essential financial resources to support adaptation (Chirisa et al., 2021; Dzingirai et al., 2023). Through this fund there are opportunities for technical support in consolidating sectoral policies, planning, and technology development and transfers, which are significant pillars for adaptation. In addition, the Infrastructure Development Bank of Zimbabwe (IDBZ) was selected as the National Implementing Entity to the Green Climate Fund (Zvirevo, 2019; Lorimer et al., 2018; Chirisa et al., 2021). The bank also formed a Climate Finance Facility (GFF) to offer a domestic financial resource base and established a Climate Finance Framework (CFF) responsible for managing the climate finance mechanism, which includes the structuring of green bonds, among others (Zvirevo, 2019; Lorimer et al., 2018; Chirisa et al., 2021). In terms of global climate finance in 2020, the Government of Zimbabwe through its Ministry of Lands, Agriculture, Water and Rural Resettlement, Climate Change Department, and UNDP Zimbabwe launched the Green Climate Fund (GCF)-financed "Building Climate Resilience of Vulnerable Agricultural Livelihoods in Southern Zimbabwe project". The innovative US$47 million 7-year project aimed at strengthening the climate resilience of vulnerable agricultural livelihoods of rural communities in southern Zimbabwe to improve food security (Dzingirai et al., 2023).

Zimbabwe's national priorities recommended for funding under the Green Climate Fund (GCF) are climate-smart agriculture, early-warning and disaster risk reduction, sustainable forestry management, renewable energy and energy efficiency, and integrated waste management as shown in Table 12.2. According to the Government of Zimbabwe and MECTHI (2020), some projects under the climate-smart agriculture have received some part of this amount.

Table 12.2 Zimbabwe national priorities per GCF result area and total amount targeted

Theme	Main national priorities	GCF result area	Total
Adaptation	Climate-smart agriculture	Health, food, and water security	US$270,430,552
		Livelihoods of people and communities	
		Ecosystems and ecosystems services	
	Early-warning and disaster risk reduction	Infrastructure and built environment	US$9,850,000
Mitigation	Sustainable forestry management	Agriculture, forestry and other land use	S$25,000,000
	Integrated waste management	Buildings, cities, industries, and appliances	US$279,180,600
	Renewable energy and energy efficiency	Energy generation and access and energy efficiency	S$154,014,000
		Transport	

Source: Authors

However, the Zimbabwe Green Climate Fund Programme does not indicate the fund that was specifically allocated to rural areas.

To reinforce climate resource mobilization, utilization, and reporting in a comprehensive way, and to fulfil GCF prospects, there is a need for a robust stakeholder coordination process. Central to this coordination process is the Climate Change Management Department under the Ministry responsible for Climate Change (Ministry of Environment, Climate, Tourism, and the Hospitality Industry (MECTHI), which is the National Designated Authority (NDA) to the GCF and plays a facilitative role in the coordination (MECTHI, 2020). The GCF Coordination Framework is building on the National Climate Policy Institutional Framework, which is the mechanism responsible for implementing the National Climate Policy (NCP, 2016). The membership of the GCF Coordination Framework includes representatives from provincial governments, UN agencies, climate-related international organizations and donor community, representatives from urban and rural community, representatives from private sector and civil society, universities, and research institutions. The country has no Direct Access Entity (DAE) with the Infrastructure Development Bank of Zimbabwe (IDBZ), FBC Bank, Steward Bank, and Environmental Management Agency (EMA) having been nominated for accreditation. The recent accreditation of Environmental Management Agency as a National Implementing Entity (NIE) to the Adaptation Fund further strengthens adaptation-focused projects in the country (Government of Zimbabwe, 2020).

Agriculture in Zambia is 95% rain-fed, and this is mainly practised in rural communities who struggle day-to-day to cope with increasing temperatures, erratic rainfall patterns, frequent floods, and protracted severe droughts (Shawoo et al., 2022). In Zambia, poverty is mostly experienced in rural areas at above 70% of population as compared to less than 30% in urban areas. Subsistence agriculture that is rain-fed is the main economic activity in the rural areas. Therefore, climate resilience building is vital for survival of these rural communities. Zambia has 12 reported climate finance projects (Funder & Dupuy, 2022; Chirambo, 2023). However, literature did not clearly specify how many projects were specifically targeting rural communities. Two of these projects account for 3% (or US$3.5 million) of the total approved climate finance to date (Chirambo, 2023). Four other projects account for 77% (or US$80.9 million). These projects are usually regionally targeted and locally piloted. Six other projects account for 20% or US$20.9 million (Mulenga et al., 2017; Funder & Dupuy, 2022; Chirambo, 2023). These projects are generally national-level activities.

The (Pilot Program for Climate Resilience) PPCR supports a National Programme in Zambia through the Climate Investment Funds of the World Bank. Zambia's PPCR goal is to "mainstream climate change into the most susceptible sectors of the economy," aligning with national plans and intentionally integrated with national processes and planning for Zambia's climate change response. The next largest amount accessed is from the Least Developed Countries Fund (LDCF) at 17% (Mulenga et al., 2017; Funder & Dupuy, 2022). PPCR is supporting three major projects for reinforcing climate resilience and accounts for 72% of all climate finance accessed to date (Mulenga et al., 2017; Funder & Dupuy, 2022). The projects are building institutional capacity of the national, provincial, and local systems to mainstream climate adaptation and resilience into development targeting vulnerable and poorest communities.

In Malawi, 85% of the population is rural, which depends on rain-fed agriculture, and poverty remains one of the most important social challenges exacerbated by climate change (Chirambo, 2016; McCauley et al., 2021). The economy of Malawi is largely agricultural-based, and the agriculture sector accounts for over 30% of the gross domestic product (GDP) and 90% of export revenues, so recurrent dry spells and floods have resulted in major drops in revenues from the sector (McCauley et al., 2021; Chirambo, 2016). Malawi has tried to solve this challenge by putting in place a legal framework and an enabling policy to govern climate change and environmental management in the country which includes the Malawi Growth and Development Strategy III, National Climate Change Management Policy, Nationally Determined Contributions, and National Resilience Strategy (Schuenemann et al., 2018; Samarakoon, 2020; Aung et al., 2021). The Government of Malawi has also established a National Designated Entity (NDE) responsible for the development and transfer of climate change technologies for adaptation and mitigation and a Designated Authority (DA) for the Adaptation Fund and Operational and Political Focal Points for the GEF. Malawi was among the first countries to nominate a GCF Focal Point (FP)/NDA in April 2014 and is also among the first eight countries to have a project approved by the GCF during the first Board's approval decision (Msowoya et al., 2016). In 2019, Malawi got approval from the GCF for a National Adaptation Plan (NAP) Readiness project titled "Advancing the NAP process: Climate resilience for sustainable development in Malawi" to the tune of about US$2.9 million. The objective of the project is derived from the National Climate Change Policy which is to reduce vulnerability of people in Malawi and to promote community and ecosystem resilience to the impacts of climate change and gender-equitable adaptive capacity for planning and implementing adaptation interventions. Further, under the Readiness support, the GCF in December 2019 approved a strategic frameworks support for Malawi through the Climate Technology Centre and Network and UNEP (Bell et al., 2018).

Malawian government largely rely on global financing for its domestic climate change action. Funds accessed to date are from the World Bank Strategic Program for Climate Resilience (SPCR), the GEF through the LDC and the Adaptation Fund. Additionally, in 2015 and 2018, Malawi received some resources for two projects on mitigation and adaptation from the GCF, respectively (Katengeza et al., 2019; Ragasa & Mazunda, 2018). The GCF also supported project on Scaling Up the Use of Modernized Climate Information and Early Warning Systems in Malawi is a $16 million investment, which ran from 2016 to 2022. The priority sectors in Malawi under the National Resilience Strategy Policy of 2018 to 2030 includes resilient agriculture growth, risk reduction, flood control and early warning and response systems, human capacity, livelihoods, and social protection and catchment protection and management (Katengeza et al., 2019; Ragasa & Mazunda, 2018).

Extensive dependence on rain-fed agriculture, widespread rural poverty, and inadequate irrigation facilities in Malawi threaten the production of maize, especially in the context of recurrent shocks that are climate-induced (Msowoya et al., 2016). In Malawi, maize is the staple crop and its production is essential for food security since it has an effect on food access and availability for many households in rural areas. Therefore, Malawian government and its development partners have been prompted to encourage a farm input subsidy programme (Katengeza et al., 2019; Ragasa & Mazunda, 2018) and practices on climate-smart agriculture such as drought-tolerant maize (Katengeza et al., 2019), conservation agriculture (Bell et al., 2018), and physical infrastructures for water and soil conservation (Soroko et al., 2018).

South Africa experiences major challenges with the sustainable provision of water since it is a water-scarce country. The country is located within Southern Africa's "drought belt" and, according to the World Bank, it ranks fifth position in sub-Saharan Africa as water-scarce country. Since the 1960s, South Africa has already experienced a rise in average annual temperatures countrywide, by 1.5 °C (Republic of South Africa, 2021; Karypidou et al., 2022). South Africa ranked among the top half of most vulnerable countries in terms of climate change vulnerability overall, and in particular water-related climate change vulnerability. Through the NWRS (National Water Resource Strategy) as well as the 2018 National Water and Sanitation Masterplan (NWSMP), the DWS (Department of Water and Sanitation) proposed a suite of initiatives to solve the future and current water challenges which are projected as being a 17% deficit by the year 2030 (DWS, 2018; Cullis & Phillips, 2019). The initiatives include lessening water demand and water losses through non-revenue water, as well as other initiatives to increase the water supply mix. These include a national water reuse strategy, and this funding proposal recommended the commitment of US$235 millions of catalytic capitals to support the establishment of a national WRP (Water Reuse Programme) and to enable the scale-up of pipeline water reuse projects that will support the improvement of water security and climate resilience at local and regional scales, where climate vulnerability is the driver (DWS, 2018).

The WRP is a national intervention to address climate-induced water insecurity in South Africa and forms part of a larger NWPP (National Water Partnerships Programme), with the establishment of the NWPP being approved in principle by the Council of the Presidential Infrastructure Co-ordinating Commission (PICC) on 10 April 2021 (Republic of South Africa, 2021). The WRP is aimed at identifying, encouraging, and assisting municipalities with the scaling of their water reuse sub-projects and programmes. Several initiatives have been implemented by national government already to avoid this anticipated water shortage with a key element of these initiatives being to develop an improved level of diversification in relation to the "mix" of water supply sources. The GCF does not directly finance the underlying water reuse sub-projects. The GCF provides loans (at a Programme Level) to the Accredited Entity (the Development Bank of Sothern Africa (DBSA), who facilitates the GCF Concessional loan(s) into the underlying sub-projects, as required (DBSA, 2021). DBSA is a development finance institution, based in South Africa, with an obligation to invest in activities by both private and public sectors at national and regional levels in Africa. The GCF entered into a Funded Activity Agreement (FAA) with the DBSA, in the form of a loan agreement, with limited recourse with respect to the repayment risk of Loans (DBSA, 2021; Republic of South Africa, 2021). The GCF provides the Concessional Loans in US$ to DBSA in two tranches: (i) a public sector tranche amounting to US$130 million and (ii) a private sector tranche amounting to US$70 million. These two tranches will be priced separately. The WRP's underlying water reuse sub-projects are financed in SA Rand (ZAR), which is overseen by the DBSA. The DBSA obtains the ZAR equivalent swap rate of GCF concessional pricing to be passed onto sub-projects.

12.4 Gaps and Challenges of Climate Finance in Rural Southern Africa

Understanding climate finance gaps and challenges in rural Southern Africa is crucial to identify possible financing opportunities and guide stakeholders to effectively mobilize and allocate financial resources required for adaptation and mitigation (Michaelowa et al., 2021). This information also supports international, regional, and national policy processes in the implementation of a collective goal on climate action as well as helps in evaluating the effectiveness of climate finance flows (Murcott, 2022). Recent statistics has shown that rural Southern Africa face significant gaps and constrains on both domestic as well as international finance flows (Ngwenya & Simatele, 2020a). Countries in Southern Africa require approximately US$1 trillion in climate finance by 2030, translating to about US$90.3 billion annually (AEO, 2023). In the context of climate finance needs for Southern Africa, adaptation requires approximately 70%, while 21.7% is for mitigation, as shown in Fig. 12.2 (Roffe et al., 2023).

Of the four case studies that were reviewed, South Africa has the highest climate needs, followed by Zambia and Malawi, while Zimbabwe has the least climate needs, as shown on Table 12.3. This may be attributed to their different Climate Vulnerability and Resilience scores.

As indicated above, Southern Africa has the highest climate finance needs, yet the region has been receiving the lowest climate finance flows compared to other African regions. Despite the pledge by the developed nations to mobilize at least US$100 billion per year as reparations to support mitigation and adaptation efforts against climate change in less developed countries, Southern Africa is only receiving approximately

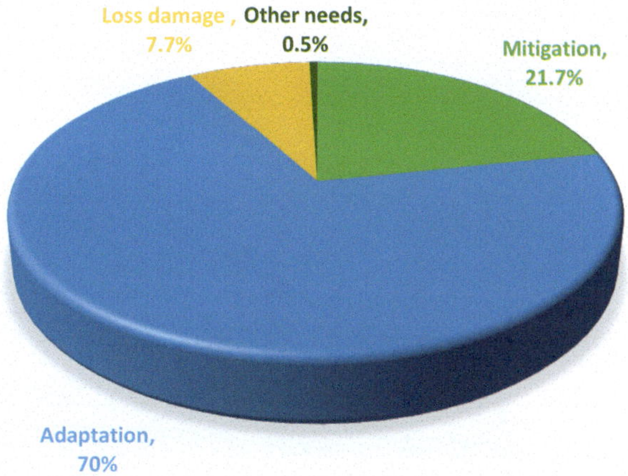

Fig. 12.2 Southern Africa's climate finance needs. (Source: Authors' own construction based on African Development Bank data, 2023)

Table 12.3 Climate finance needs for South Africa, Zambia, Malawi, and Zimbabwe (2020–2030)

Country	US$ billion
South Africa	740
Zambia	35
Malawi	15
Zimbabwe	5.2

Source: Adapted from African Development Bank data (2023)

US$6.2 billion annually, representing 6.9% of the climate finance needs for the region (Fridahl & Linnér, 2016; AEO, 2023). The climate finance flow to Southern Africa is below the regional needs and requirements to achieve the climate change goals as enshrined in the national, regional, and international targets. Further, Southern Africa's share of 6.9% in climate finance flows is less compared to what other regions get, as shown on Fig. 12.3.

While nine out of the ten highly susceptible nations to climate change are in sub-Saharan Africa, rural Southern Africa bears the brunt of climate change because of poverty and lack of adequate financial resources to cope with climatic risk (Bracking, 2015). Private climate finance share in Southern Africa remains generally low as compared with other regional counterparts in Africa making up 21% of total climate finance flows in this region (AEO, 2023). Private climate finance entails all investments made by private sector organizations, such as business organizations, financial institutions, and individual people, into projects and programmes, or institutions with the broad objective to combat climate change. Figure 12.4 shows the average private finance flows for climate mitigation and adaptation for South Africa, Zambia, Zimbabwe, and Malawi. The general low private climate finance in Southern Africa may be partly due to the poor investment climate in these countries as well as financial position of individual countries due to economic slowdowns and downturns (Thompson, 2020).

It is crucial to briefly examine the state of climate finance in each case study under the Southern African countries under review. In South Africa, the harmful effects of climate change are worsening an already dire poverty and food security situation for many people residing in the rural communities. The frequency of floods and droughts recorded across South Africa suggests the urgent need to step up funding mechanisms for climate adaptation and mitigation particularly the rural communities (Khan et al., 2023). In its efforts to tackle climate change, South Africa has embraced the private sector and was among the top African countries to receive private climate finance between 2019 and 2020 of approximately US$656 million translating to about 40% of climate finance (AEO, 2023). In 2022, the South African government indicated that it requires approximately ZAR1.5 trillion (US$84 billion) in the coming 5 years to advance its strategies to reduce carbon emissions, leverage economic prospects from transitioning to a

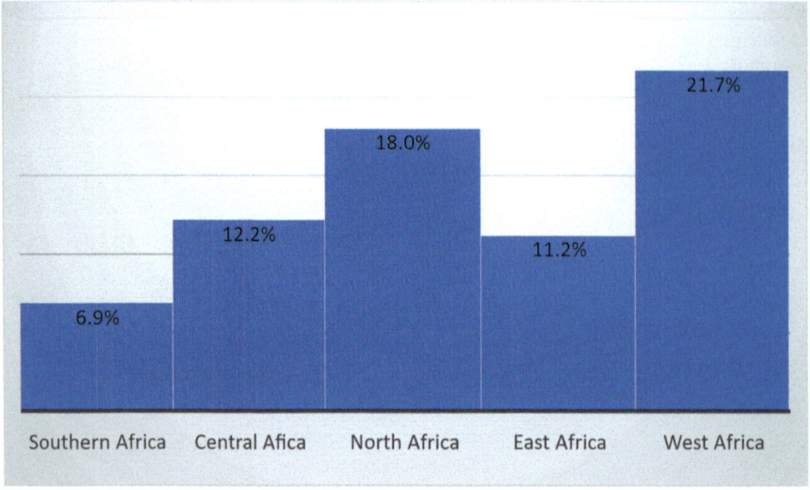

Fig. 12.3 Climate finance flow to Southern Africa. (Source: Based on AEO, 2023)

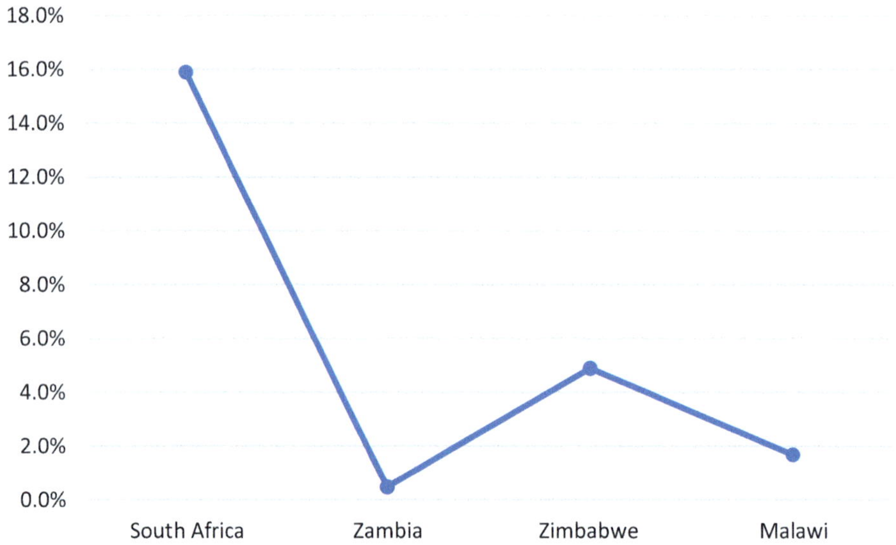

Fig. 12.4 Average annual private finance flows to Southern Africa. (Source: Adapted from African Development Bank data, 2023)

green economy, and help vulnerable communities, mainly rural areas (Winning & Kumwenda-mtambo, 2022). However, the amount of climate finance needed to attain South Africa's climate needs is considerably higher compared to the available mobilized financial resources amounting to approximately 1.7 billion or 27.7% annually (AEO, 2023). According to the Presidential Climate Finance Task Team leader of South Africa, the country has a funding gap of about 700 billion rands indicating that the country needs to strengthen its domestic and international resource mobilization efforts to fight climate change especially in rural communities where its impacts are heavily felt (Winning & Kumwenda-mtambo, 2022). Domestic sources of climate finance in South Africa accounted for 91% of tracked climate finance, while international sources accounted for 9% (de Aragão Fernandes et al., 2023). Therefore, unlike the other three reviewed case studies that need to strengthen their domestic resource mobilization, South Africa needs to strengthen its international cooperation to mobilize international climate finance which accounts for only 9% in 2022. It is estimated that South Africa requires on average R334 billion or US$17.9 billion per year to meet its net zero goal by 2050, and R535 billion or US$28.6 billion per year to meet its NDC target by 2030 (de Aragão Fernandes et al., 2023). In view of the above, it is important to ensure equitable allocation of climate finance between rural and urban areas, recognizing the unique vulnerabilities and challenges faced by rural communities. Rural communities are susceptible to the negative impacts of climate change resulting in droughts, floods, and erratic weather patterns; hence, it is imperative to prioritize rural climate finance in the budget allocation. Therefore, climate action in South Africa and elsewhere in Southern Africa needs to take a holistic approach, making investments in climate-smart infrastructure development for water management, sustainable agriculture practices, renewable energy projects and programmes tailored to rural communities, and capacity building for resilient local communities.

Similarly, by the year 2020, Zimbabwe received approximately US$990 million in climate finance, translating to about US$90 million a year. As a result, Zimbabwe currently has a climate financing gap of US$440–500 million a year, thus greatly limiting the country's ability to build climate resilience particularly for rural communities (AEO, 2023; Chirisa et al., 2021). To prioritize the achievement of its climate change targets, Zimbabwe developed the National Climate Change Fund and Climate Finance Facility to crowd in the private sector (Dzingirai et al., 2023). Zimbabwe's poor political and economic environment is currently creating uncertainty and instability, inhibiting the effective role of the private sector to make climate-related investments in renewable energy. This is largely because of challenges related to property rights, government interference, and corruption as well as the volatile hyperinflation environment and currency fluctuations which further discourage private sector investment by eroding potential returns and increasing business risks. Additionally, poor international relations with the United States of America and Britain as well as its allies following the highly contested Fast Track Land Reform Programme in 2000 has isolated and excluded Zimbabwe from being eligible for development finance.

Moreover, Zimbabwe and other countries in Southern Africa is confronted by problems related to climate finance allocation between rural and urban areas and this is mainly because of the differing needs, vulnerabilities, and capacities of these areas. Zimbabwe also lacks adequate frameworks for climate finance tracking and reporting to determine climate flows to rural areas and other parts of the country as well as capacity to mobilize scaled-up finance for technology deployment in climate action.

Zambia is one of the Southern African states where climate variability is affecting both urban and rural communities, with climate change impacts including unusual rainfall and temperature variations, increased frequency of droughts, cyclones, and other extreme weather events. Climate change in Zambia is exacerbating the entire socio-economic development landscape with the rural areas bearing the bane of drought-induced food insecurity, water shortages, floods, and cyclones (Mulenga et al., 2017). Zambia's total climate finance needs are estimated to be US$35 billion through 2030, which is anticipated to be mobilized mostly domestic and international climate mechanisms such as the Global Climate Fund and development partners (AEO, 2023). The private sector is also an emerging key player in climate finance mainly through corporate social responsibility and investment with returns (Funder & Dupuy, 2022). For instance, the Zambia National Commercial Bank, Kukula Capital, and the Worldwide Fund for Nature–Zambia recently established a US$53 million Green Outcomes Fund facility to support green projects in rural communities (AEO, 2023). Further, carbon-market financing mechanisms remain a private climate finance facility that remain untapped in Zambia as the country is endowed with many natural resources such as land, water, forest, and wildlife. In Zambia and other countries in Southern Africa, there are challenges related to capacity building beyond the injection of climate finance. There is a need to implement holistic capacity building initiatives that empower local communities to innovate, adapt, and sustainably manage their ecosystems in the face of climate change challenges. Moreover, it is crucial to develop a comprehensive capacity building strategy aimed at enhancing the resilience and ability of rural communities including women and youths to effectively plan, implement, and monitor climate change mitigation and adaptation efforts.

Malawi is susceptible to adverse climate shocks that include long dry spells, severe droughts, floods, cyclones, and other extreme weather conditions floods (Amadu et al., 2020b). Climatic shocks have increased in incidence, intensity, and scale over the last 20 years, with catastrophic results on water, food security, energy, and sustainable livelihoods of the most rural communities (Amadu et al., 2020a, b). Climate change in Malawi is now considered as one of the worst disasters leading to significant threats to meet the growing needs of most rural populations (Chirambo, 2016). The average climate finance flows for the period 2019 to 2020 was US$512 million and for 2020 alone was US$583.8 million, indicating a 16.3% increase. This is against the climate needs of approximately US$1.5 billion per year (AEO, 2023). The huge climate finance gap in Malawi is hampering efforts of climate change mitigation and adaptation particularly in rural communities because they receive little resources (Christophers et al., 2020). Malawi's climate finance landscape is dominated by both state and non-state actors which are often poorly coordinated and harmonized, resulting in a lot of duplication and conflicts (Kita, 2017). The flow of private climate finance in Malawi is generally low compared to other countries in the region accounting for approximately 1.7% of the total private climate finance to Southern Africa populations (Chirambo, 2023). In Malawi, total climate finance flows are disproportionately distributed across country and do not meet the needs of climate finance of the rural communities which bears the bane of climate change impacts. In Malawi, the Nationally Determined Contribution prioritizes the following key sectors and programmes: agriculture, water and sanitation, health, infrastructure development and transport, land planning and management, population and human settlements, and disaster risk management for adaptation. The financial resources required for mitigation and adaptation against climate change in these sectors up to the year 2040 is approximately US$46.3 billion. Credit from the private sector averaged approximately US$3.3 billion every month in 2022, of which only 15% was allocated to climate-related sectors. The public sector dominates in all climate-related sectors in Malawi, and this limits the participation of the private sector mainly due to lack of clear frameworks for their participation (Chirambo, 2023). A multi-stakeholder process is thus important where the state and development partners could develop the legal and institutional frameworks for private sector involvement in climate action.

12.5 Conceptual Framework for Promoting Climate Finance for Building Resilience in Rural Southern Africa

Systematic review of literature insights facilitated the development of a conceptual framework for improving climate finance for resilience building in rural Southern Africa, as presented in Fig. 12.5. Strengthening domestic resource mobilization including promoting a favourable environment for private sector investments is vital. Regions that have successfully sufficiently allocated financial resources and not just relied on donor support have proved to be resilient to climatic shocks (e.g. Europe and Somalia) (FAO, 2015). It is also imperative to ensure equity in allocation of finance between rural and urban areas in rural Southern Africa. Furthermore, it is

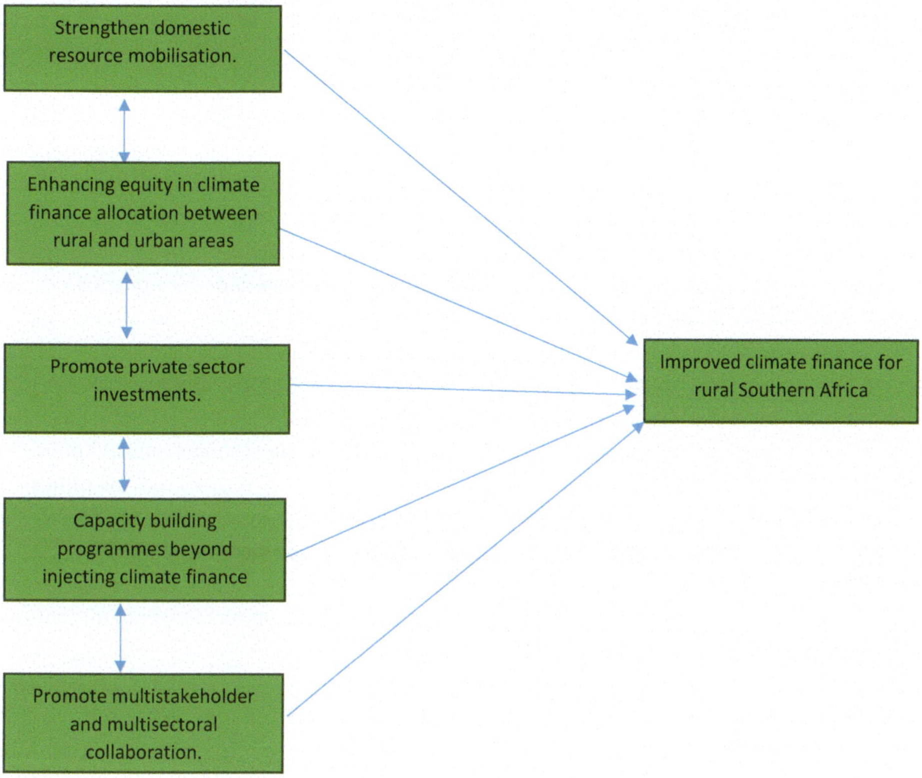

Fig. 12.5 Framework for promoting climate finance for resilience building in rural Southern Africa. (Source: Authors' own construction)

important to rollout both human and institutional capacity-building to be in a better position to advocate and lobby for equitable distribution of climate finance between rural communities and other parts of the world. Lastly, promoting multi-sectoral and multi-stakeholder partnerships is important, harnessing the effective participation of all actors including governments, philanthropic organizations, civil society, academia, and think tanks, and the private sector needs to be improved and aligned, particularly to mobilize the climate finance resources. The academics can partner governments and other stakeholders to do extensive research on the progress made so far with regards to climate finance particularly in rural areas since it was discovered, in this research, that there are scarce empirical studies that have been done to date on climate finance in rural areas of southern Africa. This will help to unravel gaps and challenges faced in different countries of southern Africa as well as disparities with regards to climate finance in rural areas. This is very critical in making well informed decisions based on empirical data to improve climate finance in rural Southern Africa for resilience building.

12.6 Conclusion and Recommendations

Climate change and development are closely linked, particularly in rural areas of Southern Africa since many people in these areas depend much on the natural environment for survival, particularly rain-fed agriculture. However, countries in Southern Africa continue to face challenges at each stage of climate finance delivery and this affect the resilient actions in this region towards sustainable development. The findings of this paper reveal that the finance gap for countries in Southern Africa is around US$1 trillion with approximately 70% going for adaptation, while 21.7% for mitigation. Although these estimates of climate finance needed for adaptation and mitigation have been the subject of great debate among experts, there is a general agreement that the climate finance gap to Southern African countries is very huge to meet the immediate adaptation needs of the region. Therefore, it is essential to put in place innovative financial and investment instruments as well as financial mechanisms that can provide regional-tailored and all-inclusive solutions to the climate finance challenges confronting Southern Africa. For instance, it is crucial to strengthen domestic resource mobilisation to better meet the climate finance needs of the region particularly rural community's adaptation which remains structurally underfunded. More so, enhancing equity in the allocation of climate finance resources between rural communities and urban areas within Southern Africa is imperative to cushion the most vulnerable groups of people and sectors. This must involve gender considerations more comprehensively and explicitly. Equity financing aligns the interests of investors and farmers, fostering a sense of shared responsibility. By sharing the risk, investors will be motivated to support farmers adopting more climate-resilient agricultural practices.

Harnessing international donor community financial support is critical in building climate resilience in rural Southern Africa. While much of the philanthropic support in the region is in rural communities, much of is for mitigation and is usually programme-based making it not sustainable in the long term. It is thus imperative for regional governments to work along and direct the efforts of international donor community towards adaptation solutions where the funding gap is huge. International donor funding anchored on a multi-stakeholder approach and informed by data-led evidence have the potential to stir investments in rural climate adaptation projects and close the adaptation finance gap. For instance, international financial institutions like the World Bank and African Development Bank can roll out low-cost loans and innovative financial mechanisms to offset the risks forecasted by the private sector. More so, donors can advance climate adaptation in rural Southern Africa through research and development, expansion, and management of infrastructure projects as well as support governments to harness indigenous knowledge systems to build climate resilience.

Promoting private sector investments in climate action can help to fill the huge gap in climate finance needed for adaptation thereby making climate-resilient rural communities in Southern Africa. Tapping into the effective role of the private sector is important to foster green investments and enhance modern climate-smart technologies critical to manage and predict future climate shocks. It is essential for policy-makers to promote a favourable and conducive environment for private sector investments in climate action in rural Southern Africa by creating an ease of doing business environment that includes removing bottlenecks and offering incentives such as tax holidays, free land, and ensuring the availability of supporting infrastructure and utilities. While public infrastructure for building climate resilience in many countries in Southern Africa is financed by public institutions, it is crucial to foster public–private partnerships through guarantees, co-financing or other strategies of risk reduction which help in attracting private capital.

Rolling out capacity building and knowledge transfer programmes beyond injecting climate finance is vital for building rural climate resilience in Southern Africa. For instance, institutional capacity strengthening programmes can be directed to international and local organizations that implement climate-reliance projects in rural areas to improve their project and financial management skills. Good governance of climate finance demands strengthening the capacity of community-based organizations in Southern Africa to positively engage in the design and implementation of climate-resilience fundable projects. Moreover, governments with support from other stakeholders can provide mentorship, technical expertise and skills, as well as networking and empowerment programmes to youths and women to adapt and effectively respond to challenges of climate change. Investors may also train the local people on how to use new climate-smart technologies to boost agriculture productivity, enhance income diversification, and increase climate resilience.

Promoting spaces for coordination, collaboration, and dialogue among different sectors and stakeholders that finance and support climate resilience in rural areas is very important to build resilience in rural Southern Africa. The lack of consensus among the different actors on how to enhance and improve climate-smart investments in the region with a particular focus on rural areas have the potential to undermine effective and timely climate action initiatives. It is therefore critical to foster effective multisectoral approaches between finance, agriculture, water, environment, and climate change as well as strengthen the multi-stakeholder partnerships to understand each other's priorities, expertise, strengths, and goals. By doing this, stakeholders will find more effective strategies to create and strengthen synergies as well as boost capacities and resources for adaptation solutions in rural Southern Africa.

References

African Economic Outlook (AEO). (2023). *Mobilizing Private Sector Financing for Climate and Green Growth in Africa*. Available at: https://www.afdb.org/en/documents/african-economic-outlook-2023. Accessed 19 Aug 2023.

Amadu, F. O., McNamara, P. E., & Miller, D. C. (2020a). Yield effects of climate-smart agriculture aid investment in southern Malawi. *Food Policy, 92*(C), 101869. https://doi.org/10.1016/j.foodpol.2020.101869

Amadu, F. O., McNamara, P. E., & Miller, D. C. (2020b). Understanding the adoption of climate-smart agriculture: A farm-level typology with empirical evidence from southern Malawi. *World Development, 126*, 104692.

Aung, T., Bailis, R., Chilongo, T., Ghilardi, A., Jumbe, C., & Jagger, P. (2021). Energy access and the ultra-poor: Do unconditional social cash transfers close the energy access gap in Malawi? *Energy for Sustainable Development, 60*, 102–112.

Bell, A. R., Cheek, J. Z., Mataya, F., & Ward, P. S. (2018). Do as they did: Peer effects explain adoption of conservation agriculture in Malawi. *Water, 10*, 51. https://doi.org/10.3390/w10010051

Bisaro, A., Roggero, M., & Villamayor-Tomas, S. (2018). Institutional analysis in climate change adaptation research: A systematic literature review. *Ecological Economics, 151*, 34–43.

Bowman, M., & Minas, S. (2019). Resilience through interlinkage: The green climate fund and climate finance governance. *Climate Policy, 19*(3), 342–353.

Bracking, S. (2015). Performativity in the Green Economy: How far does climate finance create a fictive economy? *Third World Quarterly, 36*(12), 2337–2357.

Chirambo, D. (2016). Addressing the renewable energy financing gap in Africa to promote universal energy access: Integrated renewable energy financing in Malawi. *Renewable and Sustainable Energy Reviews, 62*, 793–803.

Chirambo, D. (2023). Mitigating disaster risks and vulnerabilities through climate finance and sustainable water management: Policy considerations for Sub-Saharan Africa and Malawi. In S. Eslamian & F. Eslamian (Eds.), *Disaster risk reduction for resilience*. Springer. https://doi.org/10.1007/978-3-031-22112-5_6

Chirisa, I.; Gumbo, T.; Gundu-Jakarasi, V.N.; Zhakata,W.; Karakadzai, T.; Dipura, R. &Moyo, T. (2021). Interrogating Climate adaptation financing in Zimbabwe: Proposed direction. Sustainability, 13, 6517. https://doi.org/10.3390/su13126517.

Christophers, B., Bigger, P., & Johnson, L. (2020). Stretching scales? Risk and sociality in climate finance. *Environment and Planning A: Economy and Space, 52*(1), 88–110.

Cullis, J. D. S., & Phillips, M. (2019). *Green Book—Surface water and water supply risks for settlements: A first order assessment of the vulnerability to Climate change of surface water supply to municipalities across South Africa* (Technical report). Aurecon & CSIR.

DBSA. (2021). *Programme design and preparation of a full funding proposal to the Green Climate Fund (GCF): Market Study.*

de Aragão Fernandes, P., Gwebu, L., Johansson, L., Meattle, C., Radmore, J. V. &, Solomon, C. (2023). *South African Climate Finance Landscape 2023.* https://www.climatepolicyinitiative.org/wp-content/uploads/2023/11/The-South-African-Climate-Finance-Landscape-2023.pdf

Dionne, K. Y., & Horowitz, J. (2016). The political effects of agricultural subsidies in Africa: Evidence from Malawi. *World Development, 87*, 215–226.

DWS. (2018). *National water and sanitation masterplan. Volume2: A plan to action.* Department of Water and Sanitation, Pretoria, South Africa.

Dzingirai, M., Zvitambo, K., & Masukume, H. (2023). Green financing strategies adopted in Zimbabwe towards attainment of sustainable development goals. In S. Trivedi, R. Aggarwal, & G. Singh (Eds.), *Perspectives on blockchain technology and responsible investing* (pp. 58–84). IGI Global.

Filho, W. L., Balogun, A. L., Ayal, D. Y., Bethurem, E. M., Murambadoro, M., Mambo, J., Taddese, H., Tefera, G. W., Nagy, G. J., Fudjumdjum, H., & Mugabe, P. (2018). Strengthening climate change adaptation capacity in Africa—Case studies from six major

African cities and policy implications. *Environmental Sciences and Policy, 86*, 29–37.
Fonta, W. M., Ayuk, E. T., & van Huysen, T. (2018). Africa and the green climate fund: Current challenges and future opportunities. *Climate Policy, 18*(9), 1210–1225.
Fridahl, M., & Linnér, B. O. (2016). Perspectives on the Green Climate Fund: Possible compromises on capitalization and balanced allocation. *Climate and Development, 8*(2), 105–109.
Funder, M., & Dupuy, K. (2022). Climate finance coordination from the global to the local: Norm localization and the politics of climate finance coordination in Zambia. *The Journal of Development Studies, 58*(10), 1899–1916.
Ganti, G., Gidden, M. J., Smith, C. J., Fyson, C., Nauels, A., Riahi, K., & Schleußner, C. F. (2023). Uncompensated claims to fair emission space risk putting Paris Agreement goals out of reach. *Environmental Research Letters, 18*(2), 024040.
Government of Zimbabwe and MECTHI. (2020). *Zimbabwe Green Climate Fund Programme*.
Gray, H. B., Taraz, V., & Halliday, S. D. (2023). The impact of weather shocks on employment outcomes: Evidence from South Africa. *Environment and Development Economics, 28*(3), 285–305.
Grigoriev, L. M., Makarov, I. A., Sokolova, A. K., Pavlyushina, V. A., & Stepanov, I. A. (2020). Climate change and inequality: How to solve these problems jointly? Vestnik mezhdunarodnykh organizatsii: obrazovanie.
Hainaut, H., & Cochran, I. (2018). The landscape of domestic climate investment and finance flows: Methodological lessons from five years of application in France. *International Economics, 155*, 69–83.
Hofisi, C., Chigavazira, B., Mago, S., & Hofisi, M. (2013). "Climate finance issues": Implications for climate change adaptation for food security in Southern Africa. *Mediterranean Journal of Social Sciences, 4*(6), 47.
Intergovernmental Panel on Climate Change (IPCC). (2018). *Special Report on Global Warming of 1.5°C*. https://www.ipcc.ch/2018/10/08/summary-for-policymakers-of-ipcc-special-report-onglobal-warming-of-1-5c-approved-by-governments/. Accessed 5 May 2019.
Karypidou, M. C., Katragkou, M., & Sobolowski, S. P. (2022). Precipitation over southern Africa: Is there consensus among global climatemodels (GCMs), regional climate models (RCMs) and observational data? *Geoscientific Model Development, 15*, 3387–3404. https://doi.org/10.5194/gmd-15-3387-2022
Katengeza, S. P., Holden, S. T., & Fisher, M. (2019). Use of integrated soil fertility management technologies in Malawi: Impact of dry spells exposure. *Ecological Economics, 156*, 134–152. https://doi.org/10.1016/j.ecolecon.2018.09.018
Khan, A. R., Ratele, K., Dery, I., & Khandaker, S. (2023). Men and climate change: Some thoughts on South Africa and Bangladesh. *NORMA, 18*(2), 137–153.
Kita, S. M. (2017). "Government doesn't have the muscle": State, NGOs, local politics, and disaster risk governance in Malawi. *Risk, Hazards & Crisis in Public Policy, 8*(3), 244–267.
Lorimer, E., Zvigadza, S., Shava, C., & Parring, S. (2018). *Opportunities for accessing international Climate adaptation finance for civil society organizations in Zimbabwe*. Johannesburg, South Africa.
McCauley, D., Grant, R., & Mwathunga, E. (2021). Achieving energy justice in Malawi: from key challenges to policy recommendations. *Climatic Change (2022), 170*, 28. https://doi.org/10.1007/s10584-022-03314-1
McDowell, G., Huggel, C., Frey, H., Wang, F. M., Cramer, K., & Ricciardi, V. (2019). Adaptation action and research in glaciated mountain systems: Are they enough to meet the challenge of climate change? *Global Environmental Change, 54*, 19–30.
Michaelowa, A., Hoch, S., Weber, A. K., Kassaye, R., & Hailu, T. (2021). Mobilising private climate finance for sustainable energy access and climate change mitigation in Sub-Saharan Africa. *Climate Policy, 21*(1), 47–62.
Ministry of Environment, Climate, Tourism and Hospitality Industry (MECTHI). (2020). *Zimbabwe Green Climate Fund Programme*. Climate Change Department.
Msowoya, K., Madani, K., Davtalab, R., Mirchi, A., & Lund, J. R. (2016). Climate change impacts on maize production in the warm heart of Africa. *Water Resources Management, 30*(14), 5299–5312. https://doi.org/10.1007/s11269-016-1487-3
Mulenga, B. P., Wineman, A., & Sitko, N. J. (2017). Climate trends and farmers' perceptions of climate change in Zambia. *Environmental Management, 59*, 291–306.
Mungai, E. M., Ndiritu, S.W., & Da Silva, I. (2022). Unlocking climate finance potential and policy barriers-A case of renewable energy and energy efficiency in Sub-Saharan Africa.
Murcott, M. J. (2022). A just COP26 outcome for South Africa? *Transnational Legal Theory, 13*(2–3), 352–365.
Ngwenya, N., & Simatele, M. D. (2020a). The emergence of green bonds as an integral component of climate finance in South Africa. *South African Journal of Science, 116*(1/2), #6522, 3 pages. https://doi.org/10.17159/sajs.2020/6522
Ngwenya, N., & Simatele, M. D. (2020b). Unbundling of the green bond market in the economic hubs of Africa: Case study of Kenya. *Nigeria and South Africa, Development Southern Africa, 37*, 888. https://doi.org/10.1080/0376835X.2020.1725446
Nhamo, G., & Chapungu, L. (2024). Prospects for a sustainable and climate-resilient African economy post-COVID-19. *Global Environmental Change, 86*, 102836. https://doi.org/10.1016/j.gloenvcha.2024.102836
Ragasa, C., & Mazunda, J. (2018). The impact of agricultural extension services in the context of a heavily

subsidized input system: The case of Malawi. *World Development, 105*, 25–47.

Republic of South Africa. (2021). *South Africa First Nationally Determined Contribution Under the Paris Agreement (Updated September 2021)*. Department of Forestry, Fisheries and Environment.

Roffe, S. J., van Der Walt, A. J., & Fitchett, J. M. (2023). Spatiotemporal characteristics of human thermal comfort across southern Africa: An analysis of the Universal Thermal Climate Index for 1971–2021. *International Journal of Climatology, 43*(6), 2930–2952.

Samarakoon, S. (2020). The troubled path to ending darkness: Energy injustice encounters in Malawi's off grid solar market. *Energy Research and Social Science, 69*, 101–116.

Schuenemann, F., Msangi, S., & Zeller, M. (2018). Policies for a sustainable biomass energy sector in Malawi: Enhancing energy and food security simultaneously. *World Development, 103*, 14–26.

Shawoo, Z., Dzebo, A., Funder, M., & Dupuy, K. (2022). Country ownership in climate finance coordination: A comparative assessment of Kenya and Zambia. *Climate Policy, 22*(9–10), 1266–1280.

Skovgaard, J., Adams, K. M., Dupuy, K., Dzebo, A., Funder, M., Fejerskov, A. M., & Shawoo, Z. (2023). Multilateral climate finance coordination: Politics and depoliticization in practice. *Global Environmental Politics, 23*(2), 125–147.

Soroko, D., Mapemba, L., Phillips, J., & Jordan, M. A. (2018). *Assessment of the Wellness and Agriculture for Life Advancement (WALA) Activity, Consultancy Report*. United States Agency for International Development. Accessed 2 July 2023, at https://www.crs.org/sites/default/files/tools-research/wala_assessment.pdf

Steffen, W., Crutzen, P. J., & McNeill, J. R. (2007). The anthropocene: Are humans now overwhelming the great forces of nature AMBIO. *A Journal of the Human Environment, 36*, 614–621.

Thompson, A. (2020). Emerging powers and differentiation in global climate institutions. *Global Policy, 11*, 61–72.

UNFCCC. (2015). *Adoption of the Paris Agreement*. Available at: https://unfccc.int/files/meetings/paris_nov_2015/application/pdf/cop_auv_template_4b_new__1.pdf. Accessed 20 Aug 2023.

UNFCCC. (2019). *Climate action pathway: Resilience and adaptation*. Available at: https://unfccc.int/sites/default/files/resource/CAP_Resilience_and_Adaptation_ES.pdf . Accessed 15 Aug 2023.

UNFCCC. (2021). *United Nations Climate Change Annual Report 2021*. Available at: Https://Unfccc.Int/Sites/Default/Files/Resource/Unfccc_Annual_Report_2021.Pdf. Accessed On 18 August 2023).

Watson, C., & Schalatek, L. (2019). Climate finance regional briefing: Sub-Saharan Africa heinrich böll stiftung North America. Retrieved from http://climatefundsupdate.org/wp-content/uploads/2019/03/CFF7-2018-ENG-FINAL.pdf on 3rd 2020 at 0615 GMT.

Winning, A., & Kumwenda-mtambo, O. (2022, November 4). South Africa says it needs $84 billion for energy transition in next five years. *Routers*.

Zheng, J., Canarini, A., Fujii, K., Mmari, W. N., Kilasara, M. M., & Funakawa, S. (2023). Cropland intensification mediates the radiative balance of greenhouse gas emissions and soil carbon sequestration in maize systems of sub-Saharan Africa. *Global Change Biology, 29*(6), 1514–1529.

Zvirevo, W. (2019). *Green financing in Zimbabwe: Case of the infrastructure development Bank of Zimbabwe, Elephant Hills Hotel, Victoria Falls, Zimbabwe 6 June 2019*. Harare, Zimbabwe.

A Systematic Review of Climate Change Mitigation Policies in Southern Africa

13

Leocadia Zhou, Mark Matsa, and Admire Rukudzo Dzvene

13.1 Introduction

The impacts of global climate change are rapidly intensifying in Southern Africa (Bauer & Scholz, 2010; Seymore et al., 2014; Sithole et al., 2023). Climate change-related slow onset, as well as rapid and sudden events, continue to cause havoc on the socio-economic aspects of vulnerable populations. Such events include rising sea levels, land degradation, increased temperatures, land and deforestation, biodiversity decline, ocean acidification, floods, wildfires, heatwaves, frost, hailstorms, tropical cyclones, storm surges, salinization, pests, and droughts (Bauer & Scholz, 2010; Hoegh-Guldberg et al., 2019). Without coordinated efforts to cut greenhouse gas (GHG) emissions, temperatures in the Southern African interior could increase by more than 4 °C by 2100 (Kapwata et al., 2018).

The Intergovernmental Panel on Climate Change (IPCC, 2014) defines climate change mitigation as an effort to manage human-induced climate change factors and their overall effects, particularly greenhouse gases (GHGs). Climate change mitigation is a significant focus for human action aimed at reducing greenhouse gas sources and/or increasing their absorption in the coming decades (Panepinto et al., 2021). Consequently, there is a need to develop ambitious commitments in mitigating climate change that align with and simultaneously address key development challenges.

Approaches to climate change mitigation are typically classified into three categories, namely (1) radiative forcing geoengineering, (2) negative emissions, and (3) traditional mitigation. Some solutions, such as radiative forcing geoengineering technologies that attempt to stabilize or lower temperatures without affecting GHG concentration in the atmosphere, remain theoretical (Lawrence et al., 2018; Lockley, 2019). However, current policy frameworks incorporate the other two alternatives, which include (i) negative emissions and (ii) support for carbon dioxide (CO_2) emission reductions through traditional mitigation. Examples of the first strategy include CO_2 removal technologies such as soil carbon sequestration, biochar, afforestation or reforestation, and wetland construction and restoration (Gasser et al., 2015; Pires, 2019). The second strategy relies on traditional technologies like renewable energy, fuel switching, nuclear power, carbon capture, and storage (Østergaard et al., 2020; Su et al., 2016; Wendling, 2019).

L. Zhou (✉) · A. R. Dzvene
Centre for Global Change, Faculty of Science and Agriculture, University of Fort Hare, Alice, South Africa
e-mail: LZhou@ufh.ac.za; ADzvene@ufh.ac.za

M. Matsa
Department of Geography, Environmental Sustainability and Resilience Building, Midlands State University, Gweru, Zimbabwe
e-mail: matsam@staff.msu.ac.zw

M. Matsa et al. (eds.), *Climate Change Resilience in Rural Southern Africa*, Sustainable Development Goals Series, https://doi.org/10.1007/978-3-031-73600-1_13

Reducing carbon-based power production as a climate mitigation effort has garnered much attention in policy frameworks. The main GHGs outlined under the Kyoto protocol are CO_2, methane (CH_4), nitrous oxide (N_2O), and synthetic gases like hydrofluorocarbons (HFCs), perfluorocarbons (PFCs), and sulphur hexafluoride (SF6) (United Nations Framework Convention on Climate Change, UNFCCC, 2008). For example, South Africa, the largest CO_2 emitter in Southern Africa, implemented a Greenhouse Gas Inventory (base year 1990) Policy in 2004 (Seymore et al., 2014). The country made a political commitment to significantly reduce its GHGs at the UNFCCC Conference of the Parties (CoP) in Copenhagen (COP 15) in 2009 (UNFCCC, 2011). In 2017, Zimbabwe endorsed the Paris Agreement, obliging nations to submit their nationally determined contributions (NDCs) detailing efforts to combat climate change and restrict global temperature increases to "well below two °C" (United Nations, 2015). While there is a call for industrialized nations should take the initiative in cutting greenhouse gas (GHG) emissions and fulfil their financial and technical obligations, developing countries must also demonstrate their contributions.

Therefore, examining how different Southern African countries perform regarding climate change mitigation policy actions is critical. This is significant because it will serve as a benchmark against which we can measure, report, and verify (MRV) how the region advances regarding climate change policy. Yet, currently, literature studies on policies targeting emission reduction in Southern Africa are currently lacking. This study adopts Preferred Reporting Items for Systematic Reviews and Meta-Analyses (PRISMA) guidelines (Page et al., 2021) to synthesize the climate change mitigation policy landscape focusing on challenges and opportunities in Botswana, Malawi, Mozambique, South Africa, Zambia, and Zimbabwe. The objective is to identify and assess these policies, providing insights into their implementation barriers and recommendations within the context of each country.

The following section addresses the methodology. The third section presents results and discussions around implementing climate change policy intervention, as well as the potential and challenges for implementing climate mitigation policies in Southern Africa. To address the implications of climate change in Southern Africa, the final section provides a framework to improve the implementation of mitigation strategies.

13.2 Methodology

The methodology included extraction of data from research papers and articles, reports, and documents purification of research articles and data mining.

13.2.1 Research Papers and Articles

As highlighted earlier, the PRISMA approach was applied. The related literature was extracted by developing a retrieval query for the Scopus database. The search of relevant articles using the formula ("climate change" OR "mitigation") AND polic* AND "southern Africa") without any restriction on research type and field of knowledge. This was chosen to obtain a general overview of how climate change policies are being implemented.

The preliminary extraction yielded 258 articles. Subsequently, since the chapter focused exclusively on climate change and policy research, the search was narrowed down to 91 papers by selecting studies conducted in South Africa, Zimbabwe, Zambia, Mozambique, Botswana, and Malawi. Moreover, only research articles were included in the study to provide evidence-based information. Given the uncertainty surrounding the implementation of climate change policies in Southern Africa (Sibiya et al., 2023; Sithole et al., 2023), we considered articles from all disciplines for extraction. Therefore, the search criteria used is.

TITLE-ABS-KEY (("climate change" OR "mitigation") AND polic* AND "southern Africa") AND (LIMIT-TO (AFFILCOUNTRY, "South Africa") OR LIMIT-TO (AFFILCOUNTRY, "Zimbabwe") OR LIMIT-TO (AFFILCOUNTRY, "Botswana") OR LIMIT-TO (AFFILCOUNTRY, "Zambia") OR LIMIT-TO (AFFILCOUNTRY, "Malawi") OR LIMIT-TO (AFFILCOUNTRY, "Mozambique")) AND (LIMIT-TO (DOCTYPE, "ar")).

13.2.2 Report Documents

In this stage of the study, we concentrated on retrieving government documents such as policy documents, technical reports, and announcements related to country climate change mitigation. The policy document searches included (i) nationally determined contributions (NDCs) and (ii) selected climate change mitigation policies, as shown in Table 13.1.

13.2.3 Purification of Research Papers and Articles

After the initial online extraction, a review panel consisting of selected authors conducted a rigorous manual check following PRISMA procedures (Page et al., 2021) to refine the selection of articles relevant to the assessed domain (Fig. 13.1). This process effectively excluded false positives, which were research papers and articles containing the terms "climate change" and "policy" in their abstracts but did not align with the scope of the review. Such research documents included, for instance, research focused on climate change modelling for policy recommendations.

Table 13.1 List of selected climate change mitigation policies for extraction

Country	Selected climate change mitigation policies
Botswana	Integrated Resources Plan (IRP) (2020), National Energy Policy (NEP) (2021), Renewable Energy Strategy (RES) (2017), Nationally Appropriate Mitigation Actions (NAMA) (2016), and International Renewable Energy Agency (IRENA) Renewable Readiness Assessment (RRA) (2021)
Malawi	National Climate Change Management Policy (NCCMP) (2016), Nationally Appropriate Mitigation Actions (NAMA) (2015), National Climate Change Response (NCCR) White Paper (2012), National Climate Change Policy (NCCP) (2013), Malawi Renewable Energy Acceleration Programme (MREAP)
Mozambique	National Strategy for Climate Change (ENMC) (2013–2025), National Strategy for Adaptation and Mitigation of Climate Change (ENAMMC) (2013–2025), and National Climate Change Adaptation and Mitigation Strategy (2012)
South Africa	Framework for a Just Transition (2022), Just Energy Transition Investment Plan (JET IP) (2022), Solar Energy Technology Roadmap (SETRM) (2017), Renewable Energy Masterplan (SAREM) (2023), and National Greenhouse Gas Emissions Reporting Regulations (2017)
Zambia	National Policy on Climate Change (2016), National Climate Change Response Strategy (NCCRS) (2010), National Climate Change Learning Strategy (2020), National Strategy for Reducing Emission from Deforestation and Forest Degradation (REDD+) (2015), Low Emission Capacity Building (LECB) Programme, Renewable Energy Strategy and Action Plan (2022)
Zimbabwe	Long-term Low Greenhouse Gas Emission Development Strategy (2020–2050), The National Climate Policy (2021), National Climate Change Response Strategy (2019), Climate Action Enhancement Package (CAEP) (2021), Low Emission Development Strategy (2021), and Renewable Energy Development (2021)

Source: Authors

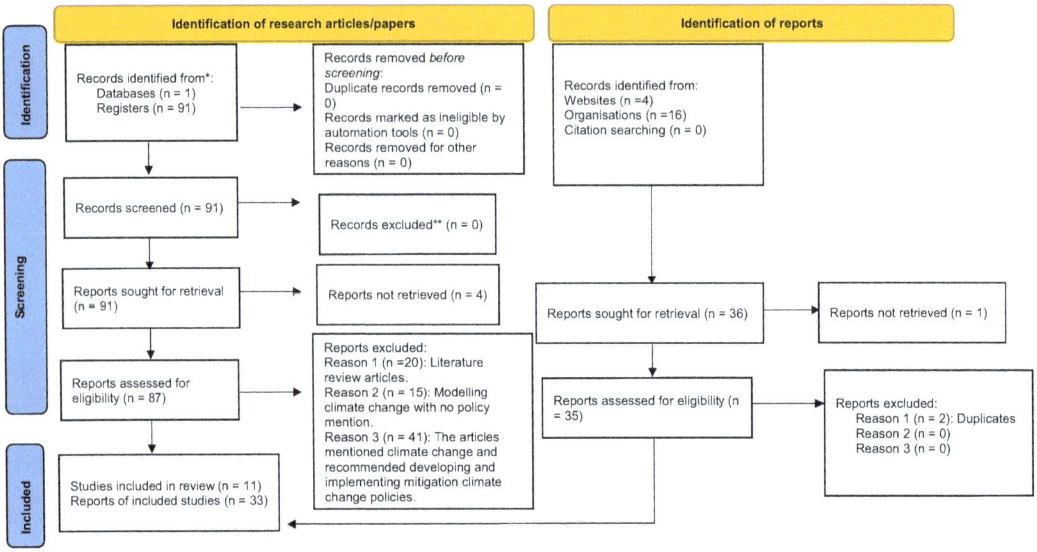

Fig. 13.1 The PRISMA statement. (Source: Authors)

During a consecutive inspection, three reviewers were assigned to read the abstract of each article and classify them as "relevant," "not relevant," or "maybe." Given that each reviewer's knowledge and personal interpretation could introduce inconsistencies in assessing articles, the panel subsequently engaged in a roundtable discussion after the initial screening. This discussion aimed to decide on the categorization of all contentious articles, ensuring consistency in the results during the refinement phase. Using the filtering criteria, the team identified 11 research papers and 33 policy documents suitable for further data extraction and analysis.

13.2.4 Data Mining

An Excel spreadsheet was used to structure the extracted data. Each row represented a document, and each column represented information regarding climate change mitigation policy interventions, barriers, and recommendations. The data were then tabulated for the corresponding southern African countries under study and presented as tables in the results section.

13.3 Results and Discussion

13.3.1 Climate Change Policy Intervention Implementation

Table 13.2 presents the spatial analysis of climate change policy implementation in five Southern African countries. It is evident that countries have robust supporting intervention policies in particular, South Africa. However, the studies reflected that policy interventions are sporadic across sectors in all five countries. While Table 13.3 pertains to barriers in Southern Africa, Botswana has the lowest interventions, particularly in the tourism sector. This can be attributed to various barriers.

13.3.2 Barriers to Implementing Climate Change Mitigation Policies

Climate mitigation policies refer to a set of activities, strategies, and measures implemented by governments and other key stakeholders to lessen GHG emissions and the effects of climate change

Table 13.2 Climate change policy interventions implemented in Southern Africa

Country	Policy interventions	References
Botswana	Most mitigation measures are targeted towards the energy sector Electricity will be generated optimally from locally available energy resources to fulfil local needs and achieve self-reliance Petroleum products will be obtained and stored in a reliable and sustainable manner to ensure the security of supply An overall GHG emissions reduction of 15% by 2030 Nationally Determined Contribution (NDC) National Energy Policy Integrated Resource Plan International Renewable Energy Agency (IRENA), Renewable Readiness Assessment (RRA)	Saarinen et al. (2020), Technology needs assessment report. Climate Change Mitigation ENERGY SECTOR September (2022), Nationally Determined Contribution (NDC) (2022 draft), National Energy Policy (2021), Integrated Resource Plan for electricity (2020), International Renewable Energy Agency (IRENA), Renewable Readiness Assessment (RRA) (2021)
Malawi	Reduction in GHG emissions relative to a business-as-usual (BAU) emissions scenario over the period to 2040 Promoting CDM activities and energy-saving technologies Providing improved energy-saving stoves to decrease the pressure on the forest reserves for the provision of firewood Low carbon energy alternatives Promoting climate-smart agriculture that upholds low emissions and is sustainable Most focus on agriculture and REDD+ programs. National Climate Change Response White Paper Nationally Determined Contribution (NDC) National Climate Change Policy Nationally Appropriate Mitigation Actions (NAMA) Climate Change Management Policy National Climate Change Policy Renewable Energy Acceleration Programme (MREAP)	Nhamo (2014), Stringer et al. (2012), England et al. (2018), Dougill et al. (2017), National Climate Change Response White Paper (2021), Nationally Determined Contribution (NDC) (2021), National Climate Change Policy (2013), Nationally Appropriate Mitigation Actions (NAMA) (2015), National Climate Change Management Policy (2016), National Climate Change Policy (2012), Malawi Renewable Energy Acceleration Programme (MREAP)
Mozambique	Focus on low-emission alternatives in energy generation, transportation, agriculture, and waste management. The country aims to expand and execute the National REDD+ strategy The National Council for Sustainable Development is mandated to coordinate environmental issues but has limited capacity for cross-sector ministerial discussions Nationally Determined Contribution (NDC) National Climate Change Adaptation and Mitigation Strategy (NCCAMS) National Strategy for Climate Change	USAID report (2017), Stringer et al. (2014), Tittonell et al. (2012), National Climate Change Adaptation and Mitigation Strategy (NCCAMS) (2013), National Strategy for Climate Change (2013), Nationally Determined Contribution (NDC) (2020)

(continued)

Table 13.2 (continued)

Country	Policy interventions	References
South Africa	Limit GHG emissions by 2025 and 2030 in a range between 398 and 614 metric tonnes of CO_2 equivalent, as defined in the National Policy Long-Term Mitigation Scenario (LTMS) has been acknowledged as a leading approach, instigating diverse interventions nationwide, especially within the energy sector, where the Western Cape Province and Gauteng stand out for their notable implementations. Despite the vigour of these mitigation policies, a noticeable discrepancy exists, as several provinces lag in their implementation efforts. Nationally Determined Contribution (NDC) South Africa's Just Energy Transition Investment Plan (JET IP) A Framework for a Just Transition South African Renewable Energy Masterplan (SAREM) National Greenhouse Gas Emission Reporting Regulations	Nhamo (2014), Dougill et al. (2017), Winkler et al. (2007), Nationally Determined Contribution (NDC) (2021), A Framework for a Just Transition (2022), National Greenhouse Gas Emission Reporting Regulations (2016), South Africa's Just Energy Transition Investment Plan (JET IP) (2023), South African Renewable Energy Masterplan (SAREM) (2022)
Zambia	Adoption of climate-smart agriculture and sustainable forest management have also resulted in substantial GHG emissions reductions Sustainable land-use planning, Nationally Appropriate Mitigation Actions (NAMAs), alternative energy sources, and energy efficiency Focus on conservation agriculture, the water sector, and UN-REDD programs UN-REDD+ program Nationally Determined Contribution (NDC) National Climate Change Policy Low Emission Capacity Building (LECB) Programme National Climate Change Response Strategy (NCCRS) Renewable Energy Strategy and Action Plan Zambia National Strategy to Reduce Emissions from Deforestation and Forest Degradation (REDD+)	Stringer et al. (2014), Tittonell et al. (2012), Stringer et al. (2012), England et al. (2018), Dougill et al. (2017), National Climate Change Policy (2016), Nationally Determined Contribution (NDC) (2021),National Climate Change Response Strategy (NCCRS) (2010), Renewable Energy Strategy and Action Plan (2022), Zambia National Strategy to Reduce Emissions from Deforestation and Forest Degradation (REDD+) (2015), Low Emission Capacity Building (LECB) Programme (2011)
Zimbabwe	Reduced energy transmission and distribution losses from 18% in 2020 to 11% in 2025 Expansion of 300 MW solar in 2025 Expansion of additional microgrids with 2.098 MW capacity by 2028 2% biodiesel in fuel by 2030 A just transition to a climate-resilient, low carbon economy The National Climate Change Response Strategy (NCCRS) 2015, coordinated through the office of the president to foresee the implementation of various sector-related policies Nationally Determined Contribution (NDC) The National Climate Change Response Strategy (NCCRS) System Development Plan Rural Energy Masterplan National Development Strategy Low Emissions Development Strategy Biofuels Policy	Stringer et al. (2014), Tittonell et al. (2012), Muhonda et al. (2014), Dougill et al. (2017), Nationally Determined Contribution (NDC) (2021), System Development Plan (2017), Rural Energy Masterplan (2021), Low Emissions Development Strategy (2020), Biofuels Policy (2019), National Development Strategy (2021), The National Climate Change Response Strategy (NCCRS) (2015)

Source: Authors

Table 13.3 Barriers to implementing climate mitigation policies in Southern Africa

Country	Policy challenges	References
Botswana	Limited community participation Insufficient awareness regarding the benefits of the tourism industry Minimal local benefits Lack of mainstreaming climate change agenda in the wildlife and biodiversity sector	Saarinen et al. (2020), Dougill et al. (2017)
Malawi	Dis-adoption due to limited socio-economic factors and physical factors Silo approach across different sectors to implement climate change intervention Policies are in place, but there is no specific reference intervention in place Current policies rely on external (international consultancy) to develop policy documents and reduce opportunities for consultation across government ministries Lack of mainstreaming climate change agenda in the wildlife and biodiversity sector Policies are in place, but there is no clarity on implementation	Tufa et al. (2023), Stringer et al. (2012), England et al. (2018), Dougill et al. (2017), Nhamo (2014)
Mozambique	Lack of strong institutional structures for implementation. Climate change development is superimposed on two existing institutional structures that rely on political will The cross-cutting nature has led to conflicts and difficulties in terms of assigning responsibility	Stringer et al. (2014), Tittonell et al. (2012), Dougill et al. (2017)
South Africa	Lack of mainstreaming climate change agenda in the wildlife and biodiversity sector Policies are in place, but there is no clarity on implementation	Dougill et al. (2017), Nhamo (2014)
Zambia	Lack of strong institutional structures for implementation Climate change development is superimposed on two existing institutional structures that rely on political will The cross-cutting nature has led to conflicts and difficulties in terms of assigning responsibility Dis-adoption due to limited socio-economic factors and physical factors Silo approach across different sectors to implement climate change intervention Policies are in place, but there is no specific reference intervention in place Current policies rely on external (international consultancy) to develop policy documents and reduce opportunities for consultation across government ministries Lack of mainstreaming climate change agenda in the wildlife and biodiversity sector	Stringer et al. (2014), Tittonell et al. (2012), Tufa et al. (2023), Stringer et al. (2012), England et al. (2018), Dougill et al. (2017)
Zimbabwe	Lack of strong institutional structures for implementation Climate change development is superimposed on two existing institutional structures that rely on political will The cross-cutting nature has led to conflicts and difficulties in terms of assigning responsibility Dis-adoption due to limited socio-economic factors and physical factors Emphasis on immediate flood response rather than a comprehensive strategy for reducing flood risks There is minimal focus on mitigation measures such as building physical structures or barriers The legislation lacks clear provisions for establishing effective funding mechanisms for the Department of Civil Protection Lack of mainstreaming climate change agenda in the wildlife and biodiversity sector	Stringer et al. (2014), Tittonell et al. (2012), Tufa et al. (2023), Muhonda et al. (2014), Dougill et al. (2017)

Source: Authors

(Dale et al., 2020; Swinburn et al., 2022). Globally, climate change has become a major concern because of different shocks and stressors that result from long-term shifts in weather patterns, which include, water pollution, poor agricultural performance, flooding, land degradation, death of domestic and wild animals, rising sea levels, droughts, extreme low and high temperatures, and the outbreak of diseases (Dube et al., 2022; Kemp et al., 2022; Liu et al., 2020; Mugambiwa & Makhubele, 2023; Mutengwa et al., 2023; Salvador et al., 2023). Energy, transport, manufacturing, agriculture, and waste handling are some of the sectors that are often targeted by climate mitigation strategies to reduce emissions (Shariffi et al., 2020; Gustafsson & Anderberg, 2021; Elavarasan et al., 2022). To reduce GHGs, these policies seek to encourage the adoption of greener, more sustainable practices, technologies, and behavioural patterns. The primary objective of climate mitigation policies is to limit the concentration of GHGs in the atmosphere, thereby curbing the rate of global warming and reducing the severity of climate change-related effects (Fawzy et al., 2020; Guo et al., 2023).

According to Nhemachena et al. (2020), Southern Africa is a region that faces unique challenges when it comes to implementing climate mitigation policies, yet climate change poses a significant threat to the countries in this region, including Botswana, Mozambique, South Africa, Zambia, and Zimbabwe. These nations are especially susceptible to climate change impacts, including severe weather events, higher temperatures, shifting rainfall patterns, and rising sea levels (Makate et al., 2019; Shobande & Asongu, 2022). Farming communities that depend mostly on rain-fed agriculture are frequently affected by climate change (Mashao et al., 2023; Mutengwa et al., 2023). Leal Filho et al. (2019) state that while there is a growing recognition of the need for climate mitigation policies in Southern Africa, several barriers hinder their effective implementation (Muchaku et al., 2023; Omer & Capaldo, 2023). These barriers arise from a combination of socio-economic, political, and institutional factors, which must be carefully addressed to achieve meaningful progress in tackling climate change in the region (Irfan et al., 2022; Singh et al., 2021; Tangonyire & Akuriba, 2021).

Firstly, countries in Southern Africa lack the necessary financial resources to invest in climate mitigation measures (Pardoe et al., 2020). According to Beg et al. (2002) and Nazir (2023), limited availability of funding makes it a challenge to allocate sufficient resources to implement comprehensive climate policies and invest in renewable energy sources, improving energy efficiency, and adopting sustainable land management practices, and other climate-friendly initiatives. Zimbabwe also faces significant financial constraints, which challenge implementing climate mitigation policies (Chanza, 2018; Manyeruke et al., 2013). Lack of financial resources also hampers capacity building, technology transfer, and infrastructure development needed to effectively mitigate climate policies (Lybbert & Sumner, 2012; Vierros & Harden-Davies, 2020). Macamo (2021) argues that limited financial resources can hinder the development of essential climate infrastructure in Mozambique. Consequently, Beira faces significant environmental challenges, particularly those associated with climate change. Located in a low-lying area, the city is highly susceptible to flooding. Climate change exacerbates these issues, with projections indicating increased vulnerability to rising sea levels and cyclones (Brimblecombe et al., 2020; Raimi et al., 2021). Moreover, Beira contends with high population density, pervasive poverty, inadequate basic infrastructure like water and sanitation, and insufficient human and financial resources. These factors pose considerable obstacles to the implementation of climate mitigation policies (Dicker et al., 2021; Macamo, 2021). The struggling economies of Southern African countries have seen the challenges of incomplete or failure in climate change mitigation policies implementation (Hussain et al., 2020; Mawonde & Togo, 2019; Mutengwa et al., 2023). Financial problems result in the failure to implement and adopt clean energy alternatives with low emissions into the atmosphere. Countries like Zambia have used cheap fossil-based energy-generating alternatives associated with high carbon emissions leading to climate change (Kapole et al., 2023).

Additionally, dependence on fossil fuels, particularly coal, for energy generation is also a barrier to implementing climate mitigation policies in Southern Africa (Schmidt et al., 2017). The region possesses significant coal reserves, which have historically played a crucial role in supporting economic growth and industrial development. Transitioning to cleaner and renewable energy sources requires substantial investments and a shift away from the long-standing dependence on fossil fuels (Ayres & Ayres, 2009; Breetz et al., 2018; Cha, 2020; Kuzemko et al., 2022). The economic implications of such a transition, including job losses in the coal sector, can create resistance to climate mitigation policies. South Africa's energy system relies heavily on coal, contributing significantly to its high emissions compared to countries with similar GDP (van Ryneveld & Islar, 2023). This coal-centric economy traces its origins to British colonization and currently displays limited progress in transitioning swiftly to mitigate climate change. The nation continues on an unsustainable trajectory of growing dependence on fossil fuels, despite its considerable untapped potential for renewable energy (Pegels, 2010; Todd & McCauley, 2021).

Countries like Zambia have opted to use diesel-powered equipment and thermal power generation because they are cheap compared to solar systems (Kapole et al., 2023). Zimbabwe and Zambia also have a high dependence on fossil fuels which makes it difficult to transition to renewable energy sources (World Bank, 2018). Zimbabwe's energy policy faced challenges during its implementation because shifting to renewable energy sources requires investment (Dzvimbo & Zhanda, 2020). According to the World Bank (2018), Zimbabwe's greenhouse gas emissions are expected to increase by 10% by 2030 due to population growth and economic activity. The Government of Botswana has pledged an NDC represents a commitment to the Paris Agreement. Regarding the power sector, it outlines that the government aims for renewable energy (RE) to satisfy 25% of peak electricity demand by 2030. However, due to the significant initial costs of RE technologies, the government intends to continue relying on a coal-based power system (Baek et al., 2019; Maswabi et al., 2021; Moyo et al., 2023).

Meadowcroft (2009) noted that political priorities and governance structures play a crucial role in determining the level of commitment to climate mitigation policies. Addressing climate change governance presents complex challenges for modern political and administrative systems (Bulkeley & Newell, 2023). In some cases, political instability, corruption, and governance challenges can divert attention and resources away from climate change issues (Doğan et al., 2023; Mahmood et al., 2021). Developing countries prioritize mobilizing funds to achieve their political interests (Zavale, 2022). In Mozambique, inadequate financial allocations to environmental sustainability issues result in the failure to implement climate change policies because of governmental funds being used for civil war resources mobilization (Mutengwa et al., 2023; Chakawa & Mangiza, 2023). Additionally, competing political interests, short-term electoral cycles, and the lack of political will also hinder the adoption and implementation of long-term climate policies. The political will to implement and enforce climate mitigation policies can be a significant barrier (Colgan et al., 2021; Di Giulio et al., 2019; Meerow & Keith, 2022). Political leaders may prioritize short-term economic growth and development over long-term climate objectives. Zimbabwe's economy has been struggling in recent years, and the government has had to prioritize economic growth over environmental concerns (Chitiyo et al., 2019).

Poor governance through misuse of funds and resources by governments and personnel involved in environmental management affects the implementation of climate change mitigation policies. Countries with poor governance characterized by corruption were mobilized resources for implementation. Climate change mitigation policies are diverted for individual benefits (Creutzig et al., 2023; Sadiq et al., 2023). In Zambia, the literature review on financial resources to environmental issues remained questionable because of corruption through misuse and unlawful diversion of cash flows and resources, leading to failure in implementing the climate change policy within the country (Sadiq et al., 2023). Poor governance is also associated with poor economic development policies, weak institutions, and unnecessary delays in the formulation and adoption of climate change mitigation policies. In South Africa, the Environmental Kuznets Curve (EKC) analytical framework revealed that the poor economic development policies facilitated direct foreign investment, escalating the carbon emissions through fossil mining and processing (Bulkeley & Newell, 2023; Udeagha & Breitenbach, 2023). These poor economic development policies mainly focus on the short-term benefits of the country's wealth success, neglecting environmental considerations such as adopting low-carbon-producing activities within the mining sector. This has led countries to have hardships in implementing sustainable practices and green energy with low carbon emissions.

Lake Malombe region in the Southern part of Malawi has been degraded because of weak laws on settlement development, which resulted in the destruction of carbon sinks, exposing the country to increased climate change and its effects (Chimonyo et al., 2023; Kosamu et al., 2022; Makwinja et al., 2021). Similarly, in Zambia, weak environmental laws enforcement means that mineral-processing industries neglect to adopt clean and renewable energy sources in production, exacerbating large quantities of carbon emissions into the atmosphere, which shows the government's failure to address greenhouse gas emissions, ultimately, failure of climate change mitigation policies implementation (Ahmed et al., 2020; Mulenga et al., 2023).

Lack of environmental knowledge and experts results in most of the Southern African countries facing challenges in implementing climate change mitigation policies. Poor environmental education affects people's capabilities to gather critical information prior to the proper formulation of climate change mitigation policies (Arif et al., 2023). Zimbabwe is not spared from countries with knowledge gaps in environmental planning because its farmers in Chiredzi and Mwenezi use indigenous knowledge (IK) and local knowledge (LK) in forecasting weather and climate issues, which is characterized by misinterpretations of the likelihood of weather events (Zvobgo et al., 2022). This results in the country facing myriad barriers in obtaining climate-related information to foster the process of implementing climate change mitigation interventions to counter the changing climate. Masinde and Thothela (2019) alluded that in countries like Malawi and Mozambique, most small-scale farmers use traditional knowledge systems to forecast weather, which provides insufficient information for climate change policy implementation.

Further, the lack of environmental experts hinders the implementation of climate change policies. The lack of environmental architects results in poor environmental research, which affects the information gathering needed to formulate and implement climate change mitigation policies (Adenle et al., 2017; Qamar et al., 2022). A study conducted in Botswana revealed that the implemented strategic environmental assessments are associated with challenges because of insufficient information gathering for environmental planning (Ehiane & Dick-Sagoe, 2023). In Southern Africa, the lack of experts and their full engagement in research and innovation results in the failure to develop and adopt renewable energy alternatives (Matekenya & Moyo, 2022). According to Zambia Climate Change Profile (2019), Zambia's ministry of lands, natural resources, and environment lacks the capacity to effectively implement climate change policies. The shortage of skilled professionals can slow down climate policy implementation and reduce the effectiveness of mitigation efforts.

Countries in Southern Africa face challenges related to weak institutional capacity, including limited technical expertise, human resources, and administrative capabilities (Bui et al., 2020). South Africa's institutions responsible for implementing climate mitigation policies lack the necessary capacities and resources to effectively implement; hence, the country's climate policy is still in stages (Lee et al., 2022; Udeagha & Breitenbach, 2023). Mogomotsi et al. (2020)

argue that although climate change poses potential risks to agricultural productivity, Botswana lacks a specific policy dedicated to addressing these challenges. Implementing effective climate policies requires institutions that can develop, enforce, and monitor regulations, as well as coordinate efforts across various sectors (André et al., 2023; Cid & Lerner, 2023; Mathetsa et al., 2023). Insufficient institutional capacity can hinder the formulation and implementation of climate policies, leading to delays, inefficiencies, and inadequate enforcement (Hope Sr, 2020; Solangi et al., 2021).

Poor international relations between nations result in most Southern African countries failing to implement climate change policies. Bad relations triggered by US sanctions on Zimbabwe have led to limited international assistance to address climate change issues. Zimbabwe is impeded by sanctions that restrict the country from freely accessing financial assistance to fund its climate mitigation policies (Marevesa & Mavengano, 2023; Mhute, 2023; Muromo et al., 2022). According to Mutandwa and Vyas-Doorgapersad (2023), in Mutoko, the Zimbabwe Energy Regulatory Authority (ZERA) failed to implement the solar project because of a lack of funding to government from developed countries as promised in the Paris Agreement of reducing carbon emissions by 2030 (Chivhenge et al., 2023).

Rapid population growth can also influence government policies to neglect environmental protection by considering settlement development and fossil-based energy provision. Zambia's gradual population increase has resulted in the country failing to adopt renewable and clean energy sources because of the high demand for coal and charcoal used for energy provision (Masud & Khan, 2023; Musa et al., 2023; Singubi et al., 2023). Moreover, in countries like Zambia, climate change is attributed to rapid population increase, which positively influences industrialization characterized by mining activities and increased greenhouse gas emissions. Settlement development significantly affects the vegetation cover through deforestation.

Overlapping mandates and unclear responsibilities among different government agencies can lead to a lack of accountability and coordination, hampering effective policy implementation. The Southern African region often lacks the necessary technical expertise and capacity to implement complex climate change mitigation policies (Dilling & Lemos, 2011; Nhemachena et al., 2020). This includes expertise in areas such as renewable energy development, carbon accounting, and climate modelling. Low levels of public awareness and comprehension of climate change issues can hinder the implementation of climate mitigation policies (Hsu et al., 2020). Without public support and engagement, it can be challenging to drive behavioural change, implement regulations, and mobilize resources for climate action.

Lastly, the geographical location of a particular country contributes to difficulties faced in the implementation of climate change mitigation policies through various constrictions. Southern African countries that are landlocked have difficulties in trading, which negatively affects their income generation capacities to mobilize funds needed to implement climate change policies. Countries such as Zimbabwe, Zambia, Malawi, and Botswana are landlocked, thereby facing challenges in trade, poor income generation, and ultimately the poor performance of their economies, resulting in failure to adopt climate change interventions that require huge capital investment (Chisale & Lee, 2023). Chisale and Lee (2023) contend that Malawi failed to facilitate the transition of renewable energy interventions because it is a landlocked country that faces difficulties in importing resources from abroad.

Addressing barriers to implementing climate mitigation policies in Southern Africa, therefore, requires a multifaceted approach that includes capacity building, financial support, institutional strengthening, public awareness campaigns, and regional cooperation. Overcoming these challenges can help Southern African countries effectively implement climate change mitigation policies and contribute to global efforts to combat climate change.

13.4 Recommendations for Implementing Climate Change Mitigation Policies

To progress, countries should pay attention to the recommendations provided in Table 13.4 and consider the proposed implementation framework outlined in Fig. 13.2.

13.5 Framework to Improve Implementation of Climate Change Mitigation Policies

The proposed climate change mitigation policy intervention framework for Southern African countries (Fig. 13.2) could assist in addressing implementing challenges at different levels of each government institutional policy landscape.

13.6 Conclusions

Southern African countries have robust national policy institutional landscapes. However, policy fragmentation, lacking responsibilities, limited resources, and expertise derail implementation. Furthermore, dependency on consultation at both national and local levels to formulate policies hampers implementation. The chapter recommends national governments provide clear roles and responsibilities. In turn, the provincial governments must lead the process, and local governments must facilitate bottom-up approaches, as proposed in Fig. 13.2. This calls for close cooperation, investment in technologies, capacity building, monitoring, and evaluation across all tiers of government, the private sector, and the international community.

Table 13.4 Recommendations for climate change mitigation policy implementation in Southern Africa

Country	Policy recommendation	References
Botswana	Information sharing Empowerment of communities to adapt to climate change impacts on the tourism industry Calls for better governance on tourism plans and development Increase the role of local participation Detailed climate change national policy plans, which are mainstream responsive strategies intended to address climate change	Saarinen et al. (2020), Dougill et al. (2017)
Malawi	Advocating for equitable representation of women and vulnerable communities in planning, decision-making, and implementation of climate change adaptation measures It is essential to integrate women's needs more extensively into regional and national climate policies to advance climate action and implementation Evaluation conducted after policy implementation to assess its effectiveness in addressing the vulnerability of women to climate change Enhancing farmers technical knowhow and enacting and enforcing by-laws regarding livestock grazing Sharing of good practices to feed into the national development policy There is a need to work across different sectors and stakeholders and to mobilize and invest in capacity and resources on the ground action Policy coordination remains weak across Southern Africa and needs strengthening to allow greater support across sectoral planning Detailed climate change national policy plans, which are mainstream responsive strategies intended to address climate change	Nhamo (2014), Tufa et al. (2023), Stringer et al. (2012), England et al. (2018), Dougill et al. (2017)

(continued)

Table 13.4 (continued)

Country	Policy recommendation	References
Mozambique	Countries must take ownership of the climate change development (CCD) approach as a conclusive development concept Countries must set their vision and develop appropriate institutions contributing to CCD outcomes State, for example, SADC and NGO-led networks to share lessons, experiences, and good practices Require a conducive policy environment in semi-arid parts of the world	Stringer et al. (2014), Tittonell et al. (2012)
South Africa	Policy on climate change awareness targeting agricultural activities Strengthening agricultural extension services in districts Improvement of access to markets Promoting fair representation of women and vulnerable communities in planning, decision-making, and implementation for climate change adaptation Further mainstreaming women's needs in regional and national climate policies is crucial in developing climate change action and implementation Post-implementation evaluation of policy implementation and efficacy in addressing women's vulnerability to climate change Detailed climate change national policy plans, which are mainstream responsive strategies intended to address climate change Countries should have effective implementation strategies to help bend the curve of the emission path downwards in South Africa	Nhamo (2014), Dougill et al. (2017), Winkler et al. (2007)
Zambia	Countries must take ownership of the climate change development (CCD) approach as a conclusive development concept Countries must set their vision and develop appropriate institutions contributing to CCD outcomes State-led, for example, SADC and NGO-led, networks to share lessons, experiences, and good practices Require a conducive policy environment in semi-arid parts of the world Enhancing farmer's technical know-how and enacting and enforcing by-laws regarding livestock grazing Detailed climate change national policy plans, which are mainstream responsive strategies intended to address climate change	Stringer et al. (2014), Tittonell et al. (2012), Tufa et al. (2023), Dougill et al. (2017)
Zimbabwe	Countries must take ownership of the climate change development (CCD) approach as a conclusive development concept Countries must set their vision and develop appropriate institutions contributing to CCD outcomes State-led, for example, SADC and NGO-led, networks to share lessons, experiences, and good practices Require a conducive policy environment in semi-arid parts of the world Enhancing farmer's technical know-how and enacting and enforcing by-laws regarding livestock grazing Community preparedness can be enhanced by the quality of information disseminated by institutions Policy coordination remains weak across Southern Africa and needs strengthening to allow greater support across sectoral planning Weak institutions lack the capacity to deal with the effective implementation of their functions There is a need to work across different sectors and stakeholders and to mobilize and invest in capacity and resources on the ground action Sharing of good practices to feed into the national development policy Detailed climate change national policy plans, which are mainstream responsive strategies intended to address climate change	Stringer et al. (2014), Tittonell et al. (2012), Tufa et al. (2023), Muhonda et al. (2014), Stringer et al. (2012), England et al. (2018), Dougill et al. (2017)

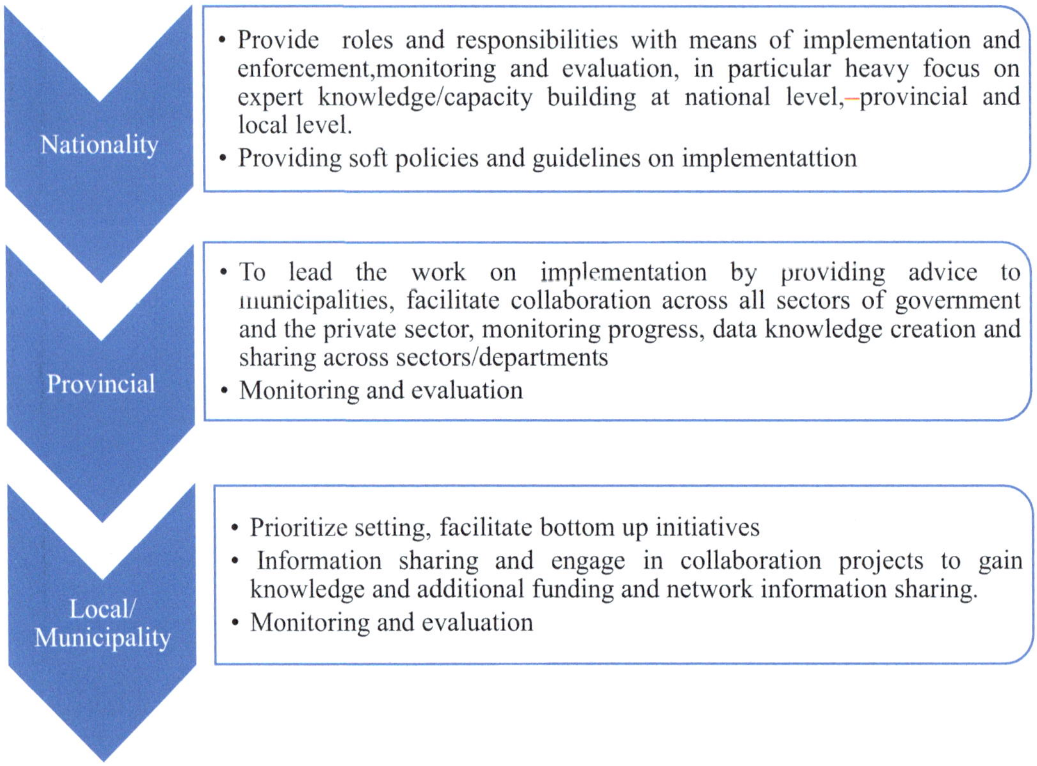

Fig. 13.2 Framework to improve implementation of climate change mitigation policies in Southern Africa. (Source: Authors)

References

Adenle, A. A., Manning, D. T., & Arbiol, J. (2017). Mitigating climate change in Africa: barriers to financing low-carbon development. *World Development, 100*, 123–132.

Aderinto, N. (2023). Tropical Cyclone Freddy exposes major health risks in the hardest-hit Southern African countries: lessons for climate change adaptation. *IJS Globalization and Health, 6*(3), e0152.9.

Ahmed, I., Parikh, P., Sianjase, G., & Coffman, D. M. (2020). The impact decades-long dependence on hydropower in El Niño impact-prone Zambia is having on carbon emissions through backup diesel generation. *Environmental Research Letters, 15*(12), 124031.

André, K., Gerger Swartling, Å., Englund, M., Petutschnig, L., Attoh, E. M., Milde, K., Lückerath, D., Cauchy, A., Botnen Holm, T., Hanssen Korsbrekke, M., & Bour, M. (2023). Improving stakeholder engagement in climate change risk assessments: insights from six co-production initiatives in Europe. *Frontiers in Climate, 5*, 1120421.

Arif, M., Jiajia, L., Tahir, M., Jie, Z., & Li, C. (2023). Environmental literacy scenarios lead to land degradation and changes in riparian zones: Implications for policy in China. *Land Degradation & Development, 34*(1), 156–172.

Armando, C. J., Yu, Z., Mavume, A. F., Ongoma, V., & Nyongesa, A. M. (2021). Formation and track of tropical cyclones Eline (2000) and Bonita (1996). *Meteorology and Atmospheric Physics, 133*, 1691–1706.

Ayres, R. U., & Ayres, E. H. (2009). *Crossing the energy divide: Moving from fossil fuel dependence to a clean-energy future*. Pearson Prentice Hall.

Baek, Y. J., Jung, T. Y., & Kang, S. J. (2019). Low carbon scenarios and policies for the power sector in Botswana. *Climate Policy, 19*(2), 219–230.

Bauer, S., & Scholz, I. (2010). Adaptation to climate change in Southern Africa: New boundaries for sustainable development? *Climate Development, 2*, 83–93. https://doi.org/10.3763/cdev.2010.0040

Beg, N., Morlot, J. C., Davidson, O., Afrane-Okesse, Y., Tyani, L., Denton, F., Sokona, Y., Thomas, J. P., La Rovere, E. L., Parikh, J. K., & Parikh, K. (2002).

Linkages between climate change and sustainable development. *Climate Policy, 2*(2–3), 129–144.

Breetz, H., Mildenberger, M., & Stokes, L. (2018). The political logics of clean energy transitions. *Business and Politics, 20*(4), 492–522.

Brimblecombe, P., Hayashi, M., & Futagami, Y. (2020). Mapping climate change, natural hazards and Tokyo's built heritage. *Atmosphere, 11*(7), 680.

Bui, T. D., Tsai, F. M., Tseng, M. L., & Ali, M. H. (2020). Identifying sustainable solid waste management barriers in practice using the fuzzy Delphi method. *Resources, Conservation and Recycling, 154*, 104625.

Bulkeley, H., & Newell, P. (2023). *Governing climate change*. Taylor & Francis.

Cha, J. M. (2020). A just transition for whom? Politics, contestation, and social identity in the disruption of coal in the Powder River Basin. *Energy Research & Social Science, 69*, 101657.

Chakawa, J., & Mangiza, O. (2023). Entangled Borderlands: Effects of the 1977–1992 Mozambican Civil War on Border Communities in Zimbabwe. In *Lived experiences of borderland communities in Zimbabwe: Livelihoods, conservation, War and Covid-19* (pp. 135–147). Springer International Publishing.

Chanza, N. (2018). Limits to climate change adaptation in Zimbabwe: Insights, experiences and lessons. In *Limits to climate change adaptation* (pp. 109–127).

Chien, F., Chau, K. Y., & Sadiq, M. (2023). Impact of climate mitigation technology and natural resource management on climate change in China. *Resources Policy, 81*, 103367.

Chikafa, M., Nejadhashemi, A. P., Moller, K., Razavi, H., & Bizimana, J. C. (2023). Multidimensional evaluation of the impacts of agricultural interventions to achieve food security in Malawi. *Food and Energy Security, 12*(4), e486.

Chimonyo, V. G. P., Mabhaudhi, T., Muthoni, F., & Chavula, G. (2023). *Context assessment for mixed farming systems in Malawi*. Ibadan.

Chisale, S. W., & Lee, H. S. (2023). Evaluation of barriers and solutions to renewable energy acceleration in Malawi, Africa, using AHP and fuzzy TOPSIS approach. *Energy for Sustainable Development, 76*, 101272.

Chitiyo, K., Dengu, C., Mbae, D., & Vandome, C. (2019). *Forging Inclusive Economic Growth in Zimbabwe*. The Royal Institute of International Affairs.

Chivhenge, E., Mabaso, A., Museva, T., Zingi, G. K., & Manatsa, P. (2023). Zimbabwe's roadmap for decarbonisation and resilience: An evaluation of policy (in) consistency. *Global Environmental Change, 82*, 102708.

Chiyangi, H. M., Mwanabute, N., Ncube, E., & Mundike, J. (2023). Assessing the Occurrence of Heavy Metals in Edible Fruits Grown around Mine Tailings Dam in Kitwe. *Journal of Environmental Protection, 14*(2), 83–95.

Cid, A., & Lerner, A. M. (2023). Local governments as key agents in climate change adaptation: Challenges and opportunities for institutional capacity-building in Mexico. *Climate Policy, 23*(5), 649–661.

Coetzee, C., Khoza, S., Nemakonde, L. D., Shoroma, L. B., Wentink, G. W., Nyirenda, M., & Van Niekerk, D. (2023). Financing disaster risk reduction: Exploring the opportunities, challenges, and threats within the Southern African Development Community Region. *International Journal of Disaster Risk Science*, 1–15. https://doi.org/10.1007/s13753-023-00499-6

Colgan, J. D., Green, J. F., & Hale, T. N. (2021). Asset revaluation and the existential politics of climate change. *International Organization, 75*(2), 586–610.

Creutzig, F., Goetzke, F., Ramakrishnan, A., Andrijevic, M., & Perkins, P. (2023). Designing a virtuous cycle: Quality of governance, effective climate change mitigation, and just outcomes support each other. *Global Environmental Change, 82*, 102726.

Dale, A., Robinson, J., King, L., Burch, S., Newell, R., Shaw, A., & Jost, F. (2020). Meeting the climate change challenge: local government climate action in British Columbia, Canada. *Climate Policy, 20*(7), 866–880.

Dechezleprêtre, A., Fabre, A., Kruse, T., Planterose, B., Chico, A. S., & Stantcheva, S. (2022). *Fighting climate change: International attitudes toward climate policies* (No. w30265). National Bureau of Economic Research.

Di Giulio, G. M., Torres, R. R., Lapola, D. M., Bedran-Martins, A. M., da Penha Vasconcellos, M., Braga, D. R., Fuck, M. P., Juk, Y., Nogueira, V., Penna, A. C., & Jacaúna, T. (2019). Bridging the gap between will and action on climate change adaptation in large cities in Brazil. *Regional Environmental Change, 19*, 2491–2502.

Dicker, S., Unsworth, S., Byrnes, R., Ward, B., Bhatt, M., & Paul, A. (2021). *Saving lives and livelihoods: the benefits of investments in climate change adaptation and resilience*. Leeds.

Dilling, L., & Lemos, M. C. (2011). Creating usable science: Opportunities and constraints for climate knowledge use and their implications for science policy. *Global Environmental Change, 21*(2), 680–689.

Doğan, B., Shahbaz, M., Bashir, M. F., Abbas, S., & Ghosh, S. (2023). Formulating energy security strategies for a sustainable environment: Evidence from the newly industrialized economies. *Renewable and Sustainable Energy Reviews, 184*, 113551.

Dougill, A. J., Whitfield, S., Stringer, L. C., Vincent, K., Wood, B. T., Chinseu, E. L., Steward, P., & Mkwambisi, D. D. (2017). Mainstreaming conservation agriculture in Malawi: Knowledge gaps and institutional barriers. *Journal of Environmental Management, 195*, 25–34. https://doi.org/10.1016/j.jenvman.2016.09.076

Dube, K., Nhamo, G., & Chikodzi, D. (2022). Climate change-induced droughts and tourism: Impacts and responses of Western Cape province, South Africa.

Journal of Outdoor Recreation and Tourism, 39, 100319.

Dzvimbo, M., & Zhanda, K. (2020). Policy-making prospects and challenges of the climate change and the urban energy sector in Zimbabwe. *Journal of Urban Systems and Innovations for Resilience in Zimbabwe-JUSIRZ, 2*(1), 203–222.

Ehiane, S. O., & Dick-Sagoe, C. (2023). Climate change and justice in Botswana's national adaptation strategy and Programming: The journey so far. In *Climate action in Southern Africa* (pp. 168–183). Routledge.

Elavarasan, R. M., Pugazhendhi, R., Irfan, M., Mihet-Popa, L., Khan, I. A., & Campana, P. E. (2022). State-of-the-art sustainable approaches for deeper decarbonization in Europe–An endowment to climate neutral vision. *Renewable and Sustainable Energy Reviews, 159*, 112204.

England, M. I., Dougill, A. J., Stringer, L. C., Vincent, K. E., Pardoe, J., Kalaba, F. K., Mkwambisi, D. D., Namaganda, E., & Afionis, S. (2018). Climate change adaptation and cross-sectoral policy coherence in southern Africa. *Regional Environmental Change, 18*, 2059–2071. https://doi.org/10.1007/s10113-018-1283-0

Fawzy, S., Osman, A. I., Doran, J., & Rooney, D. W. (2020). Strategies for mitigation of climate change: A review. *Environmental Chemistry Letters, 18*, 2069–2094.

Fuldauer, L. I., Thacker, S., Haggis, R. A., Fuso-Nerini, F., Nicholls, R. J., & Hall, J. W. (2022). Targeting climate adaptation to safeguard and advance the Sustainable Development Goals. *Nature Communications, 13*(1), 3579.

Gasana, E. M. (2021). *The effects of zonal wind on rainfall variability over East Africa during El Niño Southern Oscillation (ENSO).*

Gasser, T., Guivarch, C., Tachiiri, K., Jones, C. D., & Ciais, P. (2015). Negative emissions physically needed to keep global warming below 2 C. *Nature Communications, 6*(1), 7958. https://doi.org/10.1038/ncomms8958

Gössling, S., Balas, M., Mayer, M., & Sun, Y. Y. (2023). A review of tourism and climate change mitigation: The scales, scopes, stakeholders and strategies of carbon management. *Tourism Management, 95*, 104681.

Greer, K., King, H., & Glackin, M. (2023). The 'web of conditions' governing England's climate change education policy landscape. *Journal of Education Policy, 38*(1), 69–92.

Guo, Q., He, Z., & Wang, Z. (2023). Long-term projection of future climate change over the twenty-first century in the Sahara region in Africa under four Shared Socio-Economic Pathways scenarios. *Environmental Science and Pollution Research, 30*(9), 22319–22329.

Gustafsson, M., & Anderberg, S. (2021). Dimensions and characteristics of biogas policies–Modelling the European policy landscape. *Renewable and Sustainable Energy Reviews, 135*, 110200.

Hanlon, J. (2023). Mozambique. In *Africa yearbook volume 19* (pp. 506–517). Brill.

Harris, J., Chalemera, J., Nowa, M., Mhango, B., Lupafya, P., Museka Saidi, T., & Roschnik, N. (2023). *Malawi stories of change in nutrition: Overview*. Institute of Development Studies (IDS).

Hoegh-Guldberg, O., Jacob, D., Taylor, M., Guillén Bolaños, T., Bindi, M., Brown, S., Camilloni, I. A., Diedhiou, A., Djalante, R., Ebi, K., & Engelbrecht, F. (2019). The human imperative of stabilizing global climate change at 1.5 C. *Science, 365*, 6459, eaaw6974. https://doi.org/10.1126/science.aaw6974

Hope, K. R., Sr. (2020). Peace, justice and inclusive institutions: overcoming challenges to the implementation of Sustainable Development Goal 16. *Global Change, Peace & Security, 32*(1), 57–77.

Hsu, A., Tan, J., Ng, Y. M., Toh, W., Vanda, R., & Goyal, N. (2020). Performance determinants show European cities are delivering on climate mitigation. *Nature Climate Change, 10*(11), 1015–1022.

Hussain, M., Butt, A. R., Uzma, F., Ahmed, R., Irshad, S., Rehman, A., & Yousaf, B. (2020). A comprehensive review of climate change impacts, adaptation, and mitigation on environmental and natural calamities in Pakistan. *Environmental Monitoring and Assessment, 192*, 1–20.

Ibebuchi, C. C. (2023). Circulation patterns linked to the positive sub-tropical Indian ocean dipole. *Advances in Atmospheric Sciences, 40*(1), 110–128.

Intergovernmental Panel on Climate Change (IPCC). (2014). In T. F. Stocker et al. (Eds.), *Climate change 2013: The physical science basis. Contribution of working group I to the fifth assessment report of the intergovernmental panel on climate change.* Cambridge University Press, Cambridge.

Irfan, M., Razzaq, A., Sharif, A., & Yang, X. (2022). Influence mechanism between green finance and green innovation: Exploring regional policy intervention effects in China. *Technological Forecasting and Social Change, 182*, 121882.

Jahanger, A., Usman, M., Murshed, M., Mahmood, H., & Balsalobre-Lorente, D. (2022). The linkages between natural resources, human capital, globalization, economic growth, financial development, and ecological footprint: The moderating role of technological innovations. *Resources Policy, 76*, 102569.

Jarugula, S., & McPhaden, M. J. (2023). Indian Ocean Dipole affects eastern tropical Atlantic salinity through Congo River Basin hydrology. *Communications Earth & Environment, 4*(1), 366.

Jiang, Y., Zhou, L., Roundy, P. E., Hua, W., & Raghavendra, A. (2021). Increasing influence of Indian Ocean dipole on precipitation over central equatorial Africa. *Geophysical Research Letters, 48*(8), e2020GL092370.

Kapole, F., Mudenda, S., & Jain, P. (2023). Study of major solar energy mini-grid initiatives in Zambia. *Results in Engineering, 18*, 101095.

Kapwata, T., Gebreslasie, M. T., Mathee, A., & Wright, C. Y. (2018). Current and potential future seasonal trends of indoor dwelling temperature and likely health risks in rural Southern Africa. *International Journal of Environmental Research and Public Health, 15*(5), 952. https://doi.org/10.3390/ijerph15050952

Kemp, L., Xu, C., Depledge, J., Ebi, K. L., Gibbins, G., Kohler, T. A., et al. (2022). Climate endgame: Exploring catastrophic climate change scenarios. *Proceedings of the National Academy of Sciences, 119*(34), e2108146119.

Kosamu, I. B. M., Makwinja, R., Kaonga, C. C., Mengistou, S., Kaunda, E., Alamirew, T., & Njaya, F. (2022). Application of DPSIR and tobit models in assessing freshwater ecosystems: The case of Lake Malombe, Malawi. *Water, 14*(4), 619.

Kuzemko, C., Blondeel, M., Dupont, C., & Brisbois, M. C. (2022). Russia's war on Ukraine, European energy policy responses & implications for sustainable transformations. *Energy Research & Social Science, 93*, 102842.

Lockley, A., (2019). Distributed governance of Solar Radiation Management geoengineering: A possible solution to SRM's "free-driver" problem? *Frontiers of Engineering Management, 6*, 551–556.

Lawrence, M. G., Schäfer, S., Muri, H., Scott, V., Oschlies, A., Vaughan, N. E., Boucher, O., Schmidt, H., Haywood, J., & Scheffran, J. (2018). Evaluating climate geoengineering proposals in the context of the Paris Agreement temperature goals. *Nature Communications, 9*(1), 3734.

Leal Filho, W., Balogun, A. L., Olayide, O. E., Azeiteiro, U. M., Ayal, D. Y., Muñoz, P. D. C., Nagy, G. J., Bynoe, P., Oguge, O., Toamukum, N. Y., & Saroar, M. (2019). Assessing the impacts of climate change in cities and their adaptive capacity: Towards transformative approaches to climate change adaptation and poverty reduction in urban areas in a set of developing countries. *Science of the Total Environment, 692*, 1175–1190.

Lee, S., Paavola, J., & Dessai, S. (2022). Towards a deeper understanding of barriers to national climate change adaptation policy: A systematic review. *Climate Risk Management, 35*, 100414.

Liu, X., Hao, F., Portney, K., & Liu, Y. (2020). Examining public concern about global warming and climate change in China. *The China Quarterly, 242*, 460–486.

Lybbert, T. J., & Sumner, D. A. (2012). Agricultural technologies for climate change in developing countries: Policy options for innovation and technology diffusion. *Food Policy, 37*(1), 114–123.

Macamo, C. (2021). After Idai: Insights from Mozambique for Climate Resilient Coastal Infrastructure. *Policy Insights: African Perspectives-Global insights, 110*, 1–22.

Mahmood, H., Tanveer, M., & Furqan, M. (2021). Rule of law, corruption control, governance, and economic growth in managing renewable and nonrenewable energy consumption in South Asia. *International Journal of Environmental Research and Public Health, 18*(20), 10637.

Makate, C., Makate, M., Mango, N., & Siziba, S. (2019). Increasing resilience of smallholder farmers to climate change through multiple adoption of proven climate-smart agriculture innovations. Lessons from Southern Africa. *Journal of Environmental Management, 231*, 858–868.

Makwinja, R., Kaunda, E., Mengistou, S., & Alamirew, T. (2021). Impact of land use/land cover dynamics on ecosystem service value—A case from Lake Malombe, Southern Malawi. *Environmental Monitoring and Assessment, 193*, 1–23.

Manhique, A. J., Guirrugo, I. A., Nhantumbo, B. J., & Mavume, A. F. (2021). Seasonal to interannual variability of vertical wind shear and its relationship with tropical cyclogenesis in the Mozambique channel. *Atmosphere, 12*(6), 739.

Manyeruke, C., Hamauswa, S., & Mhandara, L. (2013). The effects of climate change and variability on food security in Zimbabwe: A socio-economic and political analysis. *International Journal of Humanities and Social Science, 3*(6), 270–286.

Marevesa, T., & Mavengano, E. (2023). Zimbabwean politics and development: An Ubuntu perspective. In *Contemporary development ethics from an african perspective: Selected readings* (pp. 243–256). Springer International Publishing.

Mashao, F. M., Mothapo, M. C., Munyai, R. B., Letsoalo, J. M., Mbokodo, I. L., Muofhe, T. P., & Chikoore, H. (2023). Extreme rainfall and flood risk prediction over the East Coast of South Africa. *Water, 15*(1), 50.

Masinde, M., & Thothela, P. N. (2019). ITIKI Plus: A mobile based application for integrating indigenous knowledge and scientific agro-climate decision support for Africa's small-scale farmers. In *2019 IEEE 2nd International Conference on Information and Computer Technologies (ICICT)* (pp. 303–309). IEEE.

Masud, S., & Khan, A. (2023). Policy implementation barriers in climate change adaptation: The case of Pakistan. *Environmental Policy and Governance*. https://doi.org/10.1002/eet.2054

Maswabi, M. G., Chun, J., & Chung, S. Y. (2021). Barriers to energy transition: A case of Botswana. *Energy Policy, 158*, 112514.

Matekenya, W., & Moyo, C. (2022). Innovation as a driver of SMME performance in South Africa: A quantile regression approach. *African Journal of Economic and Management Studies, 13*(3), 452–467.

Mathetsa, S. M., Simatele, M. D., & Rampedi, I. T. (2023). Applying the participatory approach to assess the Water-Energy-Climate Change nexus in South Africa. *Development Southern Africa, 40*(3), 696–712.

Matome, G. K., & Mulale, K. (2023). A review of the quality of strategic environmental assessment (SEA) reports in Botswana. *Impact Assessment and Project Appraisal, 41*(5), 403–415.

Mawonde, A., & Togo, M. (2019). Implementation of SDGs at the university of South Africa. *International Journal of Sustainability in Higher Education, 20*(5), 932–950.

Meadowcroft, J., 2009. Climate change governance. World Bank Policy Research Working Paper, (4941).

Meerow, S., & Keith, L. (2022). Planning for extreme heat: A national survey of US planners. *Journal of the American Planning Association, 88*(3), 319–334.

Mhute, I. (2023). A blessing or a curse: An exploration of Zimbabwe's plight in the global village. In *Contemporary development ethics from an African perspective: Selected readings* (pp. 131–142). Springer International Publishing.

Modisane, K. T. (2023). A demonstration of exceptional leadership amidst the HIV/AIDS scourge–chronicling the journey of resilience, Courage and Hope of the Former President of Botswana, Festus Gontebanye Mogae from 1998 to 2008. In *African leadership: Powerful paradigms for the 21st century* (pp. 93–108). Emerald Publishing Limited.

Mogomotsi, P. K., Sekelemani, A., & Mogomotsi, G. E. (2020). Climate change adaptation strategies of small-scale farmers in Ngamiland East, Botswana. *Climatic Change, 159*(3), 441–460.

Moyo, P., Ndlovu, T., Moyo, T. P., Mukuhlani, T., Nciizah, E., Nciizah, T., Nciizah, A. D., Ramalefane, T., Ehiane, S. O., Maphosa, F., & Maviza, G. (2023). Climate action policies, strategies and programming in Southern Africa: Missed opportunities for climate justice. In *Climate action in Southern Africa* (pp. 18–29). Routledge.

Muchaku, S., Magaiza, G., & Hamandawana, H. (2023). Translating indigenous knowledge into actionable climate-change adaption strategies: A case study of Maluti-a-Phofung local municipality, Free State Province. *South Africa. Sustainability, 15*(2), 1558.

Mugambiwa, S., & Makhubele, J. (2023). Rethinking the nexus of food insecurity, climate change and migration in the developing world: A case of Sub-Saharan Africa. *International Journal of Research in Business and Social Science, 12*(1), 249–253.

Muhonda, P., Mabiza, C., Makurira, H., Kujinga, K., Nhapi, I., Goldin, J., & Mashauri, D. A. (2014). Analysis of institutional mechanisms that support community response to impacts of floods in the middle-Zambezi river basin, Zimbabwe. *Physics and Chemistry of the Earth, Parts A/B/C, 76*, 64–71. https://doi.org/10.1016/j.pce.2014.11.013

Mulenga, E., Kabanshi, A., Mupeta, H., Ndiaye, M., Nyirenda, E., & Mulenga, K. (2023). Techno-economic analysis of off-grid PV-Diesel power generation system for rural electrification: A case study of Chilubi district in Zambia. *Renewable Energy, 203*, 601–611.

Muromo, F., Mukosi, T., Mashingaidze, T. M., & Lunga, W. (2022). *Unlocking socio-economic value to local communities from mega development projects*. Human Sciences Research Council.

Musa, M., Ibrahim, A., & Saidu, I. E. (2023). Is Financial Development Important for Zambia's Low Carbon Growth? An Extension of the STIRPAT Model. *Lapai Journal of Economics, 7*(1), 100–115.

Mutandwa, H., & Vyas-Doorgapersad, S. (2023). Public-private partnerships to improve water infrastructure in Zimbabwe. *Insights into Regional Development, 5*(2), 24–37.

Mutengwa, C. S., Mnkeni, P., & Kondwakwenda, A. (2023). Climate-smart agriculture and food security in Southern Africa: A review of the vulnerability of smallholder agriculture and food security to climate change. *Sustainability, 15*(4), 2882.

Muzerengi, F., Gandidzanwa, C. P., & Chirubvu, L. (2023). Impacts of climate change on household food security in Matande communal lands, Mwenezi district in Zimbabwe. *Jàmbá: Journal of Disaster Risk Studies, 15*(1), 1499.

Nazir, U. (2023). The winds of change are blowing: Globalization's impact on renewable energy and environmental challenges. *Archives of the Social Sciences: A Journal of Collaborative Memory, 2*(1), 78–93.

Nhamo, G. (2014). Addressing women in climate change policies: A focus on selected east and southern African countries. *Agenda, 28*(3), 156–167. https://doi.org/10.1080/10130950.2014.946734

Nhemachena, C., Nhamo, L., Matchaya, G., Nhemachena, C. R., Muchara, B., Karuaihe, S. T., & Mpandeli, S. (2020). Climate change impacts on water and agriculture sectors in Southern Africa: Threats and opportunities for sustainable development. *Water, 12*(10), 2673.

Ogwang, B. A., Ongoma, V., Shilenje, Z. W., Ramotubei, T. S., Letuma, M., & Ngaina, J. N. (2020). Influence of Indian Ocean dipole on rainfall variability and extremes over southern Africa. *Mausam, 71*(4), 637–648.

Omer, O., & Capaldo, J. (2023). The risks of the wrong climate policy for developing countries: Scenarios for South Africa. *Ecological Economics, 211*, 107874.

Østergaard, P. A., Duic, N., Noorollahi, Y., Mikulcic, H., & Kalogirou, S. (2020). Sustainable development using renewable energy technology. *Renewable Energy, 146*, 2430–2437. https://doi.org/10.1016/j.renene.2019.08.094

Page, M. J., McKenzie, J. E., Bossuyt, P. M., Boutron, I., Hoffmann, T. C., Mulrow, C. D., Shamseer, L., Tetzlaff, J. M., Akl, E. A., & Brennan, S. E. (2021). The PRISMA 2020 statement: An updated guideline for reporting systematic reviews. *Systematic Review, 10*(1), 1–11.

Panepinto, D., Riggio, V. A., & Zanetti, M. (2021). Analysis of the emergent climate change mitigation technologies. *International Journal of Environmental Research and Public Health, 18*(13), 6767. https://doi.org/10.3390/ijerph18136767

Pardoe, J., Vincent, K., Conway, D., Archer, E., Dougill, A. J., Mkwambisi, D., & Tembo-Nhlema, D. (2020). Evolution of national climate adaptation agendas in Malawi, Tanzania and Zambia: The role of national leadership and international donors. *Regional Environmental Change, 20*(4), 118.

Pegels, A. (2010). Renewable energy in South Africa: Potentials, barriers and options for support. *Energy Policy, 38*(9), 4945–4954.

Phiri, D., Mwitwa, J., Ng'andwe, P., Kanja, K., Munyaka, J., Chileshe, F., et al. (2023). Agricultural expansion into forest reserves in Zambia: A remote sensing approach. *Geocarto International, 38*(1), 2213203.

Pires, J. C. M. (2019). Negative emissions technologies: A complementary solution for climate change mitigation. *Science of the Total Environment, 672*, 502–514. https://doi.org/10.1016/j.scitotenv.2019.04.004

Qamar, S., Ahmad, M., Oryani, B., & Zhang, Q. (2022). Solar energy technology adoption and diffusion by micro, small, and medium enterprises: Sustainable energy for climate change mitigation. *Environmental Science and Pollution Research, 29*(32), 49385–49403.

Raimi, M. O., Vivien, O. T., & Oluwatoyin, O. A. (2021). Creating the healthiest nation: Climate change and environmental health impacts in Nigeria: A narrative review. *Scholink Sustainability in Environment.* https://doi.org/10.22158/se.v6n1p61

Rapolaki, R. S., Blamey, R. C., Hermes, J. C., & Reason, C. J. (2019). A classification of synoptic weather patterns linked to extreme rainfall over the Limpopo River Basin in southern Africa. *Climate Dynamics, 53*, 2265–2279.

Rej, S., Bandyopadhyay, A., Das, N., Hossain, M. E., Islam, M. S., Bera, P., & Yeediballi, T. (2023). The asymmetric influence of environmental-related technological innovation on climate change mitigation: what role do FDI and renewable energy play? *Environmental Science and Pollution Research, 30*(6), 14916–14931.

Saarinen, J., Moswete, N., Atlhopheng, J. R., & Hambira, W. L. (2020). Changing socio-ecologies of Kalahari: Local perceptions towards environmental change and tourism in Kgalagadi, Botswana. *Development Southern Africa, 37*(5), 855–870. https://doi.org/10.1080/0376835X.2020.1809997

Sadiq, M., Hassan, S. T., Khan, I., & Rahman, M. M. (2023). Policy uncertainty, renewable energy, corruption and CO2 emissions nexus in BRICS-1 countries: A panel CS-ARDL approach. *Environment, Development and Sustainability*, 1–27. https://doi.org/10.1007/s10668-023-03546-w

Salvador, C., Nieto, R., Vicente-Serrano, S. M., García-Herrera, R., Gimeno, L., & Vicedo-Cabrera, A. M. (2023). Public health implications of Drought in a Climate change context: A critical review. *Annual Review of Public Health, 44*, 213–232.

Schmidt, T. S., Matsuo, T., & Michaelowa, A. (2017). Renewable energy policy as an enabler of fossil fuel subsidy reform? Applying a socio-technical perspective to the cases of South Africa and Tunisia. *Global Environmental Change, 45*, 99–110.

Sechele, U. (2023). Botswana and the Bully: Apartheid South Africa and Southern African development coordination conference, 1978–1986. *Southern Journal for Contemporary History, 48*(1), 105–132.

Seymore, R., Inglesi-Lotz, R., & Blignaut, J. (2014). A greenhouse gas emissions inventory for South Africa: A comparative analysis. *Renewable and Sustainable Energy Reviews, 34*, 371–379. https://doi.org/10.1016/j.rser.2014.03.028

Sharifi, A. (2020). Trade-offs and conflicts between urban climate change mitigation and adaptation measures: A literature review. *Journal of Cleaner Production, 276*, 122813.

Shobande, O. A., & Asongu, S. A. (2022). The critical role of education and ICT in promoting environmental sustainability in Eastern and Southern Africa: A panel VAR approach. *Technological Forecasting and Social Change, 176*, 121480.

Sibiya, N. P., Das, D. K., Vogel, C., Mazinyo, S. P., Zhou, L., Kalumba, M. A., Sithole, M., Adom, R. K., & Simatele, M. D. (2023). Overcoming bureaucratic resistance: an analysis of barriers to climate change adaptation in South Africa. *Climate, 11*(7), 145.

Singh, N. P., Anand, B., Singh, S., Srivastava, S. K., Rao, C. S., Rao, K. V., & Bal, S. K. (2021). Synergies and trade-offs for climate-resilient agriculture in India: An agro-climatic zone assessment. *Climatic Change, 164*, 1–26.

Singubi, N., Muchanga, M., Wankie, R., & Zulu, I. (2023). Citizens science perspectives on drivers of sedimentation in the Lusitu River Catchment in Chirundu District, Zambia. *Journal of Geography, Environment and Earth Science International, 27*(7), 30–44.

Sithole, D., Tagwireyi, C., Marowa, T., Muwidzi, F., Mapanda, F., Svinurai, W., Gotore, T., Ngarize, S., Muchawona, A., Chigoverah, S., & Takavingofa, G. (2023). Climate change mitigation in Zimbabwe and links to sustainable development. *Environmental Development, 47*, 100891. https://doi.org/10.1016/j.envdev.2023.100891

Solangi, Y. A., Longsheng, C., & Shah, S. A. A. (2021). Assessing and overcoming the renewable energy barriers for sustainable development in Pakistan: An integrated AHP and fuzzy TOPSIS approach. *Renewable Energy, 173*, 209–222.

Stringer, L. C., Dougill, A. J., Mkwambisi, D. D., Dyer, J. C., Kalaba, F. K., & Mngoli, M. (2012). Challenges and opportunities for carbon management in Malawi and Zambia. *Carbon Management, 3*(2), 159–173. https://doi.org/10.4155/cmt.12.14

Stringer, L. C., Dougill, A. J., Dyer, J. C., Vincent, K., Fritzsche, F., Leventon, J., Falcão, M. P., Manyakaidze, P., Syampungani, S., Powell, P., & Kalaba, G. (2014). Advancing climate compatible development: lessons from southern Africa. *Regional Environmental Change, 14*, 713–725. https://doi.org/10.1007/s10113-013-0533-4

Su, T. M., Qin, Z. Z., Ji, H. B., Jiang, Y. X., & Huang, G. (2016). Recent advances in the photocatalytic reduction of carbon dioxide. *Environmental Chemistry Letters, 14*, 99–112. https://doi.org/10.1007/s10311-015-0528-0

Swinburn, B., Hovmand, P., Waterlander, W., & Allender, S. (2022). The global syndemic of obesity, undernutrition, and climate change. In *Clinical obesity in adults and children* (pp. 409–427).

Tangonyire, D. F., & Akuriba, G. A. (2021). Socioeconomic factors influencing farmers' specific adaptive strategies to climate change in Talensi district of the Upper East Region of Ghana. *Ecofeminism and Climate Change, 2*(2), 50–68.

Tittonell, P., Scopel, E., Andrieu, N., Posthumus, H., Mapfumo, P., Corbeels, M., van Halsema, G. E., Lahmar, R., Lugandu, S., Rakotoarisoa, J., & Mtambanengwe, F. (2012). Agroecology-based aggradation-conservation agriculture (ABACO): Targeting innovations to combat soil degradation and food insecurity in semi-arid Africa. *Field Crops Research, 132*, 168–174. https://doi.org/10.1016/j.fcr.2011.12.011

Todd, I., & McCauley, D. (2021). Assessing policy barriers to the energy transition in South Africa. *Energy Policy, 158*, 112529.

Tufa, A. H., Kanyamuka, J. S., Alene, A., Ngoma, H., Marenya, P. P., Thierfelder, C., Banda, H., & Chikoye, D. (2023). Analysis of adoption of conservation agriculture practices in southern Africa: Mixed-methods approach. *Frontiers in Sustainable Food Systems, 7*, 1151876. https://doi.org/10.3389/fsufs.2023.1151876

Udeagha, M. C., & Breitenbach, M. C. (2023). The role of financial development in climate change mitigation: Fresh policy insights from South Africa. *Biophysical Economics and Sustainability, 8*(1), 1.

UNFCCC. (2008). *Kyoto protocol reference manual on accounting of emissions and assigned amount*. https://unfccc.int/resource/docs/publications/08_unfccc_kp_ref_manual.pdf

UNFCCC. (2011). Compilation of information on nationally appropriate mitigation actions to be implemented by Parties not included in Annex I to the convention. In *AdHoc working groupon long-term cooperative action under the convention*. United Nations, FCCC/AWGLCA/2011/INF.1.

United Nations. (2015). *Paris agreement*. https://unfccc.int/process-and-meetings/the-paris-agreement/the-paris-agreement

van Ryneveld, T. N., & Islar, M. (2023). Coloniality as a barrier to climate action: Hierarchies of power in a coal-based economy. *Antipode, 55*(3), 958–981.

Vierros, M. K., & Harden-Davies, H. (2020). Capacity building and technology transfer for improving governance of marine areas both beyond and within national jurisdiction. *Marine Policy, 122*, 104158.

Wendling, Z. A. (2019). Bridges beyond renewable energy: Decarbonizing the global electricity sector under uncertainty. *Energy Research & Social Science, 48*, 235–245. https://doi.org/10.1016/j.erss.2018.09.020

Winkler, H., Howells, M., & Baumert, K. (2007). Sustainable development policies and measures: Institutional issues and electrical efficiency in South Africa. *Climate Policy, 7*(3), 212–229. https://doi.org/10.1080/14693062.2007.9685650

World Bank. (2018). Climate risk country profiles. Accessed in October 2024. Retrieved from: https://climateknowledgeportal.worldbank.org/country-profiles

Zavale, N. C. (2022). *Academic freedom in mozambique*. In *University autonomy decline* (pp. 92–118). Routledge.

Ziervogel, G., Bharwani, S., & Downing, T. E. (2006). Adapting to climate variability: pumpkins, people and policy. In *Natural resources forum* (Vol. 30, No. 4, pp. 294–305). Blackwell Publishing Ltd.

Zvobgo, L., Johnston, P., Williams, P. A., Trisos, C. H., Simpson, N. P., & Global Adaptation Mapping Initiative Team, (2022). The role of indigenous knowledge and local knowledge in water sector adaptation to climate change in Africa: a structured assessment. *Sustainability Science, 17*(5), 2077–2092.

14 Linking Gender Equality and Climate Resilience in Rural Southern Africa: Policy Gaps

Winniefridah Matsa, Mildred Chanhuwa, and Amanda T. Mugadza

14.1 Introduction

Deininger, Woodhouse, Kuriakose, Gren, and Liaqat (2023, p. 2) describe climate change (CC) as a defining challenge of our era. They highlight that addressing CC would require a fundamental shift towards green resilience and inclusive processes to capacitate individuals and communities to confront climate shocks. Shivanna (2022) also echoes that CC is a global emergency affecting not only the well-being of humans but sustainability of other life forms.

A framework that combines existing gender equality laws and policies with climate risks is needed. This chapter explores gendered impacts, the link between gender equality and climate resilience and policy gaps in implementation of CC programme and projects interventions in rural Southern Africa. It argues for integration of gender frameworks in CC actions. It emphasizes that resilience cannot be achieved without gender equality and similarly gender resilience cannot be achieved without gender empowerment. The chapter uses the lens of gender empowerment as the foundation for gender resilience. The focus is on rural Southern Africa because the people of this region experience multi-physical, economic, and social vulnerabilities. The chapter is organised according to themes premised on the objectives stated above. It begins with a discussion of gendered impacts of climate change, local policy gaps in Southern African countries, and the link between gender equality and climate resilience. Finally, the chapter discusses a gender empowerment model for climate change resilience in Southern Africa that must inform law and policy even at domestic levels. The gist of the chapter is to expose gender differentiated climate change impacts in selected rural Southern African countries and to unearth gender and climate resilience policies in local community programmes and projects. It also explores the link between gender equality, gender empowerment, and gender resilience as well as proffering a gender transformative model for climate resilience in rural Southern Africa.

14.2 Background

Pandit and Sharma (2023) view CC as a global concern with far-reaching consequences on ecosystems, cultures, and economies. CC disrupts services provision to humanity and alters the landscape for animal inhabitants. United Nations

W. Matsa (✉) · M. Chanhuwa
Gender Institute, Midlands State University, Gweru, Zimbabwe
e-mail: matsaw@staff.msu.ac.zw; chanhuwam@staff.msu.ac.zw

A. T. Mugadza
Faculty of Law, Midlands State University, Gweru, Zimbabwe
e-mail: mugadzaat@staff.msu.ac.zw

M. Matsa et al. (eds.), *Climate Change Resilience in Rural Southern Africa*, Sustainable Development Goals Series, https://doi.org/10.1007/978-3-031-73600-1_14

Convention on Climate Change UNFCCC (2016) echoes that CC presents the single biggest threat to sustainable development everywhere, with its impacts being experienced by each and every person around the world. Urgent action to halt climate change is needed to achieve all sustainable development goals (SDGs). The Post 2015 Climate Action Agenda (Paris Agreement) as well as the 2030 Agenda for Sustainable Development (SD) provide the foundation for sustainable, resilient development in the face of CC (UNFCCC, 2016). SDG 13 specifically calls for urgent action to combat climate change and its impacts. Target 13.1 focuses on strengthening resilience and adaptive capacity to climate change related hazards and natural disasters in all countries. Target 13.2 calls for the integration of climate action into national policies, strategies, and planning while target 13.3 seeks to improve educational awareness raising and institutional capacity to promote effective climate change planning and management (UNESCO, 2021). The achievement of other SDGs has ripple effects to other goals.

Urive (2016) posits that the impacts of climate change are already being experienced all over the world. The severe consequences of climate change disasters are especially felt on all the economic sectors (Shivanna, 2022). The Intergovernmental Panel on Climate Change (IPCC) also notes that the increase in average global temperature will permanently alter ecosystems as well as people's interactions with them (IPCC, 2022). The results have been increasing frequency of severe storms, floods, droughts, rising sea levels, and ocean acidification which impact on the lives of people and animals (Asian Development Bank (ADB), 2023). Therefore, the prevalence of disasters calls for urgent action to strengthen resilience against impacts of CC. According to IPCC (2007, p. 9), CC refers to any long-term change in the statistical distribution of weather patterns, including changes in average conditions (more or less rainfall, high or low temperatures) or in the distribution of events around the average (extreme weather events such as floods or droughts).

Southern Africa has been hard-hit by disasters with most of the larger and poorer populations situated in rural areas and have low levels of resilience from the disproportionate impacts of these disasters (UNHCR, Global report, 2022). Within this context, women in particular are faced with higher risks, greater vulnerability, and lower capacity to build resilience. Women in low-income countries in particular are faced with higher risks, greater vulnerability, and fewer tools for resilience (Sabaa & Peter, 2023). Yet, women and girls are the backbone in building climate resilience (International Centre for Climate Change and Development, 2023). The IPCC in Watson (1996, p. 23) defines vulnerability as "the extent to which climate change may damage or harm a system." The harm depends on the system sensitivity and its ability to adapt to new conditions.

UNICEF (2022) identifies gender equality as being at the "heart" of climate resilience. According to the Ahmed et al. (2023), climate risks and gender inequality raise barriers to women resilience. Women lack influence in environmentally related decision making resulting in the lack of focus on their vulnerabilities in the policy sphere. The ADB (2019) highlights that gender is a factor of vulnerability in CC, and CC-related events amplify existing disparities between men and women; hence, the need for gender-responsive approaches to climate change resilience—approaches taking action to redress gender inequalities in CC resilience activities.

The gender debates in development date back to 2015, a pivotal year for gender equity and empowerment of women and girls. It saw the review of the implementation status of the 1995 Beijing Declaration and platform for Action s' commitment to gender equality (UN Women, 2016). As the global community transitioned to implement the phase of the post 2015 development agenda, it became imperative that gender equality and women's empowerment contribute to influence, shape, and drive the collective and human development effort (Habtezion, 2016, p. 1–2). It came to light that there can be no genuine sustainable human development without gender equality.

Governments are increasingly acknowledging threats of CC to the poor rural people, but climate resilient programmes and projects at local levels have remained not gender-transformative. They do not tackle gender inequalities that determine CC impacts and resilience. It is against this background that this chapter seeks to determine the disproportionate impact of climate change risks by gender in rural Southern Africa and to explore the intricate link of gender equality and climate change resilience. The chapter also assesses policy gaps in gender approaches to climate change resilience as well as to proffer a gender empowerment model for climate change resilience and gender equality in rural Southern Africa.

14.3 Methodology

The chapter is an exploratory qualitative desk review of literature of primary and secondary policy documents on gender declarations, conventions and protocols, gender and climate change policies to determine gaps in policy implementation in rural Southern Africa country cases. The gender and climate resilient policy frameworks are categorized into international, regional, and national as they trickle down to community levels in programmes and projects. The gendered impact of climate change, along with the concepts of gender resilience, gender empowerment, and gender equality, and their intricate relationship, is discussed within the theoretical framework of CC justice, which focuses on CC as a cause of disproportionate harm to certain members of society or people (particularly men and women living on low income or in poverty). Country cases from literature are analyzsed and synthesized to reveal gender differentiated impacts. Incorporating CC justice means prioritizing the well-being of people or communities most exposed. It is putting communities that are at risks in the first line in pre-, current-, and post-disaster resilience planning and building. CC justice also focuses on addressing past injustices such as class, racial, or gender inequality injustice by taking socially just actions and responses (Union of concerned scientists, 2022). The paper also uses gender frameworks to disaggregate the impacts of CC risks and needs assessment by gender. The gender matrix, in particular the gender empowerment framework by Sarah Longwe (1995) model is used in this paper to empower men and women on climate change resilience building. The empowerment tool starts with gender vulnerability analysis and gender needs (practical and strategic) for inclusive assessment need, project or programme design, planning, implementation, and gender resilient indicators as outcome of monitoring and evaluation. The gender lens approach assists in the analysis of the gender dynamics of roles, access to livelihood assets, goods, services, power, decision making, control over resources, power relations, and rights as well as in analyzing climate change laws and policies to determine their blindness, neutrality, or sensitivity (Tanyag & True, 2019). The sources of data include secondary scientific papers, reports by organizations, and policy documents from websites such as Google Scholar and ResearchGate, accessed between November 2023 and February 2024. The literature review followed logical steps of exploring history of gender and climate change since 2015, gendered impact of CC in rural Southern Africa, case studies on policy gaps, and the absence of CC resilience in gender strategies, projects and programmes. The chapter also explores conceptual definitions of key terms such as climate resilience, gender empowerment, and gender equality and their linkage. The final part discusses the suggested gender and climate change resilience model for rural Southern Africa.

14.4 Findings and Discussion

14.4.1 Gendered Impact of Climate Change in Southern Africa

In Namibia, 47.9% of the population live in rural areas (HRCs, 2021). This population relies mostly on production of fish, mining exports, and farming as their sources of livelihood. The 47.9% population of those in rural areas occupy only 1% of land which is available for use because the country's agriculture has been severely affected by changes in the climatic conditions and there is great limitation on the type of crops that can be farmed. 51.57% of those in farming communities are females (HRCs, 2021). The impact of climate change is felt more by women than by men. Climate change has seen an increase in water shortages, flooding, or low rainfall. Women are affected severely as they have to travel long distance to fetch water, fail to do small-scale agricultural projects, and have to find other means of survival (Angula & Mejono, 2010). The men in the family migrate with livestock to areas that are less affected by climate change. Thus, women struggle to carry out all the tasks, and assume both gender and sex roles. The daily activities enunciated by the gender roles place women in a disadvantaged position. Women's chores, such as collection of wild fruit, medicinal plants, and harvesting natural resources of products, expose them to greater impact from the changes in the environment (Angula, 2008). Health concerns have also become alive due to climate change. Furthermore, the health risks associated with climate change, including waterborne illnesses and heat-related stress, disproportionately affect women, who often bear the burden of caring for their families (Angula, 2008).

In Zimbabwe, more than 60% of population resides in rural areas. Majority of these people depend on rain-fed agriculture for livelihoods. At the same time, about 80% of women reside in the rural areas. These women constitute about 60% of the smallholder farmers. Females mainly found in the informal sector and agro-based industries as well as subsistence cultivation dependent on rain-fed rainfall. The women are more vulnerable to climate change due to a number of factors. Some of these include culture which positions females in roles of caregivers who have no access to change information. There is also unequal participation in labour and education with females mainly found in the informal sector and agro-based industries as well as subsistence cultivation dependent on rain-fed rainfall (Awiti, 2022).

In Mozambique, 73% of the population lives in the rural areas (Rebeiro & Chaque, 2020). The main source of livelihoods is agriculture with emphasis on cashew nuts, cotton, and agriculture. There is also livestock farming and fisheries. Women are believed to suffer more from the differential impact due to the multiple roles they play (Rebeiro & Chauque, 2014). In a study conducted in the Mapi-Ngale community, it was revealed that, due to climate change, the agricultural practice of ploughing has deteriorated. In a bid to maintain livelihoods, there is an increase in the collection of fruits and worms by women. The change of climatic conditions in Mgodzwene has also seen women performing men's roles. Men migrate and women become responsible for all the tasks. They plant crops but due to the poor rains, they find alternate income generating projects such as selling alcoholic drinks and fishing to boost their finances. In addition to this, women are responsible for fetching water and firewood. The immediate consequence is that women fail to spend time with their children. As such, children, especially girls, are forced to drop out of school to assist their mothers to raise the young ones. Therefore, women's burden has increased due to climate change. Men who migrate are said to contribute very little or nothing at all (De Vletter, 2006). Clearly, the climate change conditions in the country have impacted men and women differently due to the varying perceptions of gender roles and the extreme poverty in the country.

In Zambia, it was reported that approximately 54.24 of their population live in the rural areas. Of this population, more than 60% are women who are farmers. The main source of livelihoods in Zambia is farming, and cassava is the main crop of production. The findings in literature reveal that in Zambia mortality rate from climate-related disasters are higher in women in comparison to man. Women rely on natural resources for their livelihoods in comparison to men. Thus, the decline on land and biomass productivity affects women more than men. In Zambia, female households are more likely to use less-efficient and more labour-intensive methods for collecting water such as buckets, whereas males are more likely to have access to wheelbarrows, draught, and electric pumps. The impacts depend on gender roles and available resources for men and women. Females are more likely to fetch firewood for daily use than men, leading females to travel long distances in search of firewood. Women can spend 3 h a day collecting firewood.

14.4.2 Legal and Policy Gaps, Opportunities, and Limitations in Advancing a Gendered Response to Climate Change Resilience in Southern Africa

The international laws and gender policy frameworks that have been crafted include the Convention on the Elimination of Discrimination against Women (CEDAW 1979), the Beijing Declaration and Platform for Action (1995), the African Charter on People Rights, SADC Declaration on Gender and Development (1997) (a domestic policy for Southern Africa), Maputo Protocol (2003), MDGs (2000), and SDGs (2015). CEDAW provides for equality and non-discrimination against women promoted climate resilient development (article 2, CEDAW) (CEDAW Committee 44th Session New York 2009). CEDAW specifically places an obligation on state parties to take into account the challenges of rural women and the roles they play in economic survival of their families (article 14, CEDAW, CEDAW Committee 2009). However, implementation of the provisions of the CEDAW are not married to the provisions of climate change resilient strategies. At regional level, similar provisions to CEDAW exist in the African Charter on Human and Peoples' Rights through its Protocol on the Rights of Women in Africa (The Maputo Protocol, 2003). This Protocol calls for equality and non-discrimination of women.

At global level, climate resilience is principally provided for in the United Nations Framework Convention on Climate Change of 1992, the Paris Agreement of 2015, and Sendai and Hyogo Framework for Disaster Risk Reduction 2015–2030. The Paris Agreement consolidates the commitments of state parties to the UNFCCC by providing legally binding obligations on resilience building. The main objectives of the Paris Agreement are, inter alia, to foster climate resilience in a manner that does not threaten food production and to make finance flow towards climate-resilient development (articles 2(b) and (c) and 7, Paris Agreement, 2015). To achieve this, state parties should take a country-driven, participatory, and fully transparent approach, taking into consideration vulnerable groups, communities and ecosystems (article 7(5), Paris Agreement, 2015). However, the framework does not provide for the tackling of inequality in terms of access to information, technical and scientific knowledge by rural people especially women and does not elaborate clearly how empowerment at local levels for resilience building ought to be done especially in the event of existing inequalities and gender gaps. The gender action is not clearly spelt out.

The Sendai Framework for Disaster Risk Reduction 2015–2030 calls for states to adopt people-centred and preventive disaster risk reduction, which involves engagements with stakeholders such as women and other vulnerable groups (paragraph 7, Sendai Framework, 2015). This further involves empowerment that is inclusive, accessible, and non-discriminatory, paying special attention to people disproportionately affected by disasters and promoting women leadership (paragraph 19, Sendai Framework, 2015). This Sendai is silent on the gender dynamics of climate resilience. The implementation, the framework is applied in a gender-neutral way, applying equally to males and females as if there are no regional and national gender inequalities even within the vulnerable groups as we have seen in the above projects and programmes. Gender is a cross-cutting principle in Hyogo Framework for Action 2000–2015. A gender perspective was to be integrated into resilience building of nations providing the basis upon which resilience building is undertaken by state parties at a broader level (SADC, 2020).

It is these comprehensive international and regional policy provisions on CC resilience that should work hand in glove with gender conventions, thus engendering gender in climate resilient approaches. Unfortunately, gender responsive

CC policy and action is missing at local level implementation (Rohr, 2009; FAO, 2018).

In the African regional platforms for disaster risk reduction (DRR) and Hyogo Framework for resilience building, the commitment to gender issues is unfortunately rarely stated explicitly, rather it tends to be an implicit part of a larger commitment to implement (Tanner, 2022, p. 7), and only a handful of organizations and women activists' groups show dedication. Some policy frameworks are gender-blind as we have seen in the background. Gender-blind policies and interventions exacerbate inequalities and worsen vulnerabilities (Deininger et al., 2023). Gender blindness is failure to recognize gender in development policy, that men and women are affected differently by development endeavours. It is premised on a notion of a male development actor appearing neutral by applying equally to men and women. The evidence in this paper reveals that gender-blind policies are male biased in terms of privileges, needs, interests, priorities, and in the distribution of resources (Kabeer & Subrahmanian, 1996, p. 1).

Following the SADC declaration on gender and development on mainstreaming of gender in all sectors, the expectation was to see all projects, programmes, and interventions integrating the gender aspect. However, full commitment in this endeavour has not been realized in the region. Most of the projects continue to lack the gender flair despite legal and policy commitments at international, regional, and national levels.

The SADC Transfrontier Conservation Project has set out to promote a people-centered approach, including women and youth, in its conservation efforts. However, the project's plans and implementation have been criticized for lacking inclusivity and equity. The project aims to support shared conservation areas, such as the Kgalagadi Transfrontier Park, which spans across Namibia, South Africa, Botswana, and Mozambique, as well as the Okavango Delta and Mapungubwe Park, to promote sustainable eco-cultural tourism and management of wetlands in South Africa, Mozambique, Namibia, Botswana, and Zimbabwe. The shared conservation areas are Kgalagadi between Namibia, South Africa, Botswana, Mosia-Tunya, Okavango Delta and Mapungubwe Park between (SADC Tran frontier conservation, 2022). Women have not been active in these environmental endeavours and the project plans are not transformatory to fully integrate women despite that the focus areas also affect women to a greater extent. Gender policies are far from being visible in CC and environmental resilience projects in rural Southern African Countries.

The All strategy for gender equality and women's empowerment 2018–2027 lays out the implementation plan for Agenda 2063 and the principles enshrined in Article 4(1) of the AU constitution of promotion of gender equality. There are four priority areas where rapid change is possible namely security and resilience, law policy, education for women and girls, and leadership, voice, and visibility. The challenges are that policy documents are either weak, absent, or remain as documentation for integration of gender into DRR and CC action programmes and activity as we have seen above. There are no uniform or contextual risk assessments. There is also lack of gender disaggregated data and gender-sensitive indicators as well as over reliance on externally funded models (SADC, 2020).

14.4.3 Linking Gender Equality with Climate Resilience in Rural Southern Africa

At this juncture it is critical to define the key concepts such as gender equality, gender resilience, and gender empowerment, before we explore their intricate relationship. Resilience explores how societies have thrived or are thriving in the face of dramatic changes in the natural world. It relates to how communities, individuals, or groups craft accounts of the past and learn valuable lessons for the future. Resilience is the capacity to bounce back, that is, getting back to the same or better function after a disturbance or shock, and it is not maintaining the equilibrium. In the context of climate shocks, resilience is the ability to bounce back better after being exposed to stressors and shocks (ARCAB in UN Women,

2015). It involves anticipation, exploration, uncovering, and preparation for the most alarming future warnings, as well as responding to shifting environmental circumstances (Degroot, 2020).

For resilience to take place, there must be an underlying invisible power (power within) among individual men and women, collective groups, and communities. According to Williams et al. (1994), power within comes from self-esteem, self-respect, self-acceptance, self-worth, and self-knowledge helping individuals to gain confidence to act, influence, and change their prevailing circumstances. It is about internal, intrinsic, or personal power. Resilience is intrinsic, continuous, and a sustainable venture (continuum) (UN Women, 2015). Resilience can also be externally triggered where governments and NGOs help the poor and vulnerable realise their powerlessness and create awareness of who they are or even suggest projects and programmes to build resilience. They also make individuals and communities aware of current and future uncertainties or shocks and help them identify existing resources and practices and knowledge at their disposal to enable them to get ready for current shocks and anticipate future threats (UN Women, 2015).

Resilience is driven by empowerment. In fact, empowerment is a foundation of resilience. Empowerment, according to Hannan (2003), is a process where individuals who are marginalized individually or collectively become aware of their powerlessness then gain knowledge and skills to challenge and overcome the power structures and processes that constrain them in order to reach the level of control, the highest stage of empowerment. Kabeer (1999) in Tanner (2022, p. 8) defines "empowerment as the process by which those denied the ability to make strategic choices acquire such an ability." Empowerment is premised on different types of power which individuals, groups, or communities possess or do not possess. The types of power are personal, which include inner-self, self-respect, self-confidence, or power within (Williams et al., 1994). Individuals need this type of power that enables them to have control over tangible resources such as finance, capital, land, water, and minerals, and intangible resources such as time, labour, knowledge, and skills. "Power to" is the power one possesses to do something, for example, power to act or decide. "Power with" is collective or group power also known as relational power. Not all persons or groups have all types of power. The poor, in the majority of cases, lack some or all the types of power (Van Eerdewijk et al., 2017, p. 16). Empowerment then becomes fundamental in facilitating the attainment of power by individuals and groups. However, it must be noted that empowerment is both a process and an outcome.

According to Albert (1985, p. 12), power cannot be given, but one can be manipulated, motivated, capacitated, or facilitated to change desires, needs, and view of self, thereby facilitating a change in status. It is women empowerment if it concerns females only, and gender empowerment if it is directed to both males and females. The resilience model below borrows from gender empowerment. According to Abound (2011, p. 8), gender equality refers to equal rights, opportunities, responsibilities of men and women, and girls and boys. It implies taking into consideration the interests, needs, priorities of both men and women recognizing diversity of groups. UN views gender equality as both a human right issue and a precondition for and an indicator of sustainable development (Abound, 2011).

Like empowerment, resilience occurs at various levels such as individual, group, or community levels. Both start at an individual personal level and cascade up to the community and national levels. Resilient individuals make up resilient communities and resilient nations. Like empowerment, resilience indicators are not external although they can be motivated externally through empowerment by governments and other development agencies. Both empowerment and resilience cannot be "done for," "done to someone," or imposed, they are earned during the process of empowerment (Van Eerdewijk, 2017; Albert, 1985).

Climate resilience cannot be achieved without gender empowerment and gender equality. The climate frameworks and resilient approaches that are gender-blind, neutral, and irresponsive cannot achieve climate resilience (Habtezion, 2016, Plan International, 2019). As we have noted above, some climate change frameworks are not in sync with gender equality frameworks or missing at local levels yet inequality is a barrier in achieving resilience (Rohr, 2009; UN Women, 2016; Deininger, 2023). According to Schipper and Langton (2014) in Tanner et al. (2022, p. 5), "Unless underlying gender inequality is tackled, even those policies and projects that explicitly seek to address gender concerns in climate change will not be effective." Gender equality is central and pivotal in climate-induced disasters (Babugura, 2010; UN Women, 2015).

Contreras and Roudibari (2021) posit that gender participatory programmes and projects start with the knowledge of ownership and power relations at grassroots levels of interventions. After establishing the gendered impacts of climate change and gender gaps, social structures of patriarchy are to be dealt with otherwise development programmes and projects will remain male-dominated. They did a study in Zambia to establish gender mainstreaming in climate change projects, and the results revealed that gender policies in climate change were explicit about gender but inequality were not tacked and climate action plans lacked gender-responsive strategies. What were addressed were the vulnerability of girls and women and not the social structures of inequality that caused and shaped the vulnerabilities (Abound, 2011). Djoudi et al. (2016) collude that "implementing resilient projects without tackling gender inequalities is normalising inequalities …."

A study by Ruparanganda and Kefasi (2018) in Manicaland province, Zimbabwe, on female adaptation and resilience to climate change revealed that there was no acknowledgement of gender issues. Most males were migrants but differential gender roles were ignored. Female households also lacked access to irrigation due to systematic constraints relating to access to land and resources.

South Africa through UNFCCC showed gender concerns in its action plan for climate empowerment, technology, and finance policies. However, because of inequalities, women faced social, economic, and political barriers that limited their coping capacity. Women were found to live on less than a dollar per day, owning less than 1% of the worlds' property and being 75% of the illiterate adults. Although gender mainstreaming is entrenched in its environmental programmes, inequalities render the projects ineffective.

Namibia was the first to develop and adopt CC policy and action plan (Nyahunda et al., 2020). She recognized the role played by rural women and the need for mainstreaming and empowering women in all sectors in its CC strategy and Action plan for 2030. In 2020, a drought relief programme was implemented in the Ohangwena, Omusati, Hardap, Kunence, and Omahene rural areas. The project was being funded by UN central emergence response Fund. Farmer's Unions reached 1.2 million beneficiaries increasing farmer productivity and income. Funding was through existing networks and redistributive approaches within the existing systems of inequality (Panova & Adams, 2020).

Oxfam of Southern Africa through Chinhoyi University of Technology and Zimbabwe meteorological services department conducted a study on gender and climate information dissemination systems in semi-arid areas of Zimbabwe. The areas covered were Chirumanzi, Zvishavane, Gutu in Midlands, and Masvingo Provinces. The results revealed that the gender analysis applied in research was not contextual as global impact models differed according to regions. Gender roles in SA had been altered by CC; therefore, they differ from other countries such as Namibia, Mozambique, and Botswana, yet gender analysis tools were applied equally to all four countries. The gender related dimensions on access to information were ignored. Gender baseline study on gender access to timing, delivery channels or platforms was not done before the project on SMS information on climate change. As a result, only male champion farmers were trained on the use of digital technologies, such as SMS gadgets, phones, radios, and TVs, while women were excluded from this training. Furthermore, the project failed to consider the significant gender differences in financial status, human capital, and social capital, which further exacerbated the existing inequalities. The results revealed gendered access to SMS and climate information. Women would use traditional methods of access (schools and school children) to information compared to men. As a result, women had limited access to SMS extension services in rural areas (African Community of Practice, 2023). Men and women are not a homogenous group. This is another programme in rural Southern Africa which was gender-neutral (treating men and women as if they are the same).

Another study conducted by Petrie (2010) on women as agents of change in South Africa, Botswana, Mozambique, and Namibia revealed that men and women are not the same across countries and within individual countries. Except for South Africa, all communities reported that women generally adapted better than men to changing circumstances and were repositories of knowledge on crops, environment, and food presentation techniques, although this did not make them less vulnerable. However, resilience activities did not target women. They lacked national toolkits on gender indicators. Zambia has Article 8 on principles of equality and non-discrimination, Article 257 on environmental protection and CC action, and gender action plan 2018 to ensure gender-transformative initiatives with women as beneficiaries. The CC policy 2012–2030 seeks to improve women's access to productive resources and services. However, the policies and programmes are far from engendering CC programmes and activities to achieve gender equality and equity because of the long history of vulnerability, exclusion, inequality, and marginalization from CC decision-making processes that render women passive participants in development (Nyahunda et al., 2020). Most strategies fail to address inequalities, systematic poverty, or historical imbalances. Women are treated as mere victims of CC, rendering women to mere infusion in gender policy documents.

It must be noted, however, that in South Africa, through its National Policy Framework for Women Empowerment on Gender Equality, women accounted for 53% of the environmental sector and public works programmes in 2015. The programme targeted 50% women, 20% youths, and 5% of the disabled. About 200,000 individuals (52% of whom were women) were trained and 5000 (85% of whom were women) were employed. South Africa hosts environmental conferences and dialogues creating awareness among stakeholders including women in the gender-responsive approaches. However, the environmental strategies lack gender concern which is accompanied by gender planning. The integration of gender is within the existing structures of inequality and unequal power relations and decision making. Increasing the number of men and women into projects does not necessarily mean active participation in planning, design, and decision making.

14.4.4 Opportunities and Challenges

The recognition that gender is at the heart of climate change action has created a unique opportunity to drive climate resilience efforts. In response, international, regional, and national frameworks have been established to address both gender and climate resilience. The SADC region has developed various initiatives to promote climate resilience, including regional and national strategies that incorporate gender considerations. For example, the Southern African Development Strategy has mainstreamed gender in the environmental sector, while Zimbabwe's National Gender Policy 2013 outlines strategies for gender mainstreaming across all sectors, including the environmental sector. Other countries, such as Malawi, have introduced programs and projects that empower women in rural areas, such as vocational skills training for smart agriculture among smallholder farmers. These efforts have helped women become active in local resilience initiatives, while countries like Namibia have taken a proactive approach to climate change by developing and adopting their own climate policies, making them a leader in the region.

There is also a realization that women are agents of CC resilience, hence the need for their participation, voices, and involvement in decision making (Petrie 2010; SDCC, 2020). Governments are increasingly acknowledging threats of CC on women and girls, particularly those living in rural areas. Women are being positioned as change makers in this space, and gender integration is viewed as a way of enhancing effectiveness of CC action (UNFCCC, 2022 in Deininger et al. 2023, p. 9). South Africa has developed a national policy framework for women empowerment and gender equality mandating application of gender mainstreaming in the environmental sector under UNFCCC or gender response approaches. Researches in CC have also mainstreamed gender unearthing the dynamic gendered impacts of CC (Gender vulnerability SADC, 2020).

However, the limitations are that some climate resilience frameworks and responses are not in sync with gender frameworks. Others are either blind, neutral (applying equally to men and women), and irresponsive (not taking action to redress existing gender inequalities) or remain at international, regional, and national levels not trickling to grassroots projects at local levels.

14.4.5 Gender Empowerment Model for CC Resilience in Southern Africa

The model in this chapter is adapted to the Sara Longwe, 1995 Empowerment Framework and OP2 Strategy 2030 Operational Plan to accelerate progress in gender equality and climate change discourse. Longwe (1995) cited in March et al. (1999, p. 94) Empowerment Framework has five levels of intervention: they are welfare level of material needs for survival (practical gender needs) (PGN), access, conscientization, participation (mobilization), and control as the highest stage of empowerment. This chapter takes the lens of gender empowerment as the foundation for gender resilience. With the gender perspective, the application of the model reflects the gender disaggregated impacts of disaster risks, vulnerability assessment, experiences of each gender, strategies applied to each gender, as well as gender-responsive distribution of resources.

This model has an underlying assumption that when one reaches the highest stage of empowerment, they are most likely to reach highest stages of climate resilience and gender equality. Climate resilience is withstanding and thriving in the face of global climate crisis. It is about successfully coping and managing the impacts of climate change while preventing those impacts from growing worse. It involves adjusting how people live and work to keep themselves safe or to be able to recover from the impacts of climate change risks (Union of Concerned Scientists, 2022).

The model below shows three routes: gender empowerment, gender resilience, and gender equality. An increase in one level of each root triggers an increase in the other. The basic levels of each root require baseline disaggregated information on basic needs (PGN) in case of gender empowerment, gendered impacts of climate risks in case of gender resilience, and disaggregated data on gender inequalities in case of gender equality root. The second level concerns access, focusing on whether males and females have access to strategic gender needs which allow them to be in a strategic position or resources to use to achieve empowerment resilience and equality. These are tangible and intangible resources, for example, law, policies, plans, time, skills, land, credit loans, capital, property rights, education, services, and information. This is the level of conscientization, conscious raising, being conscious of inequalities, gender and cultural constraints working against equality, empowerment, and resilience. It is about self-discovery, self-identity, self-worth, understanding concepts like gender empowerment, equality, equity, resilience, discrimination, and mainstreaming. The fourth level involves individual, household, and community mobilization for active participation in actions, programmes, and strategies meant to achieve equality, empowerment, or resilience. This relates to personal and collective empowerment emanating from power within and collective power (power with). It also has monitoring and evaluation of programmes, strategies, and policies to see the impact on men and women. This is the level of control the high-

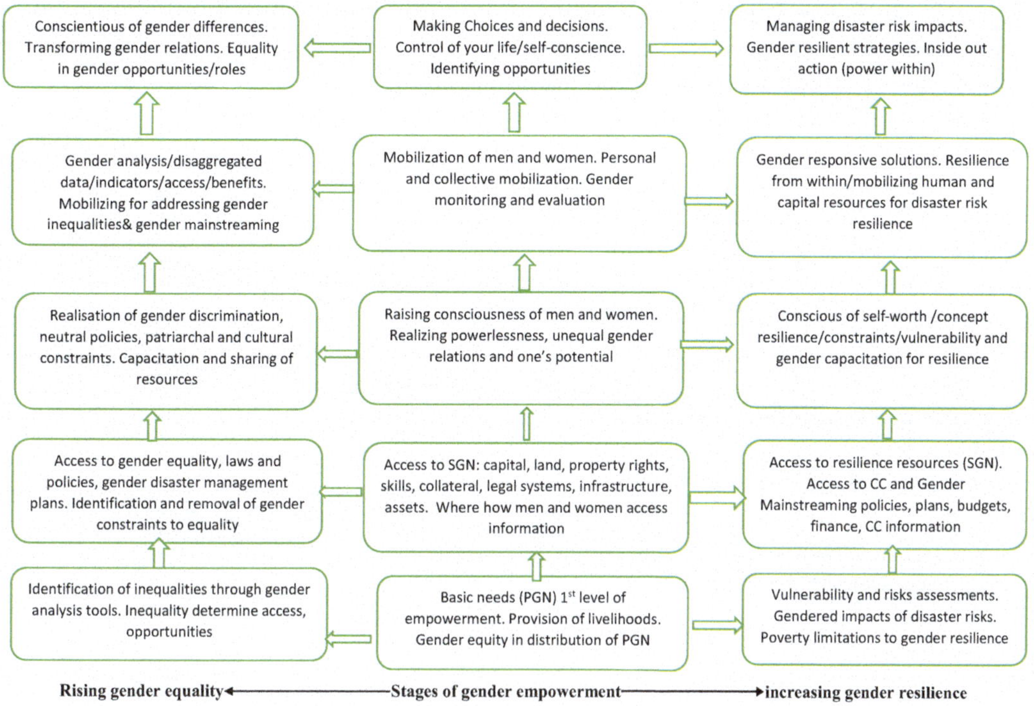

Fig. 14.1 Model for DR resilience in rural Southern Africa. [Adapted from Sara Longwe, 1995 and COP 2 Strategy (2030)]

est level of empowerment likely to help one reach equality and resilience. Individuals or communities are in control of the direction of their lives. They make choices among alternative strategies and plans. It is the stage of decision making and choices. It is an indicator of empowerment, equality, and resilience.

Below is the chain of gender empowerment, gender resilient, and gender equality model for climate change risk resilience for men and women in rural Southern Africa (Fig. 14.1).

14.5 Conclusion

The chapter explored through literature the impact of CC in Southern African women. The chapter has revealed that in Southern Africa, there are more women who live in rural areas than men. The source of livelihood of the women is agricultural in nature; therefore, CC has greatly impacted their livelihoods and their methods of survival. Evidence reveals that women experience greater impact in comparison to men because of the multiple roles they play and because of the patriarchal nature of the society. The men in the family migrate with livestock to areas that are less affected by climate change. Thus, women struggle to carry out all the tasks, and assume both gender and sex roles. Care work, on the other hand, continues to weigh heavily on women. Due to climate change women travel longer distances to fetch water, find firewood, and to gather wild fruits for their children's consumption. This chapter identified that there is also a correlation between empowerment, resilience, and equality. As empowerment increases, resilience and equality increase. Empowerment is the foundation of resilience and equality. The chapter further concludes that climate resilient policies and strategies at local levels are not engendered. The resilient strategies at national and local levels are also not contextual, are externally defined, and driven. Women are not active participants in resilience because of poverty and gender inequality barriers. Resilient projects and programmes do not transform unequal power relations of access, participation, decision making, and control. Invisible resources and the knowledge of communities is not tapped. There is an assumption that issues of power and equality are invalid yet they are core. Power is inherent in resilience building. Empowerment is a means to equality and resilience. There is correlation between economic development, men and women's legal rights, property rights, access to land, bank loans, and policies (UN Women, 2015). As such, there should be explicit laws, policies, and actions to address inequalities and gendered vulnerabilities. Global, regional, and national gender empowerment laws, policies, and frameworks should also work hand in glove with climate change, disaster risk management, and resilient frameworks (ADB, 2023).

14.6 Recommendations

The paper recommends the

- Crafting of gender-sensitive and gender-responsive climate resilient local-level projects and programmes.
- Implement the projects in a gender-responsive manner, starting with gender assessments of impacts, followed by gender planning, gender budgeting, gender-transformative implementation, and gender-sensitive indicators.
- Empowering men and women on gender and climate resilience policies.
- Creation of awareness in issues of access to tangible and intangible resources, participation, decision making, and control over the resources and their resilient courses of action.
- Empowering communities and individuals to link equality with climate resilience through a resilience model described.

References

ADB. (albert 2019). *How to use gender approaches to build climate resilience tips for redesigning ADB projects based on experiences of the Pacific.*

Abound, A. (2011). Climate change and agriculture in Africa: A review of the literature. *Journal of Climate Change and Sustainable Development, 2*(1), 1–12.

Ahmed, D. et al., (2023). Climate change, gender, and resilience: A review of the literature. *Journal of Climate Change and Human Security, 1*(1), 1–15. https://doi.org/10.1155/2023/123456

Albert, P. H. (1985). *People, power: Applying non-violence*. New Society Publishers Philadelphia.

Angula, M. N. (2008). *Gender and climate change: Namibia case study*. Heinrich Boll Stiftung.

Angula, M. N., & Menjono, E. (2010). Gender, culture and climate change in rural Namibia University of Namibia. *Journal for Studies in Humanities and Social Sciences, 3*(1 & 2), 225–238.

Asian Development Bank DB. (2023). *Women's resilience: How laws and policies promote gender equality in climate change environment and Disaster risk management in Asia and the Pacific*. ADBBrigs, adb-biref-292-women-gender-climate.

Awiti, A. O. (2022). *Climate change and gender in Africa: A review of impact and gender responsive solutions*. Institute of Development Studies.

Babugura, A. (2010). *Gender and CC: South Africa case study*. The Green political Foundation. https://www.boell.de/en/2010/04/27/gender-and-climate-change-south-africa-case-study

Contreras, A., & Roudibari, A. (2021). Climate change and migration: A review of the literature. *Journal of Climate Change and Human Security, 3*(1), 1–15. https://doi.org/10.1155/2021/123456

Degroot, D. J. (2020). Climate Change and Human Migration: A Review of the Literature. *Journal of Climate Change and Human Security, 2*(1), 1–15. https://doi.org/10.1155/2020/123456

Deininger, K., Dercon, S., & Outes, S. (2023). Climate Change, Resilience, and Poverty Reduction: A Review of the Literature. *Journal of Development Studies, 59*(2), 249–265. https://doi.org/10.1080/00220388.2022.2137953

De Vletter, F. (2006). In J. Crush (Ed.), *Migration and development in Mozambique: Poverty, inequality and survival*. Migration and Policy Series, 43. Southern African Migration Project (SAMP).

Djoudi, et al. (2016). Beyond dichotomies: Gender and intersection, inequalities and climate change studies. *Ambio, 45*(3), 248–268.

Food and Agriculture Organization of the United Nations (FAO). (2018). Climate change and agriculture: Impacts, vulnerability, and adaptation. FAO Technical Paper No. 684.

Habtezion, A. (2016). Climate Change and Women's Livelihoods: A Review of the Literature. *Journal of Climate Change and Sustainable Development, 3*(1), 1–12.

Hannan, P. (2003). Women's empowerment and microfinance: A review of the literature. *Journal of International Development, 15*(4), 465–479.

Intergovernmental Panel on Climate Change (IPCC). (2007). *Climate change 2007: Impacts, vulnerability, and adaptation. Contribution of working group II to the fourth assessment report of the intergovernmental panel on climate change*. Cambridge University Press.

IPCC. (2022). *Climate change 2022: Impacts, adaptation and vulnerability*. IPCC Sixth Assessment Report. https://www.ipcc.ch/report/ar6/wg2/

Kabeer, N., & Subrahmanian, R. (1996). Institutions, relations, and outcomes: A framework for analyzing women's empowerment. *World Development, 24*(11), 1797–1817.

Kabeer, N. (1999). Resources, agency, and achievements: Reflections on the measurement of women's empowerment. *Development and Change, 30*(3), 435–464.

Longwe, S. (1995). Gender awareness: The missing element in the third world development program. In *Candida March & Tina Wallace (ends) changingperceptions: New writings on gender and development*. Oxfam.

March, C., Smyth, I., & Mukhopadhyay, M. (1999). *A Guide to Gender Analysis Frameworks*, Oxfam GB Practical Action Publishers. https://policy-practice.oxfam.org.uk/publications/a-guide-to-gender-analysis-frameworks-115397

Nyahunda, L. et al. (2020). Climate change and agriculture in Africa: A review of the literature. *Journal of Climate Change and Sustainable Development, 6*(1), 1–12.

Pandit, J., & Sharma, A. K. (2023). A comprehensive review of CCs' impacts on ecosystems. *Journal of water and CC, 14*(11), 4273–4284.

Paris Agreement. (2015). UNFCCC on CC, DEC 12, 2015.TIAS. No 16-1104.

Petrie, H. G. C. (2010). Gender and climate change: A review of the literature. *Journal of Gender Studies, 19*(2), 147–162.

Ribeiro, N., & Chauque, A. (2014). *Gender and climate change: Mozambique case study*. Heinrich Boll Stiftung.

Rohr, A. (2009). Climate change, gender, and development: A review of the literature. *Journal of Development Studies, 45*(9), 1660–1673.

Ruparanganda, J. E., & Kefasi, J. (2018). Climate change and women's livelihoods in Southern Africa: A review of the literature. *Journal of Climate Change and Sustainable Development, 5*(1), 1–12.

Sabaa, N. J. & Peter, S. (2023). Bolstering women's climate resilience and adaptation through financial services. https://www.cgap.org/research/publication/bolstering-womens-climate-resilience-and-adaptation-through-financial-services

Sendai Framework. (2015). *Sendai framework for Disaster Risk Reduction 2015–2030*. UN. https://www.preventionweb.net/files/43291-sendaiframeworkrdrren.pdf

Shivanna, K. R. (2022). CC and its impact on biodiversity and human welfare. *Proceedings of the Indian National Science Academy, 88*, 160–171.

Southern African Development Community (SADC). (2020). SADC Regional indicative strategic development plan (RISDP) 2020–2063.

Tanner, T. (2022). Youth, gender and climate resilience voices of adolescent and young women in southern Africa. *Sustainability, 14*(14), 8797. https://doi.org/10.3390/su14148797

Tanyag, M., & True, J. C. (2019). *Gender and responsive alternatives to climate change: Monash gender peace and security institute*. Action Aid.

UN Women. (2015). *Climate resilient and empowerment livelihood for womendevelopment*. IGCCAD.

UNESCO. (2021). *Goal 13: Climate action*. Unesco/org/reports/science/2021/site/default.

UNFCCC. (2016). The Paris agreement unfcc. https://unfcc.int/sites/default/files/rescourecs/parisagreement-publication.pdf

UNICEF. (2022). *UNICEF Strategic Plan 2022–2025*. https://www.unicef.org/sites/default/files/2022-02/UNICEF-strategic-plan-2022-2025-publication-EN.pdf

Union of Concerned Scientist. (2022). *What is climate resilience?* https://www.uc.susa.org/resources/what-climateresilience

Urive, J. V. (2016). *Gender and climate change: A closer look at evidence global gender and climate change alliance*. GGCA. https://womendelivery.org/gender-climate-inforgraphic-ggca-factsheets-closer-look-existing-evidence

Van Eeldewijk, A., et al. (2017). *A conceptual model of women and girls empowerment*. KIT Gender. https://www.kit.nl/wp-content/uploads/2028/10rmcif-KIT-whirpaper-web-pdf. Royal Tropical Institute

Watson, R. T. (1996). Climate change: Implications for women. *Journal of International Affairs, 50*(1), 97–116.

Williams, et al. (1994). *Oxford gender training manual*. Oxfam.

15 Prospects and Challenges for Adopting Artificial Intelligence and Digital Tools in Climate Resilience in Rural Southern Africa: A Systematic Review

Lazarus Chapungu, Muchaiteyi Togo, and Ines Raimundo

15.1 Introduction

The impact of climate change on environmental and human systems is undeniably a global concern (Milán-garcía et al., 2021; Warner et al., 2010). The frequency and severity of extreme weather occurrences have increased over the last 10 years, providing evidence that the repercussions are becoming more pronounced (Singh & Goyal, 2023). The dramatic increase in extreme weather events such as floods, heat waves, droughts, and cyclones caused by a changing climate in recent decades has increased the vulnerability of several global industries, including agriculture, energy, and transportation (Kumar et al., 2021). Climate projections indicate that there will be significantly more extreme events in the future, raising the stakes for the socioeconomic system (Kumar et al., 2021; Smirnov et al., 2023). Thus, sectors across the global economy are increasingly becoming more vulnerable to climatic vagaries.

The groups who are most at risk from climate change are also the ones that contribute the least to its causes and depend most on the weather for their way of life. These communities lack access to resources, information, and opportunity for adaptation and resilience building related to climate change (Toplic, 2022). There is consensus among development practitioners that if people's lives are to be safeguarded, their means of subsistence and ecosystems must adapt and become resilient to climate change. Rural areas in Southern Africa have not been spared from the enfeeblements authored by climate change (Nhemachena et al., 2020; Thomas et al., 2022), and the need to build resilience cannot be overemphasized.

To effectively address climate resilience in rural Southern Africa, there is a need to integrate digital technologies into conventional decision-making processes (Singh & Goyal, 2023). This enables rural-based sectors to assess their sensitivity to climate change, anticipate the future impacts, and plan to improve resilience. Digitalization has great potential to assist humanity to build climate resilience, enhance adaptation, and foster sustainability. Digitalization, which includes the application of artificial intelligence (AI) has proven to be effective in a variety

L. Chapungu (✉)
Institute for Corporate Citizenship, University of South Africa, Pretoria, South Africa
e-mail: tchapul@unisa.ac.za

M. Togo
Department of Environmental Sciences, University of South Africa, Pretoria, South Africa
e-mail: togom@unisa.ac.za

I. Raimundo
Eduardo Mondlane University, Maputo, Mozambique

M. Matsa et al. (eds.), *Climate Change Resilience in Rural Southern Africa*, Sustainable Development Goals Series, https://doi.org/10.1007/978-3-031-73600-1_15

of domains. Its applications in climate change can be widely defined to include the use of information technologies to disseminate, process, analyze, and interpret data that can be used by people to make climate-related decisions (Papadopoulos & Balta, 2022). The incorporation of AI and deep learning techniques in risk management may be a viable approach for increasing resilience to climate change. At corporate level, valuable insights can be gained from massive climate datasets by utilizing big data analytics and machine learning algorithms (Alzubaidi et al., 2021). This enables them to evaluate vulnerability and recognize potential risks and repercussions (Perera et al., 2020).

Digitalization is a crucial enabler of the socio-economic dynamics, with the ability to build climate-friendly landscapes and communities (Balogun et al., 2020; Lee & Mwebaza, 2022; Parida et al., 2019). With the arrival of the fourth industrial revolution, there has been a greater use of digital technologies in rural areas, especially within the agricultural sector (Gebresenbet et al., 2023). Digitalization is expected to have a greater impact on rural areas in the future, transforming occupations and causing lifestyle changes with far-reaching consequences that will eventually affect rural areas' resilience and adaptive capacities. Resilience building will enable protection of existing infrastructure, economies, and food security and cultures. Digital technologies (DTs) are emerging as powerful tools that can assist communities, policy makers, climate experts, and researchers to make rural Southern Africa more resilient to climatic change. They have the potential to effectively assess, predict, and mitigate climate risks through efficient data use, sensing devices, and learning algorithms. DT enables stakeholders within the climate change space to understand the trends and dynamics of climate variables from which strategies for resilience building can be designed.

The use of modern technology in agriculture nowadays includes the Internet of Things (IoT), big data analysis, artificial intelligence, cloud computing, remote sensing, and more. By creating low-cost sensor and network platforms aimed at the optimization of production efficiency, coupled with decreases in the utilization of water resources and energy with minimal environmental effects, the adoption of new technologies has considerably enhanced agricultural activities and bolstered climate resilience (Howden et al., 2007; The World Bank, 2019; Ward, 2022; World Bank, 2020). Big data in smart farming enables farmers to make informed decisions by providing extrapolative overviews of current agricultural conditions. Real-time programming is created using AI principles and integrated into IoT devices to assist farmers in making the best decisions.

Significant progress has been made to understand the climate change-DT applications in cities (ADB, 2021; De Coss-Corzo & Luque-Ayala, 2021; Dwivedi et al., 2022). However, a lot still needs to be done for rural areas in Southern Africa. While it has already been shown that there is convenient scope and opportunity to use DTs, including artificial intelligence and Internet of Things (IoTs) to build the adaptive and resilience capacity in cities, there is still need to evaluate the prospects and challenges for rural Southern Africa. Rural Southern Africa has been dangerously exposed to climate risks which have exacerbated their vulnerability. Exploring the prospects and challenges of DT applications to understand climate-related hazards in these areas provides a platform for launching a framework for climate resilience building in Southern Africa. This chapter reviews recent research and applications of AI in climate change resilience building, with an emphasis on identifying opportunities and barriers in the process of integrating DT in the frameworks, strategies, and policies for climate resilience building in Southern African rural areas.

15.2 Methodology

15.2.1 Study Area Description

The study focused on the Southern African region, which is located in the continent's southernmost section. South Africa, Lesotho,

Zimbabwe, Malawi, Eswatini, Botswana, Namibia, Angola, Zambia, and Mozambique are among its members (Fig. 15.1). The interior of Southern Africa is comprised of a series of undulating plateaus that extend into central Angola and encompass the majority of Namibia, Botswana, and South Africa. This is bounded by uplands in Zambia and Zimbabwe. Coastal mountains and escarpments edge the high ground in South Africa, Namibia, Angola, northern Mozambique, and the Mozambique–Zimbabwe border. Coastal lowlands border the Atlantic and Indian Oceans in Mozambique and Namibia, respectively. Figure 15.1 depicts Southern Africa's geographical location.

Climates in the region include seasonal, semi-arid, temperate, and tropical. Droughts are becoming more frequent and severe. The IPCC anticipates that the region will grow drier as the climate system changes. Climate change is also hastening the occurrence of weather extremes like tropical cyclones, heat waves, heavy rains, cold snaps, floods, and dust storms.

15.2.2 Document Analysis Protocol and Screening

The study is premised on the document analysis approach. Only peer-reviewed articles were considered. The documents were screened following the PRISMA framework. The PRISMA approach was chosen for its completeness and broad application. Numerous studies from various academic domains have employed it (Chapungu et al., 2022; Janjua et al., 2021). The procedure included criteria for selecting literature, search phrases, and databases (Cascini et al., 2021). Literature focusing on climate change resilience, artificial intelligence, and digitalization tools was sought. Figure 15.2 shows an overview of the processes followed to come up with a final sample of journal articles that were used in the study.

There are numerous databases with publications that might be used for this investigation. However, only two databases, namely Scopus and ScienceDirect, were selected. The reasons for their selection include that they index double-blind peer-reviewed publications and have been

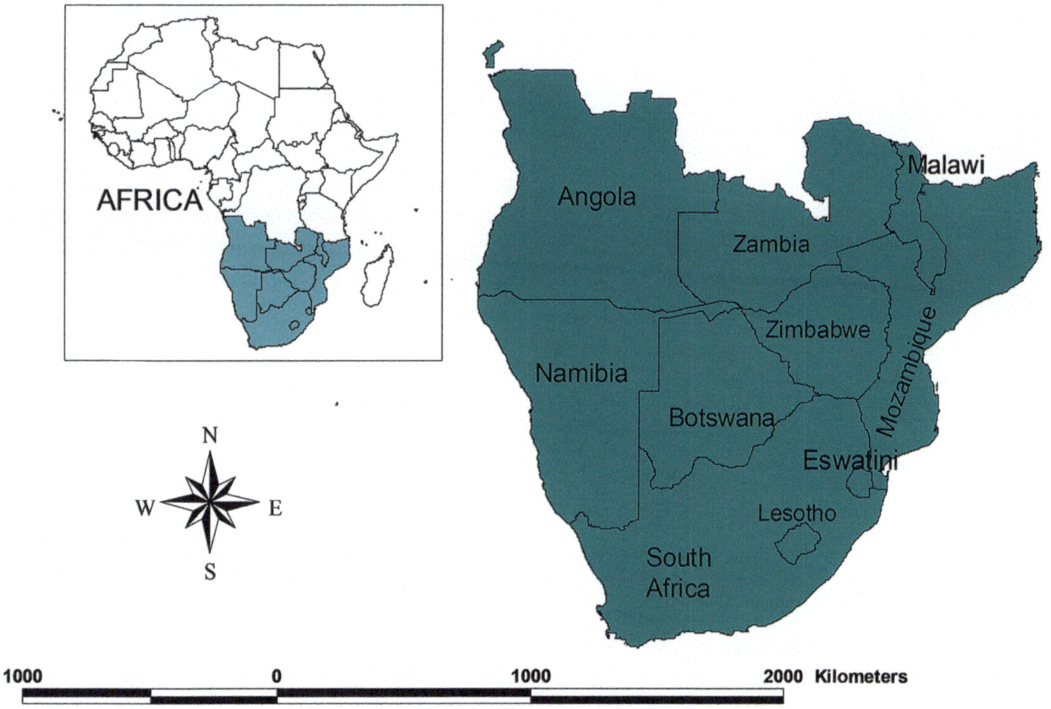

Fig. 15.1 Map of study area, the Southern African region. (Source: Authors)

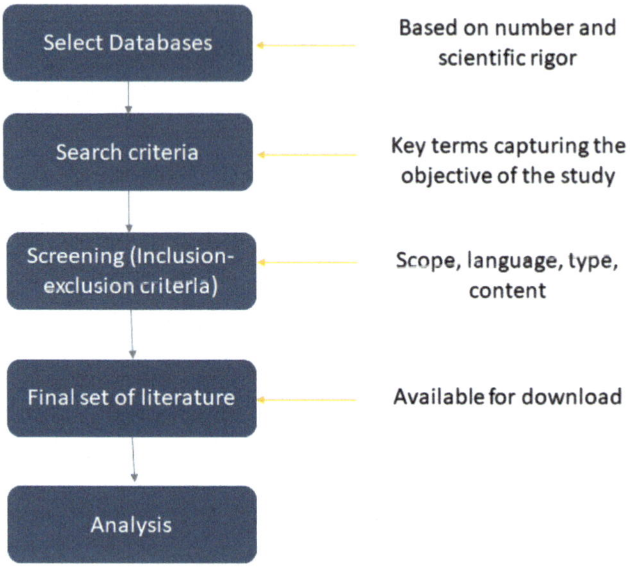

Fig. 15.2 Database selection and article screening process. (Source: Authors)

used widely in other systematic scientific reviews. The following word schemes and Boolean operators were employed:

> TITLE-ABS-KEY ("Climate Change" AND (ict OR it OR "Artificial intelligence" OR digital AND technology) southern AND Africa) AND (LIMIT-TO (DOCTYPE , "ar") OR LIMIT-TO (DOCTYPE , "re")) AND (LIMIT-TO (LANGUAGE , "English"))

The following inclusion and exclusion criteria were used to identify relevant sources of information from the selected databases:

- The geographic scope is confined to the Southern African region.
- The content focuses on artificial intelligence and digitalization in the context of climate change.
- Include only content addressing rural areas and populations.
- Exclude books and book chapters.
- Include only research articles and reviewed organizational reports.

15.2.2.1 Screening and Computation of the Coefficient of Agreement

The screening of the research was done in two steps. Title and abstract were used as the basis for the initial screening process, and studies that passed this stage were exposed to a second round of screening using the full texts. Data were taken from the study and evaluated for the quality of the evidence and extracted in accordance with a predesigned coding instrument. The screening was done independently by three researchers (one author and two hired experts in the field of climate Science). Following independent screening, the results were compared and the coefficient of agreement was computed. The level of agreement was 0.82%, which reflects high levels of agreement. All the documents that the assessors agreed qualified were further read to extract and analyze the content.

15.2.2.2 Data Extraction

A predesigned template was used to capture themes, content, and statistics of importance from each qualifying document. The major themes included prospects of climate-smart digital technologies, challenges and barriers to the implementation of climate-smart digital technologies, climate change mitigation digital technologies, climate change adaptation digital technologies, and applications of artificial intelligence to understand climate-related phenomena.

15.3 Results and Discussions

From the systematic literature review process, a total of 300 research articles were identified through the use of a predefined search criteria and 12 other records were identified through random google search. The articles were initially screened based on a predetermined inclusion and exclusion criteria, resulting in 231 studies deemed unsuitable for further analysis due to duplications, automated ineligibility, and failure to meet the set criteria. This resulted in 90 documents undergoing further screening using refined screening criteria. Additional 10 articles were removed after repeated screening. Of the 80 remaining studies, 7 were not accessible and 73 went through further screening focusing on the geographic and conceptual scope. This resulted in 52 articles being removed and 21 qualified for final validation and analysis. Figure 15.3 shows the flow diagram with details on how the screening process was done.

From the 12 reports identified from random Google search, 3 reports were irretrievable and 5 were excluded due to irrelevant conceptual scope. This resulted in 4 of the reports qualifying for final validation and analysis. Thus, the total number of documents included in the final analysis was 24. The small number of qualifying articles is an indication that artificial intelligence and digitalization tools in the context of climate change resilience in rural Southern Africa is an area that still requires academic attention. Most of the qualifying studies considered in this study were not very explicit about the role of digital and AI tools in climate resilience building in Southern African rural areas but implicitly, the analysis extrapolated the challenges and prospects for using AI and digital tools.

15.3.1 Digital and AI Technology Applications in Climate Resilience in Southern Africa

It is projected that Southern African nations would experience greater damage as a result of climate change. Agriculture, water, mining, and tourism are just a few of the crucial economic sectors in Southern Africa that heavily rely on natural resources and are very sensitive to the effects of climate change. In rural areas,

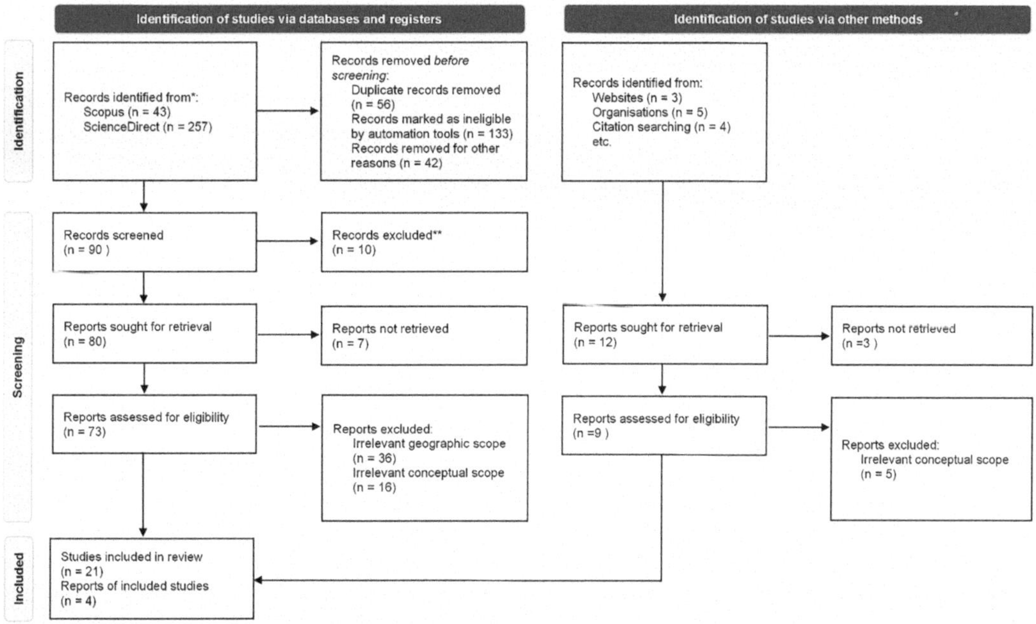

Fig. 15.3 Prisma flow diagram for the study. (Source: Authors)

development issues such as food insecurity, disease outbreaks, water scarcity, humanitarian disasters, and poverty are inextricably linked to climate change. Therefore, strengthening climate change adaptive capacity and overall resilience is essential. Digital technologies and AI tools have a pivotal role to play in enabling to resilience status for rural communities. They have the capability to improve the efficiency of climate communication, better decision making using real-time environmental information, enable access of historically marginalized societies and groups to climate information, and enhance collaboration and effective coordination among partners working on climate issues.

An in-depth analysis of the studies and reports produced by non-governmental organizations working in rural Southern Africa shows that there are three main approaches in which digital and AI technologies are applied in climate resilience (Aruleba & Jere, 2022; Mayoyo et al., 2023; Toplic, 2022). This includes application in anticipatory action for climate resilience, climate advisory services, and climate mitigation efforts (Toplic, 2022). For effective climate resilience building, the applications consider community-led initiatives, gender balance, race balance, poverty alleviation, collaboration among stakeholders, and sustainability of initiatives (Toplic, 2022). Figure 15.4 shows the digital applications in climate resilience building as reflected in most of the literature.

Anticipatory action is increasingly viewed as essential for mitigating and lowering the severe effects, such as loss of life and damage to property, assets, and the environment, as climate change continues to threaten human lives, nature, and economy. Communities can save lives, stop loss and damage, and maintain livelihoods when they have the knowledge to prepare for climate shocks and the means to put adaptation measures into place.

One of the main applications of digital technologies is climate risk assessments (International Telecommunication Union (ITU), 2019; Kumar et al., 2021; Lee & Mwebaza, 2022). Climate risk assessments make it possible to estimate the potential effects of climate change and improvise ways of building resilience. Digital tools are used to evaluate the possible effects of climate change on rural farmers and design strategies for building resilience. Some scholars in Southern Africa have used climate modelling and simulation to inform anticipatory climate action, including resilience building. They help to give decision makers crucial knowledge about the potential effects of climate change. Vast amounts of environmental data collected from ground-based observations and remote sensing, including precipitation, temperature, and other meteorological variables, are analyzed using artificial intelligence and machine learning algorithms to determine the temporal and spatial trends of susceptibility to climate change.

The Mozambican National Institute of Health is collaborating with climate scientists from the Future Climate for Africa programme (FCFA) to create a digital tool that will assist the local administration in better tracking and anticipating disease hotspots by utilizing a vast database of weather and climate-related data (Rael, 2019). In Zimbabwe, the Zimbabwe Resilience Building Fund (ZRBF) programme has promoted digital weather stations to assist farmers to monitor weather patterns and determine agriculture dynamics in their areas. The same programme has made significant strides in promoting e-extension through smartphone app, an e-marketing app, and a finance app. These initiatives have seen progress in the deployment of digital technologies to track groundwater extraction and prevent over-abstraction. Another digital application introduced by the Food and Agriculture Organization (FAO), the Kurima Mari App, is used by a group of smallholder farmers for group savings and loans, as part of the Livelihoods and Food Security Programme (LFSP). Members need to invest in Internet and data to use the app (Mayoyo et al., 2023).

Using Unstructured Supplementary Service Data (USSD) technology, a patient monitoring system was developed in South Africa for sparsely inhabited rural areas in order to support the model. The suggested system is driven by decision support. It assists medical professionals (nurses, doctors) in determining the best course

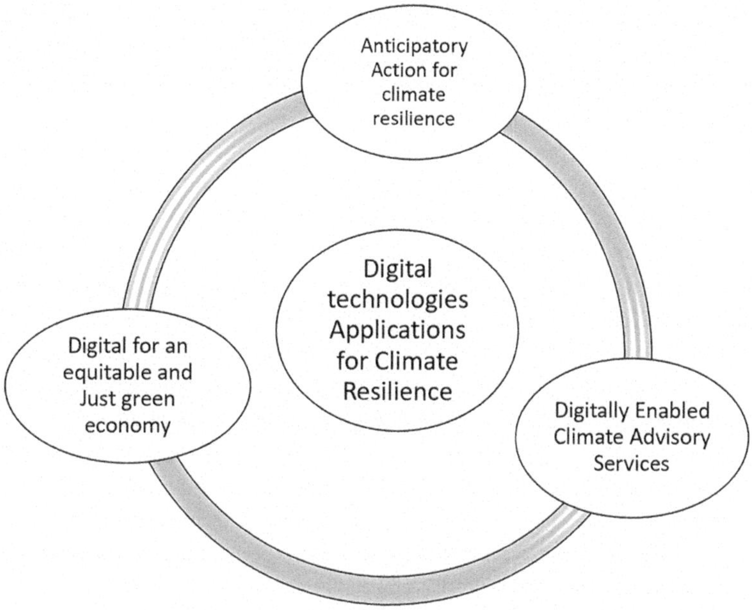

Fig. 15.4 Digital and AI tools applications in climate resilience building in Southern Africa. (Source: Authors, adapted from Toplic (2022))

of action or additional care based on the vital signs of patients who get routine telemonitoring. Such technology help reduce exposure to climate-related hazards as patients reduce travel to distant medical centres (Barjis et al., 2013). In the Eastern Cape province of South Africa, digital technologies have proved key in providing telemedicine services (Cilliers & Flowerday, 2014), a service which contributes in reducing vulnerability to climate-induced hazards and spillover effects.

In Namibia, the establishment of AI and digital technology infrastructure in rural areas has been the government's top priority to close the digital divide between the country's 52% rural and 48% urban populations (Government of Namibia, 2020). The country has a 120% mobile phone penetration rate, more than that of the world (67%) and Africa (81%) (Kamutuezu et al., 2021). The Internet connection rate is 51% (Government of Namibia, 2020), making it feasible to develop digitalization initiatives to build resilience against climate change impacts in rural areas.

In Mozambique, more frequent and severe rainstorms, flash floods, and water stagnation have caused outbreaks of malaria and cholera, particularly in areas with poor planning and high population density (Rael, 2019). This poses a serious health threat to communities. However, digital technologies and AI tools have proved to be instrumental in reducing vulnerability of communities to climate related hazards. A computerized dashboard that uses weather forecasts as a basis has been created to facilitate the planning and preparation for the spread of vector-borne and waterborne illnesses. Even more, it can help users locate the locations of suggested medical facilities and predict illness outbreaks in certain high-risk areas by 10–20 days. The digitization of meteorological data is rapidly growing in Mozambique. The National Institute of Meteorology (INAM) has successfully archived a significant amount of historical climate data and has assisted technical staff members in honing their digital archiving tool skills. The goal of the initiative is to improve climate data quality, availability, and accessibility to support risk management and adaptation in many industries and geographical areas.

In Zimbabwe, South Africa, Botswana, and Mozambique, farmers have largely relied on digital tools to support prompt, informed decision making in order to increase preparedness for climatic hazards. Early warning systems and other data-generating tools have been used to map upcoming calamities, such as the drought, and digital tools like mobile phones, radios, TVs, newspapers, and social media platforms have been used to disseminate information and increase disaster preparedness (Kareem et al., 2020; Makate et al., 2019). Farmers have also been receiving updates on prices and weather. The use of community radios that broadcast in their local tongues has increased farmers' com prehension of the messages and willingness to heed guidance (Mayoyo et al., 2023). Table 15.1 shows some of the types of AI and digital infrastructure and their applications in Southern Africa.

Numerous studies have demonstrated how the Internet of Things (IoT) has sped up innovation and disruptive ideas that are redefining and transforming sectors (Balogun et al., 2020; Dhanaraju et al., 2022; Ferrández-Pastor et al., 2016; Lee & Mwebaza, 2022; Ngcakani, 2023; Parida et al., 2019). The Southern African rural communities have also benefited from this wave of innovation, albeit at a pace slower than in urban areas and most parts of the world. Its ability to seamlessly merge the digital and physical realms to generate intelligent solutions keeps it at the forefront of the digital trend landscape. Energy conservation, natural catastrophe monitoring, yield measurement and optimization, and other applications are just a few of the ways that IoT is significantly improving corporate operations and agricultural ecosystems. The industry still needs to overcome a few obstacles before the IoT can completely fulfil its promise in the near and long term.

15.3.2 Digitalization Challenges in Rural Southern Africa

Several studies (e.g. ADB, 2021; Argyroudis et al., 2022; Balogun et al., 2020; George et al., 2021; International Telecommunication Union (ITU), 2019; Mayoyo et al., 2023; Toplic, 2022) have demonstrated how the use of digital technology has become crucial for promoting climate change adaptation, assisting businesses and communities in understanding and preparing for climate change threats. However, analysis of the existing literature shows that the Southern African region is saddled with unyielding challenges that inhibit effective use and implementation of AI and digitalization tools in climate resilience building. The challenges include lack of requisite infrastructure and lack of knowledge and skills.

For Zimbabwe, Mayoyo et al. (2023) illustrated that adaptation and resilience building efforts have been made in the agricultural sector using digital technologies, but the country's agricultural population is still saddled with challenges which include lack of ICT proficiency, lack of access to smart gadgets, inaccessibility of some areas, lack of access to energy to power AI and digital technologies, and unreliable network services from the network providers.

Since the majority of digital technologies rely on energy to function, resolving energy-related concerns is necessary to create an atmosphere that makes it easier for these technologies to be adopted. Through enhanced information access, lowering the cost of instruments like phones and sim cards presents prospects for accelerating digitalization. To upgrade the ICT infrastructure and increase coverage in underserved areas, however, wise investment is still required. Funding for research and development of locally created digital tools is required; doing so could lower the price of technology, making them more affordable for more people, particularly women. Zimbabwe is promoting the use of both conventional and scientific early warning systems in an effort to increase community awareness of climate change and decrease its effects. Though they are the custodians of indigenous knowledge systems (IKS), women in particular in communities have limited access to and proficiency integrating ICTs for digital innovations and interventions within the IKS.

15.3.3 Prospects for Digital and AI Tools for Climate Resilience in Rural Southern Africa

Several studies have indicated that Southern African rural areas are generally ripe for advancement in digital technologies and AI applications (Aruleba & Jere, 2022; Mayoyo

Table 15.1 Types of AI and digital technology infrastructures and their applications in rural southern Africa

Infrastructure	Countries	Applications	Enabling factors	Disenabling factors	Source
Remote sensing technologies, for example, drones, satellites, satellite-receiving stations	Zimbabwe, South Africa, Namibia, Botswana, Mozambique	Examining environmental variables for climate resilience strategies Climate vulnerability assessments Climate impact assessment Veld fire monitoring	Digital literacy Training of farmers by NGOs International support Availability of drone infrastructure	No access to reliable sources of energy High cost to connect to the Internet Limited access to high-speed Internet Ineffective telecommunication links	(Mayoyo et al., 2023) (World Bank, 2023) (Matarira et al., 1996)
Mobile phones	Malawi, Zimbabwe, South Africa, Namibia, Botswana, Mozambique, Angola, Lesotho, Eswatini	Health delivery services Climate and weather information dissemination Early warning Agricultural information dissemination	Digital literacy Training of farmers by NGOs International support Government policy	High cost to connect to the Internet Limited access to high-speed Internet	(Aruleba & Jere, 2022) (Mokwena, 2021) (Kyobe, 2011)
Machine learning	Malawi, Zimbabwe, South Africa, Namibia, Botswana, Mozambique, Angola, Lesotho, Eswatini	Climate projections Analyzing satellite imagery to detect changes in land use and vegetation	Digital literacy International support Government policy Universities participation		(World Bank, 2023) (Kyobe, 2011) (Mayoyo et al., 2023)
Deep learning	Malawi, Zimbabwe, South Africa, Namibia, Botswana, Angola, Mozambique, Lesotho, Eswatini	Monitoring deforestation, guiding land-use planning			(Mayoyo et al., 2023) (World Bank, 2023)
The Internet of Things	Malawi, Zimbabwe, South Africa, Namibia, Botswana, Mozambique, Angola, Lesotho, Eswatini	Data collection, communication, processing, and actionable intelligence by farmers, agricultural extension officers, climate disaster experts, and other stakeholders	Increased penetration of ICT infrastructure including mobile devices and mobile data in rural areas	Inadequate access to ICT infrastructure Sociocultural norms, unemployment, and lack of digital skills	(Aruleba & Jere, 2022) (World Bank, 2023)

(continued)

Table 15.1 (continued)

Infrastructure	Countries	Applications	Enabling factors	Disenabling factors	Source
Global positioning systems	Zimbabwe, South Africa, Botswana	Monitor food and water resources			(Mayoyo et al., 2023) (World Bank, 2023)
PCs, servers, mainframes, and network databases	Malawi, Zimbabwe, South Africa, Namibia, Botswana, Mozambique, Angola, Lesotho, Eswatini	Food security mapping			(World Bank, 2023)
Natural language processing	Zimbabwe, South Africa, Botswana	Social media monitoring, for disaster response, Tweets, posts, and messages related to disasters, analysis of public sentiment, identification of people in need, and coordination of rescue efforts	Expertise still limited		(Jain et al., 2023) (Mayoyo et al., 2023)

Source: Authors

et al., 2023; Mokwena, 2021) . The increase in network coverage, number of households with mobile phones, rural electrification programmes, increased literacy rates, extension work by agriculture departments, and the general increase in awareness on digital applications create enormous opportunities for the use of digital platforms and AI tools.

There are prospects for leveraging digital tools and AI technologies in increasing access to climate information. This enhances proactive strategies that increase resilience to climate change. There are currently mobile applications in use that offer weather forecasts, trends, and notifications to communities about possible hazards. Real-time environmental data collection is being facilitated by the development of community-based monitoring systems, which enable communities to get ready for any climate-related hazards.

Southern African countries have policies that promote the use of ICT and these can be leveraged to support climate resilience initiatives. There has been a drive towards digital governance, with South Africa having the most advanced digital governance infrastructure which has also contributed to building climate resilience among rural population (CCARDSA, 2022). Other areas enhanced by the policy environment include innovation-driven entrepreneurship, digital skills, G5 digital economy, ICT infrastructure, and digital business. Box 15.1 presents an example of ICT policy objectives that are linked to climate change resilience building in rural areas. Although the policies were put in place to digitalize all facets of the economy, rural population greatly benefits from them in building their resilience to climate change. For example, Zambia's ICT policy sets out to achieve goals that enable ICT to support rural farmers in increasing their outputs and improve the flow of information to enhance awareness regarding extreme climatic events (CCARDSA, 2022) (Box 15.1).

Box 15.1 Zambia's ICT Policy Objectives Linked to Climate Resilience Building

The policy sets out the following strategies to achieve those goals:

- Create an integrated agricultural information system on agro-technologies and techniques, pricing, and market information for all agro-products to provide strategic information for farmers, government authorities, and other stakeholder at national, provincial, and district levels.
- Undertake intensive ICT awareness campaigns for all types of farmers in the use of traditional and new ICT tools at all levels.
- Intensify the use of radio and TV programmes and integrate new technologies to reach extension workers and farmers alike.
- Develop weather and agrometeorology early warning systems to support agricultural production and predict as well as prevent disasters.
- Develop incentives for deployment of affordable ICT solutions to support rural connectivity of farmers especially those within the catchments of existing ICT infrastructure.
- Develop and promote ICT skills development among agricultural extension workers and farmers.
- Develop database systems and applications including GIS to support agricultural input resource management as well as to support land and water resource management, environmental monitoring and impact assessment, crop yield assessment, and livestock management.
- Develop a monitoring and evaluation system for the conservation and sustainable utilization of natural resources in the agricultural production process.
- Promote two-way information dissemination to support the physical and socio-economic planning process in the agricultural systems.

Source: CCARDSA (2022)

15.3.4 Discussion

The existing literature has clearly shown that there are inordinate prospects for artificial intelligence and digital tools in rural Southern Africa to boost resilience to climate change. The Internet of Things (IoT) is a general future development in rural Southern Africa. Through local and worldwide, frequently wireless network infrastructures, various smart gadgets are connected and interact with one another. Precision agriculture can be seen as an example of this evolution, and it is frequently cited as a key factor driving Big Data (Wolfert et al., 2017), with the ultimate goal of attaining food security to boost resilience to climate change. The anticipated outcome of this is a significant shift in the management of rural livelihoods due to the availability of clear information and the ability to make decisions that were previously unattainable from a technical or financial standpoint.

A massive wave of digitally enabled climate adaptation programmes is underway, primarily spearheaded by non-governmental organizations (Toplic, 2022). These programmes aim to improve decision making with real-time information, reach more communities more quickly and affordably, give historically marginalized groups access to resources, information, and opportunities, and foster more effective and inclusive partnerships and harmonization of interventions.

The push for digitalization of systems throughout the region has been advantageous for Southern African rural communities. Hazard maps, which offer details on significant risks like droughts and floods and can assist farmers in better addressing such risks, are made using digitalization data. The availability of scientific data on weather and climate helps to improve the forecasting of natural disasters, which helps to improve early warning systems (EWS) and emergency response. Digitally preserved local data can also contribute to the expanding body of knowledge on climate change, helping to build more efficient national, regional, and global climate information networks. This has also made it possible to plan and execute adaptation programmes based on scientific evidence as well as to formulate climate policy.

Digital technologies have already increased the capacity of humanitarian actors, enhanced service delivery, and increased access to resources and information. The digital revolution is bringing about a number of benefits, including the removal of uncertainty brought about by climate variability, the growth of resilience through data-driven recommendations, the delivery of life-saving information, and the mapping of hazardous regions and alerting about approaching natural disasters. It is evident that when used responsibly and appropriately in the context of the community, digital technology can have a transformative effect in aiding communities in preparing for and enduring climate shocks as well as constructing longer-term resilience. However, more research is required to fully understand the role of digital technology in climate adaptation of particular vulnerable communities.

There are opportunities for developing information modelling for climate change adaptation given the advancements in infrastructure, rural electricity, and wireless access in several rural Southern African locations. This makes it more likely that infrastructure and buildings will be constructed to withstand increasingly common extreme climate change-related disasters like heat waves and flooding. Strategies to avert the effects of extreme weather events include the use of cutting-edge materials, the creation of green spaces, and the application of certain construction methods. Digital copies of physical assets can help assess the consequences of concurrent occurrences and determine the most effective ways to improve resilience.

It emerged from this study that a plethora of challenges still exist, inhibiting the development of AI and digital tools for climate resilience building in rural Southern Africa. However, solutions exist for each of these challenges, but it takes political will and leadership commitment to enable quick adoption of digital technologies and AI tools to bolster the resilience of communities to climate change. Since the majority of digital technologies rely on energy to function, resolving energy-related concerns is necessary to create an atmosphere that makes it easier for these technologies to be adopted. Through enhanced information access, lowering the cost of instruments like phones and sim cards presents prospects for accelerating digitalization. To upgrade the ICT infrastructure and increase coverage in underserved areas, however, wise investment is still required. Funding for research and development of locally created digital tools is required; doing so could lower the price of technology, making them more affordable for more people, particularly women.

15.3.5 Conclusions

This study sought to examine the prospects and challenges for using artificial intelligence and digital tools for climate resilience building in rural Southern Africa. It has shown that digital technologies and AI tools are widely used to increase farmers' resilience directly and indirectly to climate change induced challenges. Digitalisation has facilitated climate information flows among stakeholders and rural-based populations, predicting weather conditions, performing risk assessments and enabled evacuations following climate induced disasters. Several conditions make this possible and there are prospects for improvement. Thus, the increase in network coverage, number of households with mobile phones, rural electrification programmes, increased literacy rates, extension work by agriculture departments and the general increase in awareness on digital applications create enormous opportunities for the use of digital platforms and AI tools. The policy environment has created an opportunity for farmers to improve their digital skills and increase the use of ICT to overcome challenges posed by climate change.

However, a number of drawbacks for using digital and AI tools have been noted. These include lack of access to energy, lack of digital literacy in some communities, lack of ICT proficiency, lack of access to smart gadgets, inaccessibility of some areas, lack of access to energy to power AI and digital technologies, and unreliable network services from the network providers. The emerging digital technologies have been noted as a threat to citizens' security and issues of misinformation have been associated with digitalization. Instead of boosting the communities' resilience, digitalization can be a driver of vulnerability through misinformation and cyberattacks. These technologies remain heavily reliant on the power supply and the interdependencies between their operators. Additionally, there is an involved tension in human–machine interaction during decision making. This is due to the distrust that the technologies have users' best interests at heart, which often stems from the challenges in explaining and interpreting digital data into actionable and human terms.

Despite the challenges that exist in the digital space, there is a general increase in the use of digital technologies in rural Southern Africa with further prospects for expanding the use of ICT in enhancing the capacity of rural farmers to withstand the threats of climate change. There is need for continued government support through infrastructure development. The role of non-governmental organizations cannot be underestimated as they have contributed significantly in enhancing the digitalization of rural areas especially in supporting agricultural initiatives that build resilience to climate change. There is a need for governments in Southern Africa to enhance cooperation among stakeholders through enacting supportive sector-wide policies and providing the necessary infrastructure to increase resilience of rural populations to climate-induced phenomena and reduce damage and loss associated with climate change.

References

ADB. (2021). *Digital technologies for climate action, disaster resilience, and environmental sustainability.* https://www.adb.org/sites/default/files/publication/700396/digital-technologies-climate-action.pdf

Alzubaidi, L., Zhang, J., Humaidi, A. J., Al-Dujaili, A., Duan, Y., Al-Shamma, O., Santamaría, J., Fadhel, M. A., Al-Amidie, M., & Farhan, L. (2021). Review of deep learning: concepts, CNN architectures, challenges, applications, future directions. *Journal of Big Data, 8*(1). Springer International Publishing. https://doi.org/10.1186/s40537-021-00444-8

Argyroudis, S. A., Mitoulis, S. A., Chatzi, E., Baker, J. W., Brilakis, I., Gkoumas, K., Vousdoukas, M., Hynes, W., Carluccio, S., Keou, O., Frangopol, D. M., & Linkov, I. (2022). Digital technologies can enhance climate resilience of critical infrastructure. *Climate Risk Management, 35*(November 2021), 100387. https://doi.org/10.1016/j.crm.2021.100387

Aruleba, K., & Jere, N. (2022). Exploring digital transforming challenges in rural areas of South Africa through a systematic review of empirical studies. *Scientific African, 16*, e01190. https://doi.org/10.1016/j.sciaf.2022.e01190

Balogun, A.-L., Marks, D., Sharma, R., Shekhar, H., Balmes, C., Maheng, D., Arshad, A., & Salehi, P. (2020). Assessing the Potentials of Digitalization as a Tool for Climate Change Adaptation and Sustainable Development in Urban Centres. *Sustainable Cities and Society, 53*, 101888. https://doi.org/10.1016/j.scs.2019.101888

Barjis, J., Kolfschoten, G., & Maritz, J. (2013). A sustainable and affordable support system for rural healthcare delivery. *Decision Support Systems, 56*, 223–233. https://doi.org/10.1016/j.dss.2013.06.005

Cascini, F., Pantovic, A., Al-Ajlouni, Y., Failla, G., & Ricciardi, W. (2021). Attitudes, acceptance and hesitancy among the general population worldwide to receive the COVID-19 vaccines and their contributing factors: A systematic review. *EClinicalMedicine, 40*, 101113. https://doi.org/10.1016/j.eclinm.2021.101113

CCARDSA. (2022). Assessment of Digitalization in the Agricultural Systems of the SADC Region, Supplement to the Situational Analysis Report. Digital Agriculture Country Study Annex: Zambia. https://www.ccardesa.org.knowledge-products

Chapungu, L., Nhamo, G., Chikodzi, D., & Maoela, M. A. (2022). BRICS and the Race to Net-Zero Emissions by 2050: Is COVID-19 a Barrier or an Opportunity? *Journal of Open Innovation: Technology, Market, and Complexity, 8*(4). https://doi.org/10.3390/joitmc8040172

Cilliers, L., & Flowerday, S. (2014). User acceptance of telemedicine by health care workers a case of the Eastern cape province, South Africa. *Electronic Journal of Information Systems in Developing Countries, 65*(1), 1–10. https://doi.org/10.1002/j.1681-4835.2014.tb00467.x

De Coss-Corzo, A., & Luque-Ayala, A. (2021, May). Academic Literature Review: Smart cities and digital technologies in DRR. pp. 1–60.

Dhanaraju, M., Chenniappan, P., Ramalingam, K., Pazhanivelan, S., & Kaliaperumal, R. (2022). Smart Farming: Internet of Things (IoT)-Based Sustainable Agriculture. *Agriculture (Switzerland), 12*(10), 1–26. https://doi.org/10.3390/agriculture12101745

Dwivedi, Y. K., Hughes, L., Kar, A. K., Baabdullah, A. M., Grover, P., Abbas, R., Andreini, D., Abumoghli, I., Barlette, Y., Bunker, D., Chandra Kruse, L., Constantiou, I., Davison, R. M., De', R., Dubey, R., Fenby-Taylor, H., Gupta, B., He, W., Kodama, M., et al. (2022). Climate change and COP26: Are digital technologies and information management part of the problem or the solution? An editorial reflection and call to action. *International Journal of Information Management, 63*, 102456. https://doi.org/10.1016/j.ijinfomgt.2021.102456

Ferrández-Pastor, F. J., García-Chamizo, J. M., Nieto-Hidalgo, M., Mora-Pascual, J., & Mora-Martínez, J. (2016). Developing ubiquitous sensor network platform using internet of things: Application in precision agriculture. *Sensors (Switzerland), 16*(7). https://doi.org/10.3390/s16071141

Gebresenbet, G., Bosona, T., Patterson, D., Persson, H., Fischer, B., Mandaluniz, N., Chirici, G., Zacepins, A., Komasilovs, V., Pitulac, T., & Nasirahmadi, A. (2023). A concept for application of integrated digital technologies to enhance future smart agricultural systems. *Smart Agricultural Technology, 5*, 100255. https://doi.org/10.1016/j.atech.2023.100255

George, G., Merrill, R. K., & Schillebeeckx, S. J. D. (2021). Digital sustainability and entrepreneurship: How digital innovations are helping tackle climate change and sustainable development. *Entrepreneurship: Theory and Practice, 45*(5), 999–1027. https://doi.org/10.1177/1042258719899425

Government of Namibia. (2020). *Republic of Namibia Broadband Policy*. chrome-extension://efaidnbmnnnibpcajpcglchttps//www.npc.gov.na/wp-content/uploads/2022/06/NamibiaBroadband-Policy-2018.pdf

Howden, S. M., Soussana, J. F., Tubiello, F. N., Chhetri, N., Dunlop, M., & Meinke, H. (2007). Adapting agriculture to climate change. *Proceedings of the National Academy of Sciences of the United States of America, 104*(50), 19691–19696. https://doi.org/10.1073/pnas.0701890104

International Telecommunication Union (ITU). (2019). Turning digital innovation into climate action. In https://www.itu.int/en/publications/Documents/tsb/2019-Turning-digital-technology-innovation-into-climate-action/mobile/index.html#p=1

Jain, H., Dhupper, R., Shrivastava, A., Kumar, D., & Kumari, M. (2023). AI-enabled strategies for climate change adaptation: protecting communities, infrastructure, and businesses from the impacts of climate change. *Computational Urban Science, 3*(1). https://doi.org/10.1007/s43762-023-00100-2

Janjua, Z., Krishnapillai, G., & Rahman, M. (2021). A systematic literature review of rural homestays and sustainability in tourism. *SAGE Open, 11*(2). https://doi.org/10.1177/21582440211007117

Kamutuezu, E. U., Winschiers-Theophilus, H., & Peters, A. (2021). *An exploration of factors influencing the adoption of ICT enabled entrepreneurship applications in Namibian rural communities*. http://arxiv.org/abs/2108.09789.

Kareem, B., Lwasa, S., Tugume, D., Mukwaya, P., Walubwa, J., Owuor, S., Kasaija, P., Sseviiri, H., Nsangi, G., & Byarugaba, D. (2020). Pathways for resilience to climate change in African cities. *Environmental Research Letters, 15*(7). https://doi.org/10.1088/1748-9326/ab7951

Kumar, N., Poonia, V., Gupta, B. B., & Goyal, M. K. (2021). A novel framework for risk assessment and resilience of critical infrastructure towards climate change. *Technological Forecasting and Social Change, 165*(December 2020). https://doi.org/10.1016/j.techfore.2020.120532

Kyobe, M. (2011). Investigating the key factors influencing ICT adoption in South Africa. *Journal of Systems and Information Technology, 13*(3), 255–267. https://doi.org/10.1108/13287261111164844

Lee, W. J., & Mwebaza, R. (2022). Digitalization to achieve technology innovation in climate technology transfer. *Sustainability (Switzerland), 14*(1). https://doi.org/10.3390/su14010063

Makate, C., Makate, M., Mango, N., & Siziba, S. (2019). Increasing resilience of smallholder farmers to climate change through multiple adoption of proven climate-smart agriculture innovations. Lessons from Southern Africa. *Journal of Environmental Management, 231*, 858–868. https://doi.org/10.1016/j.jenvman.2018.10.069

Matarira, C. H., Kamukondiwa, W., Mwamuka, F. C., Makadho, J. M., & Unganai, L. S. (1996). Vulnerability and Adaptation Assessments for Zimbabwe. In J. B. Smith, S. Huq, S. Lenhart, L. J. Mata, I. Nemešová, & S. Toure (Eds.), *Vulnerability and adaptation to climate change. Environmental Science and Technology Library* (Vol. 8). Springer. https://doi.org/10.1007/978-94-017-3653-4_6

Mayoyo, A., Chapoto, A., Matchaya, G., Aheeyar, M., & Chiwunze, G. (2023). Digital Climate Adaptation in Agriculture Profile for Zimbabwe June 2023 Acknowledgement. *June*.

Milán-garcía, J., Caparrós-martínez, J. L., Rueda-lópez, N., & Valenciano, J. D. P. (2021). Climate change-induced migration: A bibliometric review. *Globalization and Health, 3*, 1–10.

Mokwena, T. (2021). More South African SMES are going digital due to COVID-19. *The Brave Ones*. https://ewn.co.za/2021/12/20/more-south-african-smes-are-going-digital-due-to-covid-19

Ngcakani, N. (2023). *How the Internet of Things is transforming agri*. Food For Mzanzi. https://www.foodformzansi.co.za/how-the-internet-of-things-is-transforming-agri/

Nhemachena, C., Nhamo, L., Matchaya, G., Nhemachena, C. R., Muchara, B., Karuaihe, S. T., & Mpandeli, S. (2020). Climate change impacts on water and agriculture sectors in southern africa: Threats and opportunities for sustainable development. *Water (Switzerland), 12*(10), 1–17. https://doi.org/10.3390/w12102673

Papadopoulos, T., & Balta, M. E. (2022). Climate Change and big data analytics: Challenges and opportunities. *International Journal of Information Management, 63*. https://doi.org/10.1016/j.ijinfomgt.2021.102448

Parida, V., Sjodin, D., & Reim, W. (2019). Reviewing literature on digitalization, business model innovation, and sustainable industry: Past achievements and future promises. *Sustainability, 11*(2). https://doi.org/10.3390/su11020391

Perera, D., Agnihotri, J., Seidou, O., & Djalante, R. (2020). Identifying societal challenges in flood early warning systems. *International Journal of Disaster Risk Reduction, 51*. https://doi.org/10.1016/j.ijdrr.2020.101794

Rael, L. (2019). *Future climate for Africa. Digital dashboard brings early-warning tool for disease outbreaks in Maputo (blog)*. https://futureclimateafrica.org/news/digitaldashboard-%0Abrings-early-warning-tool-for-disease-outbreaks-in-maputo/

Singh, S., & Goyal, M. K. (2023). Enhancing climate resilience in businesses: The role of artificial intelligence. *Journal of Cleaner Production, 418*(April), 138228. https://doi.org/10.1016/j.jclepro.2023.138228

Smirnov, O., Lahav, G., Orbell, J., Zhang, M., & Xiao, T. (2023). Climate change, drought, and potential environmental migration flows under different policy scenarios. *International Migration Review, 57*(1), 36–67. https://doi.org/10.1177/01979183221079850

The World Bank. (2019). *Climate -smart agriculture investment plan, Zimbabwe*. www.worldbank.org

Thomas, T. S., Robertson, R. D., Strzepek, K., & Arndt, C. (2022). Extreme events and production shocks for key crops in Southern Africa under climate change. *Frontiers in Climate, 4*(May). https://doi.org/10.3389/fclim.2022.787582

Toplic, L. (2022). *Digital for climate adaptation and resilience of vulnerable communities first installment*.

Ward, F. A. (2022). Enhancing climate resilience of irrigated agriculture: A review. *Journal of Environmental Management, 302*(PA), 114032. https://doi.org/10.1016/j.jenvman.2021.114032

Warner, K., Hamza, M., Oliver-Smith, A., Renaud, F., & Julca, A. (2010). Climate change, environmental degradation and migration. *Natural Hazards, 55*(3), 689–715. https://doi.org/10.1007/s11069-009-9419-7

Wolfert, S., Ge, L., Verdouw, C., & Bogaardt, M. J. (2017). Big data in smart farming—A review. *Agricultural Systems, 153*, 69–80. https://doi.org/10.1016/j.agsy.2017.01.023

World Bank. (2020). *Climate-smart agriculture implementation brief climate-smart agriculture implementation brief*.

World Bank. (2023). *Zimbabwe: Risk, historical hazards*. Climate Change Knowledge Portal for Development Practitioners and Policy Makers. https://climateknowledgeportal.worldbank.org/country/zimbabwe/vulnerability

16 Livelihood Diversification for Climate Change Adaptation Among Smallholder Farmers in Gokwe North District of Zimbabwe

Tapiwa Musasa

16.1 Introduction

Climate change is an alteration of the known and recorded patterns of the weather over a long period of time, while environmental change has to do with observed changes in vegetative cover, soil fertility, water availability, and many more (Nwankwoala, 2015). According to Joshi and Maharjan (2013), climate change cannot be disputed anymore and it is affecting the developing countries more than developed countries. According to Onoja et al. (2011), natural causes of climate change include ocean currents, volcanic activities, the earth's orbital changes, and solar variations, while anthropogenic causes like greenhouse gases, burning fossil fuel land use, and deforestation remain the most outstanding with very high carbon footprints. Global temperatures are rising (Jithitikulchai, 2023), sea levels are rising, heatwave frequency increasing, floods, intense drought, and storms have become the new norm, harming people and animals as well as destroying ecosystems and causing people to relocate (Feris, 2015).

In Zimbabwe, the economy is agro-based (FAO, 2020), and 70% of the people are in the rural areas relying on agriculture as a livelihood (Zimstats, 2012; UNDP, 2012; Runganga & Mhaka, 2021). The statistics have remained the same with slight variations (Zimstats, 2012). Rural communities are more vulnerable to any slight change in temperature and rainfall patterns because their livelihoods like farming and fisheries depend largely on the climate patterns which are not certain (Onoja et al., 2011; Dube & Phiri 2013; Duru et al., 2022). According to UNDP (2012), 75% of the world's rural poor live in rural areas, and highly depend on farming and fishing. Runganga and Mhaka (2021) proffer that agriculture has not been having good returns of late due to climate change. This chapter therefore seeks to establish how the poor rural farmers in marginalized areas like Gokwe North district are coping in the face of climate change. The rural farmers are not only exposed to vagaries of climate change, but to other underlying factors like poverty, lack of education, skills, and absence of a strong resource base. The sustainability of most rural livelihoods has been threatened by factors like seasonal drought, lack of climate change adaptation knowledge (Dube & Phiri, 2013), lack of credit and capital base, lack of improved marketing strategies, and high poverty incidence levels. In some cases, the farming environments are so marginal that there is need to abandon them (Cannon, 2014). Climate change adaptation knowledge (Dube & Phiri, 2013) therefore becomes essential to enable effective

T. Musasa (✉)
Department of Social Sciences and humanities,
Catholic University of Zimbabwe, Harare, Zimbabwe

M. Matsa et al. (eds.), *Climate Change Resilience in Rural Southern Africa*, Sustainable Development Goals Series, https://doi.org/10.1007/978-3-031-73600-1_16

livelihoods diversification strategies as a credible entry point for enhancing the resilience of farmers in the face of climate change (Rodriguez, 2019).

According to Ellis (2012) livelihood diversification is the process by which rural families construct a diverse portfolio of activities and social support capabilities in order to survive and to improve their standards of living. In concurrence, Israr et al. (2014) who also define it as a new way of raising income and reducing risk by using alternative ways of survival, or an attempt by individuals and households to find new ways of survival which add to the already existing strategies in a bid to cushion them in case of disasters. In order to enable the livelihood diversification approach to be effective for resource-poor rural farmers, Baharu (2016) maintains that rural infrastructure, road connectivity, market creation and accessibility, credit facility, and increase of input supply have to be improved. A well-maintained road network, for example, enables rural produce to be taken to the market within good time, particularly for highly perishable produce like vegetables. The good road networks give the farmers an urge over others when they can easily get access to the market. Consumers in urban areas can also easily travel to the rural areas to purchase the produce directly from the farmers reducing transport costs challenges for the farmers. Livelihood diversification has always been there in the rural areas through a mixed farming approach, but there is a need for new ways of enhancing it to cater for new challenges posed by climate change, and contemporary pandemics like COVID-19, thus there has to be a mixture of on-farm and off-farm livelihood diversification activities (Israr et al., 2014). This chapter aims to explore the experiences of the rural smallholder farmers in Gokwe North district in livelihood diversification strategies. In addition, the chapter intends to find out the levels of awareness in terms of climate change. Lastly, the chapter intends to explore adaptation and livelihood diversification strategies being employed by the rural farmers.

16.2 Conceptual Framework

This chapter makes reference to climate-smart agriculture (CSA) as the major guiding framework alongside the sustainable livelihoods approach (SLA). Climate-smart agriculture was defined by FAO (2010), as an approach that helps to guide actions needed to transform and reorient agricultural systems to effectively support development and ensure food security in a changing climate. Makate et al. (2016) note that hundreds of practices, approaches, and technologies belong to the smart climate agriculture title including agroforestry, cultivation of drought-resistant crops, conservation agriculture, water harvesting, and integrated soil fertility management. Most smallholder farmers have adopted CSA pillars, which are specific and responding to their own environments and needs. CSA aims to improve productivity, livelihood, and environmental outcomes, thus policy makers and concerned stakeholders always recommend it for food and nutrition security for the benefit of the more vulnerable sectors of the population. Rosenstock et al. (2016) posit that climate-smart agriculture has the following main principles, which FAO (2010) presents as pillars: addressing climate-related risk while improving food, income, and nutrition security; achieving productivity and livelihoods benefits; and, where possible, reducing GHG emission by enhancing the carbon sink through vegetative cover.

These pillars are key to improving the livelihoods of the resource poor smallholder farmers especially if the farmers are involved through the participatory rural approaches. While farmers are aware that the situation is changing around them and there is need to practice agriculture differently, resources become the major limiting factor. Most of the CSA technologies are knowledge-intensive and their benefits might take time to accrue or notice. In some instances, smallholder farmers might continue practising agriculture in the same way they have been doing because levels of technology adoption differ depending on many factors which include, age, levels of education, and available resources. It is therefore crucial to develop multi-stakeholder approaches where smallholder rural farmers, institutions, and the private sector are included from policy planning, adoption, and implementation so that technologies, policies, and investments are tabled (FAO, 2010).

The chapter also adopted the sustainable livelihoods approach (SLA) as a complementary framework since it focuses so much on the poor and vulnerable people (Morse et al., 2009). It is a way of thinking, planning, implementing, and monitoring development strategies targeting the needs of the vulnerable poor people (Serat, 2017). The approach seeks pragmatic priorities, views, and interests of the concerned, searching for livelihood strategies that link the people with enabling environments and referring to their skills, social networks, physical and financial resources, as well as the ability to influence core institutions. The sustainable livelihoods approach taps from the following types of capital: human, financial, natural, physical, and social capital. Khuzwayo (2016) posits that human capital involves the skills and good health which enable people to work, while natural capital involves available natural resources like land, water, vegetation, and conservation strategies. Financial capital means financial resources people have and physical capital refers to the infrastructure at hand which includes roads, secure shelter, adequate water supply, sanitation, and clean affordable energy. Lastly, social capital refers to the networks and connectedness in society that enable people to trust each other and work together towards the common good of the community (Serat, 2017; Khuzwayo, 2016). This will lead to the formulation of development policies that are people-centred and inclusive of the recipients of the development programmes, in a bid to identify and reduce vulnerabilities of the poor. In the context of this paper, the poor are the rural women and their families. However, critics of the SLA argue that the framework concentrates too much on the microlevel ignoring macrolevel factors like the global environment, international wars, national policy-making factors, political stability, and many more which control market prices and availability of raw materials. It is also criticized for ignoring power relations, particularly gender relations which have a lot of influence on how resources are shared, and who get which assets, how and when. The two theoretical frameworks will complement each other in this paper in explaining the extent to which smallholder farmers in Gokwe have been able to diversify their livelihoods in response to climate change as well explaining the challenges they are facing.

16.3 Literature Review

The Inter-Governmental Panel on Climate Change report (IPCC, 2023) states that human induced climate change is unequivocal, causing damage to people and ecosystems, and the evidence of climate change includes rising global temperatures (A.1.1 and A1.2), increased greenhouse gas (GHG) emissions (A.1.3 and A.1.4), historical variations of CO2 emissions (A.1.5) and widespread changes in the atmosphere, hydrosphere, cryosphere, and biosphere (A.2). This calls for comprehensive strategies which includes all stakeholders. However, challenges are that some of the mitigation measures are fragmented and maladapted (A.3.3). The major challenges to effective climate change adaptation (CCA), include limited resources, lack of private sector and citizen engagement, insufficient mobilization of finance (including for research), low climate literacy, lack of political commitment, limited research, slow and low uptake of adaptation science, and low sense of urgency (A.3.5). According to Joshi and Maharjan (2013), some remain negative to the call by the international community to participate effectively and change consumerism patterns, practice irrigation, sustainable land management approaches, ecosystem-based strategies like urban greening and wetland restoration (AR6-A.3.2), thinking mother nature has own ways of reversing the situation at hand (Oka, 2016). The situation at hand indicates an urgent need to make pragmatic rather than theoretical approaches to the guidelines given by IPCC (2023). These challenges affect smallholder farmers more than other groups, since they have always relied on the known patterns for survival and mitigation measure, thus Indigenous knowledge systems will be challenged because of the changes. Rural farmers in particular have always tried to mitigate the impacts of climate change as perceived at community level through Indigenous knowledge systems, but the intensity and frequency of the quagmires are getting beyond their capabilities. As expressed by Duru et al. (2020), environmental changes caused by climate change have more negative impacts on the poor smallholder farmers, particularly the women who rely so much on the environment for survival on a day-to-day basis. One example of the challenge in areas like Gokwe North district is the increased distances travelled by women and girls to fetch firewood and water, while more time is spent queuing for water at boreholes (Brazier, 2022).

In sub-Saharan Africa (SSA), the impacts of climate change are clearly visible on smallholder farmers because they are the most vulnerable to any changes in climate (Serdeczny et al., 2017; Jithitikulchai, 2023). Most (75%) of the farmers are women (Alliance for Green Revolution Agriculture, 2014) and they depend on rain-fed agriculture. McKinsey (2020) posits that it is difficult for African smallholder farmers to find diversification strategies because they lack integrated and coordinated support from stakeholders. According to Inoni et al. (2021), smallholder farmers in SSA cannot easily practice crop diversification because of challenges which include farm size, financial resources, access to productive land, farmers' age, level of education, and access to information. Farmers also fear risk factors involved in venturing into new activities (Inoni et al., 2021). In Zimbabwe, smallholder farmers still face the same challenges as their counterparts in sub-Saharan Africa (Serdeczny et al., 2017). Mutami (2015) posits that smallholder agricultural is the mainstay of Zimbabwe's economy although the agriculture is done under very difficult conditions due to shortage of inputs, markets, and options in case of climate-induced hazards and disasters like droughts and floods. The farmers' challenges include lack of food, malnutrition, low incomes, lack of livelihood diversification, and stunted development due to lack of capital (Dube & Phiri, 2013; Runganga & Mhaka, 2021). The low crop yields are caused by midseason moisture stress, lack of irrigation facilities, and lack of effective water harvesting and conservation techniques to keep fields moist (Tsee, 2020; Dyke et al., 2020 and FAO, 2022). In addition, some farmers continue growing unsuitable crops despite the increased agricultural risks which include diminishing soil fertility, low annual rainfall totals, and lack of agricultural inputs are the most hazards to smallholder farmers because they are threatening food security and incomes (Oka, 2016; Makate et al., 2016; Mawonike & Mandonga, 2017; Dyke et al., 2020). Any policy for sustainable development should be inclusive of the rights, challenges, threats, and opportunities for the majority rural smallholder farmers. According to FAO (2010), governments should enforce climate-smart agriculture so that farmers build resilience to climate change and diversify their livelihoods in a changing climate. In Zimbabwe, Livelihood and Food Security Program (LFSP) is working hard to improve the farmers' resilience to shocks through the introduction of climate-smart agricultural practices, which include subsidies to enable farmers to engage into alternative projects during times of drought and floods. The growing of drought resilient crops like cowpeas, sorghum, and rapoko is also being promoted as a crop diversification strategy (FAO, 2022). These strategies are part of diversification in a changing climate (Onoja et al., 2011; Runganga & Mhaka, 2021), since some farmers have continued to grow crops unsuitable for their agricultural regions (Dube & Phiri, 2013). Malnutrition is being addressed through the introduction of vitamin A, maize, and zinc soyabean which is important in fighting vitamin and nutrient deficiency (FAO, 2022).

Smallholder rural farmers have always faced challenges on how to diversify their livelihoods in the face of a changing climate. Although Indigenous knowledge systems are there (Dube & Phiri, 2013), the magnitude of the change is too big and their efforts are not reaping much yield (Oka, 2016; Mawonike & Mandonga, 2017; Tsee, 2020). The challenges of the farmers in

Zimbabwe include lack of equipment, pests and disease attacks, lack of expertise, and lack of local processing enterprises (Tsee, 2020). According to Dyke et al. (2020), rural farmers perceive climate change differently and sometimes it is very difficult to adopt technology and new strategies of livelihood diversification, which makes it necessary for stakeholders to continue with awareness raising, education, and training (Cannon, 2014). Following Ellis (2012), definition of livelihood diversification which emphasizes the increase of activities to improve standards of living, stakeholders should encourage farmers to change their usual ways of survival and adopt new ways like increasing rural non-farm activities like buying and selling, abandoning non-productive areas, and buying food instead of growing it (Cannon, 2014). Another important challenge to be addressed is post-harvest management is very critical for smallholder farmers since some farmers lose the little harvest they manage to produce under the difficult climate conditions due to lack of storage and preservation facilities (McKinsey, 2020). Guided by FAO (2010), definition of climate-smart agriculture which emphasizes the need to reorient agricultural systems to efficiently support development and ensure food security in a changing climate, there is need to assist all rural smallholder farmers with post-harvest handling so that all grain harvested is kept in safe standards for consumption (Cannon, 2014; Tsee, 2020). In most parts of Zimbabwe, lack of storage and preservation facilities has led to food wastage during times of excess and food shortages during times of scarcity. Mushonga and Scoones (2012) traced livelihood changes in Mazvihwa communal area in Zimbabwe and noted that livelihoods change with time and the factors contributing to change range from the size and age of family members, while weather pattern changes have also contributed to the increase of off-farm activities as mitigation measures. This is an indication that change is inevitable, but livelihoods can be more sustainable when smallholder farmers are cushioned from the impacts of human-induced climate change through skills provision, small loans, and irrigation equipment supply in the driest areas. According to Tsee (2020), providing smallholder farmers with skills is the first step to ensure that they are self-reliant and are able to come up with improved strategies to promote food security for themselves and their livestock in times of drought.

Crop diversification is one of the strategies used by smallholder farmers to improve their yields and ensure food and nutrition security. It involves growing more than one variety of crops in the form of intercropping or crop rotation (Makate et al., 2016). Crop diversification is an aspect of climate-smart agriculture, contributing to all three pillars of CSA which include improving productivity, resilience of farming ecosystems, and reducing carbon dioxide emissions (FAO, 2010; Ellis, 2012; Makate et al., 2016). Wider adoption of crop diversification is therefore recommended since it has positive impacts on climate change through ability of local flora to generate less carbon dioxide. The inclusion of nitrogen-fixing plants is highly recommended, as shown by a study done in three districts of Zimbabwe by Makate et al. (2016).

16.4 Methodology

16.4.1 Location of Study Area

The study was conducted in Gokwe North district (Fig. 16.1). The district was established in 1993 after the amalgamation of Rural councils and District councils, with Nembudziya Growth Point as the administrative centre. The district has 36 wards in four constituencies (Brazier, 2022) which are Nembudziya, Gumunyu, Chireya, and Kabuyuni (Chiguye, 2016). According to Brazier (2022), Gokwe North has a population of 300,000 people of which 48% are male. The main crops grown are maize, cotton, groundnuts, millet, sunflower, tobacco, and horticulture, making crop production the main source of food and income.

The Growth Point is located about 310 km from the Provincial Capital Gweru, 72 km from Gokwe town, 165 km from Kadoma, and 317 km from the capital city of Zimbabwe, Harare (Gwimbi, 2009). The district lies in

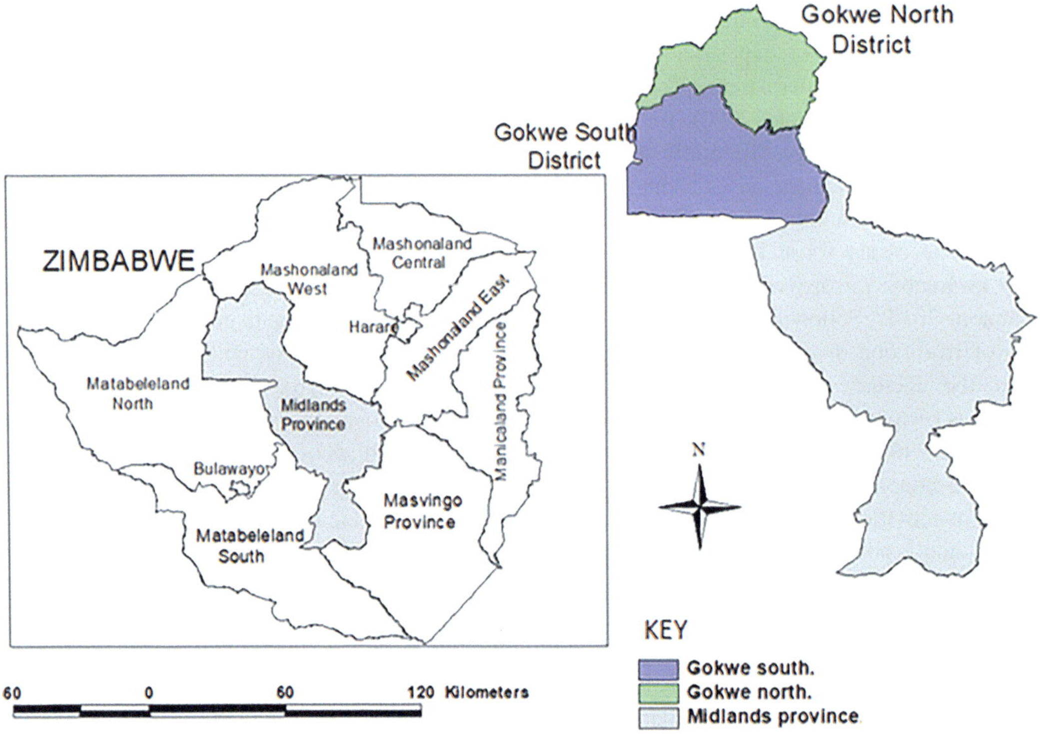

Fig. 16.1 Location of study area—Gokwe North. (Source: Author)

agro-ecological region 4 with arid and semi-arid conditions (Mawonike & Mandonga, 2017), experiencing low rainfall which vacillates at 250–500 mm per anum, presenting a severe shortage of agricultural, domestic, and clean drinking water due to the absence of perennial rivers (Chiguye, 2016). Women and girls travel long distances to fetch water, and spend long hours queuing for clean drinking water at boreholes (Brazier, 2022), thus affecting their reading time if they are in school. In Gokwe North, girl's education is not a priority, so when need arises, girls drop out of school to attend basic needs like fetching water at home or caregiving duties. This further aggravates the situation faced by girlchild in accessing education, as lamented by Rodriguez (2019), Reliefweb (2017), and Brazier (2022). According to Rise Against Hunger (2023), Gokwe North has been affected by El Nino-induced droughts since 2016 and school dropouts increased due to food shortages, only for an increase to be noted when feeding schemes were introduced at five schools in the district.

16.4.2 Research Methods and Strategies

The paper made use of purely qualitative methodologies in the data collection process with an exploratory design. According to Dyke et al. (2020), the exploratory research paradigm puts emphasis on understanding, analyzing, and describing phenomenon without necessarily having to rely on quantitative measurements and statistics. This paper relied on exploratory research design for the convenience of the researcher based on the time available during the data-gathering period. In terms of sample size, the paper relied on arguments from Rusu (2020), who indicate that there is no consensus on the sample size to be used in qualitative research because there are no rules, and research designs are not universal. It also depends on the researcher's resources, time, depth, and size of the research. Boddy (2016) posits that 12 respondents can be enough in qualitative research, while Aguboshini (2021) points out that respondents between 15 and 50 are enough for qualitative research. This chapter used a total 76 participants (50 female and 26 male) purposively drawn from 4 villages in ward 13 and ward 25 of Gokwe North. The respondents participated in Focus Group Discussions (35 participants) and Interviews (30 participants), which were the main data collection methods. An additional 11 key informants were identified as village heads, extension officers and program coordinators from the wards under study The respondents participated in Gokwe North has a total population of 300,000 of which 52% is female (Brazier, 2022). Constant interviews were conducted over the phone from July to August 2023 to verify the responses from the focus group discussions and individual interviews. The results were analyzed qualitatively through thematic analysis. The results of the research are presented in Tables 1 and 2, which show the responses from the participants and the emerging themes from the focus group discussions. Table 2 shows observations done during the data collection period in the two wards, which were also confirmed by the key informants and the rest of the interviewees.

16.5 Results

The study discovered that the livelihoods of smallholder farmers in Gokwe North have been affected so much by human-induced climate change. Farmers no longer rely totally on rain-fed agriculture; instead, they engage in other off-farm activities such as small-scale artisanal mining, fruit and vegetable vending, small-scale irrigated gardens, and paid employment. The findings from the focus group discussions indicate that farmers have diversified their livelihoods into their on-farm activities like market gardening, and growing drought-resistant small grains like rapoko and sorghum, instead of the usual cotton and maize. They have also increased the permanence of water sources through deep well and water harvesting, while continuing to farm the usual crops but reducing the areas under cultivation. Results from the four focus group discussions and the interviews are shown by Table 16.1 below.

Table 16.1 shows that smallholder farmers in Gokwe are aware of their quagmire in terms of human-induced climate change and how their livelihoods have changed significantly over a period of 30–40 years. The smallholder farmers in Gokwe North district have no option but to opt for alternative livelihoods to enable themselves a decent meal and an education for their children. The key informants in the study also validated the results of the focus groups as they added more information.

One of the key informants, a 70-year-old village head, had this to say:

> "Life has changed in Gokwe North as compared when we were young. During our days as young people, the forests were so thick and we never went far away from home to search for grazing land for our family herds. You would stay closer home because either you were too scared to go deep into the forests or there was just so much grass and thickets for the goats and cattle that there was no need at all to go far away. But nowadays you cannot even find water nearby for the cattle because dry riverbeds are common in the area. We do not have any large inland dams as well so water is now a problem. We have to did wells in the riverbeds (Mufuku) and keep maize stocks after harvest for food especially during the times of scarcity."

This is an indication that the smallholder farmers in Gokwe North are aware that there have been changes in their environment due to climate-induced activities which have affected their livelihoods and there is no other option for them, except to look for alternative sources of food and water for themselves and their livestock.

Another female participant in the study also said:

> "The problem is there are too many people and too many homes everywhere leading to the clearance of forests for construction purposes. We now travel long distances as women to fetch water and firewood, responsibilities which are mainly left for women. This affects our work schedules because we have to be in the fields to till the land with others. Some who can afford are now using paraffin stoves, solar lights for lighting and pumping water from deep wells which makes life easier for women. However, not everyone can afford that because of the hard times everywhere so, the problems continue."

Table 16.1 Indicators of climate change and livelihood diversification strategies in Gokwe North

Group and number of participants	Indicators of climate change	Livelihood diversification strategies	Themes emerging
1 (10 participants)	Less rainfall intensity as compared to previous years Long distances travelled to fetch water and firewood	Small-scale mining Small poultry projects Beer brewing for cash Buying and selling clothes	The environment has changed and people in the district are opting for other sources of income
2 (8 participants)	Less vegetation cover as compared to 30 years ago	Construction work	Lifestyles have changed and people are aware of the impacts of climate change to the environment
	Soil degradation less grazing land for our animals	Fruit and vegetable vending	
	Dry riverbeds evidence of less rainfall	Goats and cattle projects	
3 (7 participants)	Wetlands drying Less vegetation covers Firewood not easily available anymore Deeper wells More frequent droughts	Water harvesting project at the homestead Paid employment Solar project at the homestead to pump water for gardening Buying and selling clothes	Residents acknowledge the environment as a source of resources, which is dwindling and they have to find alternatives
4 (10 participants)	More frequent disasters like floods Shorter wet periods Deaths of livestock due to drought Crop failure but resources lacking for irrigation	Venturing into new crops Planting early Growing drought resistant crops Increase small grains	Need for increased crop diversification but financial resources remain the major limitation

Source: Author

This response is an indication that smallholder farmers realize the changes taking place around them and the need to find alternatives. The challenges are not easy to rectify considering the hard economic conditions in the country as indicated by the respondent. The good thing is the farmers have options and ideas for diversification. For this reason, it is key to consult smallholder farmers during any development intervention in their areas.

A key informant in the study, a 65-year-old traditional leader, had this to say:

> "Our people are trying their luck all over; in urban areas, growth points and mining areas as a way to find food and schools fees for their families. Harvests have declined and people face starvation if they sit and fold their hands. I recommend the growth of drought resistant small grains and the Zunde ramambo (King's Granary) to be resuscitated so that grain stored can cushion vulnerable families in times of hardships."

This response shows that the leader is concerned about the welfare of the people and proposes some solutions to mitigate the current challenges being faced. The solutions involve the use of Indigenous knowledge systems which have been known to work.

Another respondent, a 40-year-old young farmer, had this to say:

> "We have resorted to cooperative market gardening. We dig deep wells to be able to water our vegetables and cattle because water is scarce particularly during the dry season. This makes work easier when we dig the deep wells together for our animals and vegetables. Life was going to be better if we had solar powered boreholes, irriga-

tion schemes and more inland dams. We supply vegetables to the food vendors at the growth point and we find some cash for food and stationery for school children. It is better than growing cotton because the prices of cotton have fallen. You work for nothing."

Cotton growing has been the white gold in Gokwe North for a very long time, but things have changed now due to exogenic factors like fall in international prices so the farmers have resorted to diversify into other survival strategies.

The findings from this study indicate that smallholder farmers in Gokwe North district have been affected by climate-induced environmental change. Through observation, it could be seen in the area that there is a lot of environmental degradation, evident in uncontrolled run-off, un-reclaimed land after artisanal mining, and overgrazing. Eroded bare ground, gullies, and dongas could be seen in Chireya area of Gokwe North. The respondents to the study expressed their awareness and concern of the hazardous situation facing them due to the effects of climate-induced environmental change, indicating that they have resorted to other ways of survival diversifying from their usual farming activities because of low harvests and low prices of cash crops on the market. The most common crops grown in Gokwe North are shown in Table 16.2 below.

Table 16.2 shows a few crops chosen from ward 7 and ward 36 of Gokwe North district to represent the crop diversification being practised. The cereals and vegetables shown were observed during data collections, and the respondents emphasized that crop diversification has been done for years as a food security measure.

One of the key informants interviewed said:

"Crop diversification is a good strategy so that if one crop fails, at least you get something from the next crop. However, new crops like chillis can only be grown on a small scale due to inputs and knowledge challenges. There is also the fear of risk because no one has experience on the performance of such new crops. With enough training, inputs, irrigation schemes and extension services supervision, Gokwe North can adopt a variety of crops to enhance crop diversification in a changing climate."

Table 16.2 Crops and vegetables grown in Gokwe North

Crops	Influencing Factors	Challenges Encountered
(A) Cereals		
Maize	The need for food, staple food	Low harvests due to drought
Sorghum	Its drought-resistant characteristics, beer brewing	Lack of inputs
Finger millet	Drought-resistant characteristics, beer brewing, and food needs	Lack of land as priority is given to other crops
Cotton	Government initiative, financial gains	Declining market prices, high inputs, and less returns
Sweet potatoes	Income generation and consumption	Decrease in yields due to less rain
Cowpeas	Additional food supplement	Less land allocated to women crops
Groundnuts	Nutritional benefits, income generation	Less land
Traditional rice	Traditional food source, income generation	Less water as most wetlands are drying
Sunflower	Stock feeds, oil pressing, and income	Area under cultivation has become low due to increased competition from other crops
(B) Vegetables		
Leafy vegetables (covo, tsunga)	Daily food consumption, income	Water challenges
Onions and tomatoes	Food and income generation	Water challenges
Cabbage	Income generation, food needs	Marketing strategies and water challenges
Chilli	NGO initiative, income generation	Low adoption by farmers
Pumpkins	Intercropping benefits, food needs	May not grow well with some crops, less marketable for income

Source: Author

16.6 Discussion

The diversification strategies employed by the smallholder farmers in Gokwe North district are both on-farm and off-farms as emphasized by Israr et al. (2014), who maintains that a balance should be struck between on-farm and off-farm activities for sustainability. The change was necessitated by shortage of rainfall due to drought. This shows that people are making efforts to diversify but agriculture has remained the main source of food. Even the new activities can still be affected by drought if less rainfalls annually to affect underground reservoirs which feed into the deep wells or boreholes being used for the gardens. This shows that the smallholder farmers continue to be affected by climate change and their livelihoods are not sustainable. Dyke et al. (2020) posited that under the hazardous impacts of climate change, harvests have been eroded and farmers will continue to engage into unsustainable farming activities unless there is intervention from stakeholders through monitored loans and skills training for realistic diversification. When small-scale loans are monitored, farmers are assisted to get maximum yields through extension services to enable them to repay the loans and remain income and food secure. While the diversification into drought-resistant crops was recommended by the Food and Agricultural Organization of the United Nations (FAO, 2022) as part of climate-smart agriculture strategies, still more needs to be done in terms of improving soil fertility, prevention of deforestation, and reclamation of degraded range lands so that future generations can also benefit from the same pieces of land. Without assistance from extension service workers on environmental management activities, it is difficult for farmers to achieve all the benefits of climate-smart agriculture. Makate et al. (2016) argued that income, food, and nutrition security continue to be threatened because of declining crop yields unless deliberate efforts are made to assist farmers with alternative skills or small loans to enable the purchase of inputs, as well as water harvesting and irrigation kits. The respondents in the study who managed to moot market gardening projects with water harvesting tanks and solar-powered boreholes got remittances from family members in South Africa and other European countries. This excludes other families without remittances and they remain vulnerable and in need of assistance. This is where the government, non-governmental organizations, and the private investors have to come into play in a multi-stakeholder approach. For instance, Catholic Relief Services in Beitbridge is capacitating smallholder farmers through skills training, small loans provision, and nutritious seed and input provision to grow nutritious grains for nutrition security are being helped to diversify their livelihoods (Tsee, 2020). All these factors should be known by legislators who should take an interest in the areas they represent in parliament and advocate for composite sustainable livelihood diversification projects not small self-generated projects which continue exposing the rural smallholder farmers to vulnerability.

In addition, the smallholder farmers in Gokwe North district are also diversifying into off-farm activities for survival (Chiguye, 2016; Brazier, 2022). These include artisanal mining, beer brewing, fruits and vegetable vending, buying and selling clothes, and paid employment, and projects like poultry projects, goats, and cattle projects. The off-farm activities are small in nature but they are essential as fallback plans for cash and food provisions. In the case of artisanal mining, smallholder farmers especially the young people have resorted to mining as an alternative because farming returns have declined (Brazier, 2022). Such uncertainties threaten the education security of their children, one of the reasons why school dropouts are on the increase in remote rural areas like Gokwe North (Rise Against Hunger, 2023; Musasa, 2024). Many challenges in Gokwe district have led to school dropouts for boys and girls and some of them have joined small-scale mining. In this respect, the livelihoods diversification strategies for the smallholder farmers in Gokwe North district cannot be deemed sustainable because the artisanal miners are leaving open pits everywhere to the detriment of the environment and future generations. Even arable land in some cases is being deemed useless as people search for the valuable gold in the fight for survival. Climate Smart agriculture as a concept and framework emphasizes the need to reclaim degraded land and promote safe environments for future generations (FAO, 2022). Instead of being alternative options for livelihood diversification, some of these approaches appear to be desperate efforts by the smallholder farmers to survive for the moment from hand to mouth, without any considerations for saving or for keeping resources for future generations. This is in agreement with what Serat (2017) pointed out that sustainable livelihoods approach should target all, and include skills, knowledge and abilities of the poor, and efforts done to improve the livelihoods of everyone. When they leave open pits everywhere, they are being selfish and inconsiderate about the common good of the community, as arable land will be destroyed, and cattle and children may drown in the open pits during the rainy season. However, there is nothing much Gokwe North residents, particularly the youths can do because there are no employment opportunities for them (Musasa, 2024), or capital injection for the youth to start sustainable well-planned projects for livelihood diversification. The SLA also emphasizes the need to engage physical, financial, and social capital together for comprehensive survival strategies which are environmentally friendly (Khuzwayo, 2016). Rural infrastructure should be improved rather than destroyed, and income-saving concepts can be promoted through Income and Saving and Lending Schemes (ISALS). Through the social capital concept, people can work together in cooperatives to offset labour challenges and lend each other money in ISALs, an advantage of community life in the rural area where people know each other and they have trusted each other for years to be able to work together through work beer parties (Nhimbes) and they can lend money to each other unlike in the urban area where individualism is the order of the day.

Findings from the interviews with key informants and other participants were in agreement with those findings from the focus group discussions and this helped to validate these responses through triangulation. The interviewees raised issues on the sustainability of their livelihoods which leave a lot to be desired because they cannot leave much to their children as inheritance when they are gone. The key informants highlighted challenges facing smallholder farmers in Gokwe North district and these include declining soil fertility, lack of markets, increasing population densities, deforestation, land degradation, environmental degradation through mining, and limited support for vulnerable families like child-headed families, those living with disabilities, and female-headed households. Limited access to markets for their produce is the greatest challenge, and they lose out on their inputs and efforts. Mutami (2015) also noted lack of markets as a biggest challenge for rural farmer. Only a few farmers have remained in cotton growing, since the prices of the crop are no longer attractive. In addition the soils have been overused, and need more fertisers for high productivity, while more chemicals are needed due to pest resistance thus it has become costly tpo grow the crop (AGRA, 2014), for them to be able to harvest anything, thus most of the rural farmers do not afford the inputs, thus need for well-controlled inputs destruction from the Government to eliminate corruption so that everyone can benefit.

16.7 Conclusions and Recommendations

This study concludes that smallholder farmers in Chireya-Gokwe North district are aware of the challenges related to climate change. They have diversified their livelihoods in on-farm and off-farm activities. On-farm activities include market gardening and growing traditional small grains which are drought-resistant, while off-farm activities include beer brewing and selling for cash, fruit and vegetables vending, buying and selling clothes, poultry and goats projects as shown by the responses in the findings section. The environment itself is pausing problem of poor soil fertility, increased frequency of droughts, declining river flow, and low yields. From these conclusions and results, the study makes the following recommendations:

- There is a need for continued environmental management practices in Gokwe North to curb land degradation and deforestation and reduction in soil fertility. This can be done by many stakeholders through the coordination of the government.
- There is a need to increase access to climate finance to enable smallholder farmers to start sustainable and viable projects for sustainable livelihood diversification.
- There is a need for skills and knowledge transfer from different stakeholders through climate-smart agriculture techniques.
- There is a need to promote increased water harvesting and solar equipment in Gokwe district and other rural areas so that the water harvested can be used in midseason drought for sustainable irrigation projects. At the moment only those who are better placed financially are using these renewable sources of energy. The readily available solar energy can be used to power projects for livelihoods diversification in a climate and environmental smart way. Solar energy use saves vegetation destruction and the environment will be smart.

References

Aguboshim, F. C. (2021). Adequacy of sample size in a qualitative case study and the dilemma of data saturation: A narrative review. *World Journal of Advanced Research and Reviews, 10*(03), 180–187.

Baharu, G. (2016). Determinants of livelihood diversification: The case of Kembata Zone, Southern Ethiopia. *Journal of Poverty, Investment and Development, 23*, 1–10. Available online: https://iiste.org/Journals/index.php/JPID/article/view/3020

Boddy, C. R. (2016). Sample size for qualitative research. *Qualitative Market Research, 19*(4), 426–432.

Brazier, A. (2022). *Gokwe food system study report*. Available online: https://www.researchgate.net/publication/364071242_Gokwe_Food_System_Study_Report

Cannon, T. (2014). Livelihood diversification and adaptation to climate change. Chapter 4. In J. Ensor, R. Berger, & S. Huq (Eds.), *Community-based adaptation to climate change*.

Chiguye, L. T. (2016). *Flood risk prediction using remotely sensed data in the Sanyati catchment area: A case of Gokwe North District*. [Online]. Available from: r.msu.ac.zw:8080/jspui/bitstream/11408/2124/1/R125036J.pdf

Dube, T., & Phiri, K. (2013). Rural livelihoods under stress: The impact of climate change on livelihoods in southwestern Zimbabwe. *American International Journal of Contemporary Research, 3*(5).

Duru, C., Olanipekun, G., Odili, V., Kocmich, N., Rezac, A., Ajose, T.O., et al. (2020). *Molecular Characterization of Invasive Enterobacteriaceae from Pediatric Patients in Central and Northwestern Nigeria*. PLOS ONE, 15, e0230037. https://doi.org/10.1371/journal.pone.0230037

Duru, J., Aro, J., & Ebunoluwa Oladipo, R. (2022). The effects of climate change on the livelihoods of rural women: A case study of Ilorin South Nigeria. *Bulletin of the National Research Center, 46*, 165.

Dyke, T., Tirivangasi, H. M., & Rarkoana, S. A. (2020). Rural community perceptions on the impact of climate change on subsistence farming: Mutoko Community in Zimbabwe. *Journal of Social Sciences and Humanities, 17*(7), 89–104.

Ellis, F. (2012). Household strategies and rural livelihood diversification. *The Journal of Development Studies, 35*(1). Published online 2007. Available online www.tandfonline.com/action/showCitFormats?doi=10.1080%2F00220389808422553

FAO. (2010). *Climate Smart Agriculture*. Source Book. Available online at https://www.fao.org/climatechange/37491-0c425f2caa2f5e6f3b9162d39c8507fa3.pdf

FAO (Food and Agriculture Organization). (2020). *Zimbabwe at glance*. Online. Available at http://www.fao.org/zimbabwe/fao-in-zimbabwe/zimbabwe-at-a-glance/en/

Food and Agricultural Organization (FAO). (2022). *Income security for smallholder farmers in Zimbabwe: Revamping rural livelihoods in the face of climate change*. Available online: www.fao.org/in-action/income-security-farmers-zimbabwe/en/

Feris, E. (2015). Environmental change and the planned relocation of communities. *SAIS Review of International Affairs, 35*(1), 109–117. Published by Johns Hopkins University Press. https://doi.org/10.1353/sais.2015.0001

Gwimbi, P. (2009). Linking rural community livelihoods to resilience building in flood risk reduction in Zimbabwe. *Jàmbá: Journal of Disaster Risk Studies, 2*(1), 80–89. https://doi.org/10.4102/jamba.v2i1.16

Inoni, O., Gani, B., & Sabo, E. (2021). *Drivers of crop diversification: Evidence from Smallholder Farmers in Delta State Nigeria*. Available online: https://www.researchgate.net/publication/349887812_Drivers_of_Crop_Diversification_Evidence_from_Smallholder_Farmers_in_Delta_State_Nigeria

IPCC. (2023). Summary for policymakers. In Core Writing Team, H. Lee, & J. Romero (Eds.), *Climate change 2023: Synthesis report. Contribution of Working Groups I, II and III to the Sixth Assessment Report of the Intergovernmental Panel on Climate Change* (pp. 1–34). IPCC. https://doi.org/10.59327/IPCC/AR6-9789291691647.001

Israr, M., Khan, H., Jan, D., & Ahmad, N. (2014). *Livelihood diversification: A strategy for Rural Income Enhancement in Pakistan*. Available online: http://pubs.sciepub.com/jfe/2/5/10/index.html. Retrieved 9 October 2021.

Jithitikulchai, T. (2023). The effects of climate change and agricultural diversification on the total value of agricultural outputs of farm households in Sub-Saharan Africa. *African Journal of Agricultural and Resource Economics, 2*(18), 152–170. https://doi.org/10.53936/afjare.2023.18(2).10

Joshi, N. P., & Maharjan, K. L. (2013). *Climate change, agriculture and rural, livelihoods in developing countries*. Springer Japan. Available online. SBN: 978-4-431-54342-8 (Print) 978-4-431-54343-5.

Khuzwayo, H. A. (2016). *A sustainable livelihood approach to poverty reduction: Participatory experiences of women involved in art and craft cooperative in Bhambayi, Kwa Zulu Natal*. Available online: https://researchspace.ukzn.ac.za/bitstream/handle/

Makate, C., Wang, R., Makate, M., & Mango, N. (2016). *Crop diversification and livelihoods of smallholder farmers in Zimbabwe: Adaptive management for environmental change*. Springer Plus. Available online: https://pubmed.ncbi.nlm.nih.gov/27478752

Mawonike, R., & Mandonga, G. (2017). The effect of temperature and relative humidity on rainfall in Gokwe Region Zimbabwe: A factorial design perspective. *International Journal of Multidisciplinary Academic Research, 5*(2)., ISSN 2309–3218. Available online: https://www.academia.edu/34535297/

McKinsey and Company. (2020). *How will African Farmers Adjust to Changing Patterns of Precipitation?* McKinsey Global Institute. Available online: https://www.mckinsey.com/~/media/mckinsey/business%20functions/sustainability/our%20insights/how%20will%20african%20farmers%20adjust%20to%20changing%20patterns%20of%20precipitation/mgi-how-will-african-farmers-adjust-to-changing-patterns-of-precipitation.pdf

Morse, S., McNamara, N., & Acholo, M. (2009). *Sustainable livelihood approach: A critique of theory and practice*. Available online: https://doi.org/10.1007/978-94-007-6268-8

Musasa, T. (2024). Youths rights to quality education in rural areas with reference to Gokwe North District, Zimbabwe: Challenges and opportunities. In *Human Rights in Southern Africa* (pp. 67–76). Domuni Press.

Mushonga, J., & Scoones, I. (2012). Livelihood change in rural Zimbabwe over 20 years. *Journal of Development Studies, 48*(9), 1241–1257. https://doi.org/10.1080/00220388.2012.671474

Mutami, C. (2015). Smallholder agriculture production in Zimbabwe: A survey. *The Journal of Sustainable Development, 14*(2), 140–157. Available online: https://academiccommons.columbia. Edu/Doi/10.7916/D.

Nwankwoala, A. H. N. (2015). Causes of climate and environmental change: The need for environmentally friendly education policy in Nigeria. *Journal of Education and Practice, 6*(30), 224–234. Available online: https://files.eric.ed.gov/fulltext/EJ1081366.pd

Oka, N. O. (2016). *Climate change and its impact on smallholder farmers*. Available online and on LinkedIn.

Onoja, U. S., Dibua, U. M. E., & Anete, A. A. (2011). Climate change, causes, effects and mitigation measures. *Global Journal of Pure and Applied Sciences, 17*(4), 469–479. Copyright© Bachudo Sciences Co. Ltd Printed in Nigeria ISSN 1118–0579 www.globaljournalseries.com, Email: info@globaljournalseries.com

Relief web. (2017). 13 reasons why girls are not in school on international day of the girl child. *News and Press Release*. Available online https://reliefweb.int/report/world/13reasons-why-girls-are-not-school-international-day-girl-child. Accesses on 16 February 2021.

Rise Against Hunger. (2023). *Attendance increased 35% at Five Schools in Gokwe North Zimbabwe, Find Out Why*. A Relief Web Publication. Available online: https://reliefweb.int/report/zimbabwe/attendance-increased-35-five-schools-gokwe-north-zimbabwe-find-out-why.Retrived,03/02/24

Rodriguez, L. (2019). Girls' education: 7 obstacles and how to overcome them. *Global Citizen*. Available online: https://www.globalcitizen.org/en/content/barriers-to-girlseducation-around-the-world/. Accessed 16 February 2021.

Rosenstock, T. S., Lamanna, C., Chesterman, S., Bell, P., Arslan, A., Richards, M., Rioux, J., Akinleye, A. O., Cheng, Z., Champalle, C., Corner-Dolloff, C., Dohn, J., English, W., Eyrich, A. S., Girvetz, E. H., Kerr, A., Madalinska, A., Lizarazo, M., McFatridge, S., Morris, K. S., Namoi, N., Poultouchidou, N., Ravina da Silvaz, M., Ström, H., Rayess, S., Tully, K. L., & Zhou, W. (2016). The scientific basis of climate-smart agriculture: A systematic review protocol. In *CCAFS working paper no. 138*. CGIAR Research Program on Climate Change, Agriculture and Food Security (CCAFS), Copenhagen, Denmark.

Runganga, R., & Mhaka, S. (2021). *Impacts of agricultural production on economic growth in Zimbabwe*. Available Online at https://mpra.ub.uni-muenchen.de/106988/

Rusu, M. D. (2020). *Determining the sample size in qualitative research*. ResearchGate Publications. https://doi.org/10.26520/mcdsare.2020.4.181-187

Serat, O. (2017). *The sustainable livelihoods approach*. Springer Link. Available online: https://link.springer.com/chapter/10.1007/978-981-10-0983-9_5. Accessed on 31 July 2022. 1.

Serdeczny, O., Adams, S., Baarsch, F., & Couman, D. (2017). *Climate change impacts in Sub-Saharan Africa: From physical change to their social repercussions*. Springer. Available online: https://www.researchgate.net/publication/290194107_Climate_change_impacts_in_Sub-Saharan_Africa_from_physical_changes_to_their_social_repercussions

The Alliance for a Green Revolution in Africa (AGRA). (2014). "The Africa Agriculture Status Report.", contributed by Angela Okune, *STS Infrastructures, Platform for Experimental Collaborative Ethnography*, accessed 19 February 2024. https://stsinfrastructures.org/content/agra-2014-africa-agriculture-status-report

Tsee, D. (2020). *Zimbabwe farmers adapt to climate change*. Catholic Relief Services. Available online.

United Nations Development Programme. (2012). *Africa human development report 2012*. Towards a Food Secure Future.

Zimbabwe Statistical Office (Zim Stats 2012 & 2022). *Women and men summary report*. UNICEF Publications. Available online: https://www.citalivres.com/pdf/women-and-men-in-zimbabwe-report-2012

Emerging Adaptation Constrains in Mount Darwin District, Zimbabwe

17

Vincent Chenzi, Mohau J. Mateyisi, Shingirai S. Nangombe, and Malebajoa Anicia Maoela

17.1 Introduction and Background

This chapter examines climate change adaptation constrains with specific focus on Mount Darwin district in Zimbabwe. It argues that there are a number of constrains hindering climate change adaptation in rural Southern Africa in general, and Mount Darwin district in particular. In this chapter, adaptation refers to the process of adjusting to actual or anticipated climate conditions and its effects (Klein et al., 2014). In this chapter, adaptation constrains refer to barriers, obstacles, and limits that make adaptation planning and implementation more difficult. These may be institutional, financial, environmental, technical, political, and cultural. Mount Darwin district was used in this study because it is located in the region of Mashonaland central which is the agricultural mainstay of Zimbabwe. Hence, the climate change effects have broader implications beyond the district (Musemwa et al., 2013). The main economic activities in the district include smallholder and/subsistence farming and artisanal gold mining (AGM). The chapter further demonstrates that the rural community of Mount Darwin is shifting from agriculture to ASM as an adaptation mechanism against the ravages of climate change on the environment.

Contemporary climate projections of drought conditions indicate that northern Zimbabwe which houses Mount Darwin district is expected to experience some of the high temperatures in the Southern Africa region evoked by climate change which is also influencing the shift from agriculture to AGM (see Chap. 2). Furthermore, Mashonaland central is one of Zimbabwe's agricultural hubs especially tobacco and maize farming. Hence, the changes in climate have wider implications beyond the Mt Darwin district (Musemwa et al., 2013). In addition, research on adaptation in Zimbabwe has mainly focused on historically climate-prone regions, whilst Mashonaland central has not been given adequate attention because of its historically favourable climatic conditions despite emerging climate shocks affecting the region and its inhabitants (Chanza, 2018; Chanza et al., 2019). Likewise, the Global Climate Risk Index (GCRI) which

V. Chenzi (✉)
Department of Peace, Security and Society, Faculty of Arts and Humanities, University of Zimbabwe, Harare, Zimbabwe

M. J. Mateyisi
Smart Places Cluster, Council for Scientific and Industrial Research, Pretoria, South Africa
e-mail: mmateyisi@csir.co.za

S. S. Nangombe
Danish Meteorological Institute, Copenhagen Ø, Denmark
e-mail: snangombe@csir.co.za

M. A. Maoela
Institute for Corporate Citizenship, University of South Africa, Pretoria, South Africa
e-mail: maoelma@unisa.ac.za

M. Matsa et al. (eds.), *Climate Change Resilience in Rural Southern Africa*, Sustainable Development Goals Series, https://doi.org/10.1007/978-3-031-73600-1_17

tracks climate change impacts listed Zimbabwe as one of the countries most affected by extreme weather events in 2019, including droughts, floods, and elevated temperatures (Global Climate Risk Index, 2021). These emerging threats posed by climate change are also a direct consequence of unsustainable agricultural practices in the Mount Darwin district which have resulted in land degradation, deforestation, and siltation of water bodies, which heighten climate change effects. Therefore, using the district as a case study is critical towards identifying useful data that can be used in addressing the districts' adaptation constrains.

Rural communities in Southern Africa are on the frontlines of the impact of climate change, facing challenges such as shifting weather patterns, increased temperatures, rise in sea levels, and the increase in frequency and intensity of extreme weather events including heatwaves, droughts, heavy rainfall, and tropical storms (Allan et al., 2021; Epule et al., 2021). In Southern Africa, these climate change-driven disasters have acted as risk multipliers resulting in widespread food insecurity and displacement of millions of the region's inhabitants (Epule et al., 2021). For example, in 2019, Cyclone Idai was one of the deadliest storms to hit Mozambique, Zimbabwe, and Malawi killing over 1200 while displacing 2.2 million and destroying 800,000 hectares of cropland, crops, and seed stock (WFP, 2021). Again, the 2016, El Nino-induced drought was the worst in 35 years. Zambia also encountered one of its worst droughts in decades following the erratic 2018–2019 rainfall (WFP, 2021). Also, between 2021 and 2022, Southern Madagascar came across severe drought conditions that drove 1.35 million Malagasies to the brink of starvation (Rabezanahary Tanteliniaina & Andrianarimanana, 2023).

These climate changes undermine economic development, efforts to eliminate poverty, and endanger lives and livelihoods of the poorest and most vulnerable communities (Dicker et al., 2021). In particular, the most vulnerable communities situated in rural Southern Africa are disproportionately affected by climate stressors due to limited adaptive capacities and high sensitivity. Moreover, low technological development, limited access to information and dearth of skills, weak infrastructure, weak institutions, and unbalanced empowerment and access to resources also characterize the region (Thomas et al., 2021). Smallholder farming is the backbone of Southern Africa's food supply as it accounts for 90% of the total agricultural output (WFP, 2021). However, 30% of smallholder production is at risk from climate hazards especially heat stress, denoted by high temperatures during the growing season which lowers both livestock and crop productivity (Menghistu et al., 2020; Thinda et al., 2020; Khanal et al., 2021). Furthermore, the Southern Africa region is heating up at twice the rate of the global average which exacerbates the effects of pre-existing vulnerabilities in the region such as hunger, poverty, and poor health thus drastically affecting the attainment of the sustainable development goals (SDGs). Furthermore, between 2011 and 2021, cereal production fell drastically with deficits ranging between 0.1 and 8.9 million metric tonnes (Kumar et al., 2021). By 2050, approximately 30% of the region will be exposed to various climate hazards (WFP, 2021). Also, the number of undernourished children below 5 years is anticipated to increase by 2.4 million due to lack of proper adaptation by 2050. The United Nations Office for Disaster Risk Reduction (UNDRR) reported that half a million people are killed annually, whilst close to four billion people have been affected by disasters between 2000 and 2019 (UNDRR, 2019).

Due to industrialization, urbanization, and climate change, there has been a rapid drop in the world population engaged in agriculture (The World Bank, 2019). Nevertheless, 65% of the poor working adults still survive on agriculture especially in the Southern Africa region (The World Bank, 2019). Apart from the climate change effects, food production in the African continent in general and rural Southern Africa in particular is still low because of the limited use of technology, inadequate supply of agrochemicals, and other inputs and low remuneration of contracted labour by mostly subsistence-oriented farmers (Savo et al., 2016). In terms of livelihoods, climate change is destroying agriculture-based rural livelihoods through crop failure associated with heat stress and persistent droughts, thus exacerbating environmentally induced migration patterns especially rural-urban migration. This is further complicated by unemployment which is at 32.2% in Southern Africa leading many to embrace artisanal mining which has increased from 6 million miners in 1993 to 40.5 million in 2017 globally (Statista, 2023; Ofosu et al., 2020; Fritz et al., 2018). Employment and artisanal mining are some of the region's domestic copying mechanisms to enhance rural livelihood opportunities and address climate change-induced migration (Kumah et al., 2020).

Therefore, adaptation strategies are essential for these communities to build resilience and secure their livelihoods since regions with greater adaptive capability will be able to manage the climate change shocks compared to those with weak adaptive mechanisms. That is, investing in climate change adaptation and resilience is not only a cost-effective measure towards mitigating climate-induced shocks but it also defends communities, livelihoods and businesses, whilst nurturing economic development and growth. For example, climate financing towards Kenya's arid and semi-arid regions not only increased water availability which addressed food insecurity challenges resulting in fewer livestock fatalities, but it also strengthened and diversified livelihoods (Dicker et al., 2021).

This chapter is broadly categorized into four broad sections excluding the background and introduction. The global status on the green development agenda and climate adaptation capital distribution form the second segment. The literature review and methodology segments form the third and fourth parts, respectively. The findings section and the discussion of findings make up the third and fourth segments. This study is of importance because it provides emerging treads on climate change domestic adaptation mechanisms in rural Southern Africa which is a neglected area of research. Furthermore, the research's findings are critical towards identifying and addressing adaptation contains in rural communities of the developing world in general and rural Southern African in particular.

17.2 Global Climate Adaptation and Capital Distribution: Challenges and Prospects

The green development agenda spearheaded by the UN which seeks to transition towards low-carbon, climate-resilient societies and economies in order to establish a sustainable future is intricately related to climate change adaptation. It was espoused by the Paris Agreement, the sustainable development goals (SDGs) within the framework of the 2030 Agenda for sustainable development, and the Sendai Framework for Disaster Risk Reduction (SFDRR) 2015–2030 (UNCCS, 2017). The goals of the green development agenda include reduction in greenhouse gas emissions to pre-industrial levels, poverty reduction, economic growth and human development, ecosystems services, enhanced biodiversity, and climate change resilience. However, these potential target outcomes closely resemble the aims of sustainable development (DEA, 2010; Deloitte, 2012; Nhamo, 2013).

Governments and organizations that have the capability to advance the green development agenda are those that have invested their institutions with considerable resources to address the key pillars of sustainable development. These include social, environmental, skills development, biodiversity, water, air pollution policies and acts among other key legislations to foster first-order green economy readiness. The second-order green economy readiness manifests when governments and stakeholders institute measures to address explicit injunctions of the new green economy, in the context of sustainable development and poverty eradication (Nhamo, 2013). However, institutional readiness is vital towards providing frameworks for policy and legislative action. The aptitude of an institution to satisfy its mandate rests not only on power relationships, but also on the representativeness of its configuration and its human resource capacities (Chagutah, 2010). The Southern Africa region is populated by countries with diverse social, economic, and political features, nevertheless, these states possess the potential to invent new pathways towards low carbon and climate resilient development (Nhemachena et al., 2020).

The realization of a green development agenda is intricately linked with the distribution of capital resources, which significantly shapes the adaptation capacity of rural areas. However, apart from the fact that capital distribution of resources (financial, human, and infrastructural) is not evenly distributed across the region, climate change adaptation has generally suffered from a lack of investment. Developing nations alone, will require US$140–300 billion per annum by 2030, and US$280–500 billion per year by 2050 towards adapting to climate change (Pauw, 2021; Bechtoldt et al., 2021). However, between 2015 and 2016, an annual average of just US$22 billion was invested globally towards adaptation (Pauw et al., 2021). Targeted countries are receiving these funds but they are inadequate to meet the current adaptation needs. In Zimbabwe, for example, the Zimbabwe's Climate Change Department in partnership with the United Nations Development Programme (UNDP) provided US$26.6 million for the 2.3 million vulnerable smallholder farmers in Manicaland, Masvingo, and Matabeleland South provinces. However, the funding did not cover the other seven provinces including the case study area despite evidence of climate change effects (UNDP, 2020). Moreover, more funding is spent on mitigating climate change shocks rather than investing towards adaptation. The Organisation for Economic Co-operation and Development (OECD) found that out of the US$78.9 billion climate finance which was mobilized through the United Nations Fram lljework Convention on Climate Change (UNFCCC) for developing nations in 2018, 55.23 billion, representing 70% was reserved for mitigation whilst only US$16.8 billion, representing 21% was channelled towards adaptation (Pauw et al., 2021). These inadequate global capital distribution funds towards climate change resilience also explains the low adaptation capacity of Southern Africa in general and Mount Darwin area in particular.

In Southern Africa, rural regions are disproportionately affected by climate change, due to dependency on climate sensitive sectors such as agriculture and natural resources, high levels of poverty, and high exposure to climate shocks. Rural Southern Africa contains 56% of the region's population (Helen Suzman Foundation, 2024). Southern Africa requires a total of US$1 trillion, with an annual demand of US$90.3 billion for the 2020–2030 period (Seyuba & Garcia, 2022). Nonetheless, the average annual climate finance flows to the region are pegged at US$6.2 billion, which translates to a mere 6.9% of the required funding (African Development Bank Group, 2023). Consequently, this lack of financing has resulted in low adaptation capacity for the region (Seyuba & Garcia, 2022).

Notwithstanding, it should be noted that the deficiency of adequate investment towards adaptation and resilience reflects the absence of investment in disaster risk reduction (DRR). One of the goals targeted by the SFDRR 2015–2030 is to "Prevent new and reduce existing disaster risk through the implementation of integrated and inclusive economic, structural, legal, social, health, cultural, educational, environmental, technological, political and institutional measures that prevent and reduce hazard exposure and vulnerability to disaster, increase preparedness for response and recovery, and thus strengthen resilience" (Kelman, 2015). Among its four priorities is the aim to "investing in disaster risk reduction for resilience" (Aitsi-Selmi et al., 2015). However, financing for DRR has remained highly volatile, ex post, and marginal such that only 3.8% or US$5.2 billion of the total humanitarian financing was spent on DRR between 2005 and 2017 (WFP, 2021).

Financial investment towards DRR is very critical towards resilience building. The Multi-hazard Mitigation Council (MMC) also posited that for every US$1 spent on disaster risk reduction (DRR) saves between US$4 and US$11 in losses from a range of disaster incidences, ranging from floods, storms, and wildfires (MMC, 2018). It should be noted that the lack of investment in adaptation and resilience mirrors the lack of investment in disaster risk reduction (DRR). The Sendai Framework for Disaster Risk Reduction 2015–2030 (United Nations, 2015) set the following goal: "Prevent new and reduce existing disaster risk through the implementation of integrated and inclusive economic, structural, legal, social, health, cultural, educational, environmental, technological, political and institutional measures that prevent and reduce hazard exposure and vulnerability to disaster, increase preparedness for response and recovery, and thus strengthen resilience." It set four priorities, including "investing in disaster risk reduction for resilience" (Dicker et al., 2021).

17.3 Climate Change Adaptation in Rural Southern Africa: A Literature Review

There is a steadily growing body of literature on climate adaptation in rural Southern Africa (Ofosu et al., 2020; Dicker et al., 2021). In order for rural populations to adapt to climate change, these studies generally concur that climate change effects have influenced a shift from agrarian rural livelihoods towards mineral extraction-based livelihoods especially in mineral-rich parts of rural Southern Africa (Tschakert, 2009; Mkodzongi & Spiegel, 2019). However, the rapid proliferation of artisanal mining as an adaptation measure in parts of rural Southern Africa has intruded on contested terrains where competing economic interests are pursued (Esdaile & Chalker, 2018). Climate change in Southern Africa has created "agricultural poverty" which has had a "push" effect on the agrarian economy's labour force by driving it out towards the artisanal and small-scale mining (ASM) sector (Mkodzongi & Spiegel, 2020). ASM is generally considered to be a low-tech, low-capital, and labour-intensive resource extraction occupation of comparatively high value alluvial minerals such as diamonds and gold; it is also labelled as a poverty-driven livelihood with low-entry requirements and minimum mechanization (Ofosu et al., 2020). Therefore, both subsistence agriculture and artisanal mining inhabit the same geographic space whilst sharing and/or competing for the same inputs and exogenous factors.

This shift from agricultural-based livelihood frameworks towards ASM especially gold and diamond mining because of the emerging threat posed by global climate change on agricultural-based livelihoods has been referred to as deagrarianization (Bansah et al., 2023). Countries such as Zimbabwe, Malawi, South Africa, Zambia, and Botswana have recorded a marked increase in ASM activities from rural population in formerly agriculture-dominated spaces as a copying mechanism (Mkodzongi & Spiegel, 2019; Bansah et al., 2023). This failure to synchronize climate change livelihood adaptation mechanisms is one of the drivers of vulnerabilities in the rural agrarian economy.

However, there are also other studies that argue that there exists a mutually reinforcing relationship between climate change, agriculture, and ASM which can be either positive or negative (Cartier & Bürge, 2011; Kamlongera, 2011). ASM affects agriculture by destroying the land, whilst agriculture contributes towards climate change via greenhouse gas emissions. Likewise, climate change affects agriculture production, leading to food insecurity. On a positive note, ASM and agriculture can be complimentary livelihood activities such that revenue from the latter is used to support the former and vice versa (Mkodzongi & Spiegel, 2019; Persaud et al., 2017). Although this literature is critical in understanding the emerging adaptation mechanisms in the Southern Africa region, it nonetheless does not capture the effectiveness of ASM as an adaptation measure or its implications on the environment and the struggle to address climate change. Therefore, this study uses Mount Darwin district to trenchantly analyze the effectiveness of the shift from agriculture to ASM and its implications on the environment, livelihoods, and on climate change.

Literature on adaptation limitations is very scant. Available literature posits that there are several adaptation constrains which are technological, cultural, financial, environmental, economic, informational, social, attitudinal, and behavioural (Klein et al., 2014). These constrains are said to be evident in both the developed and developing world. Other studies further state that these adaptation constrains are primarily driven by path-dependent rigid institutional systems governing natural resources and public goods which operate at a much slower pace than changes in climate (Barnett et al., 2015). This bureaucratic lassitude acts as a catalyst to existing constrains whilst also obstructing adaptation initiatives.

In Zimbabwe, adaptation is fragmented at best and this is further compounded by a dearth in literature. The major constrain being the institutional predisposition that the prevailing drivers of vulnerability are non-climatic (Chanza, 2018). However, empirical evidence points towards increasing climate-induced vulnerabilities in the health, agriculture, water, human settlements, energy, industrial, and biodiversity sectors (Chanza, 2018). Furthermore, Chanza (2018) also posits that adaptation constrains in Zimbabwe also emanate from institutional generalization of data by the state and its stakeholders especially information on why adaptation is failing. A homogeneous approach is adopted despite the fact that climate change and adaptation approaches are not homogeneous across the country. This is further worsened by the fact that adaptation across the country is localized and is determined by the magnitude and severity of climatic events in specific regions (Chanza et al., 2019). This literature on Zimbabwe is very critical especially in identifying knowledge gaps on adaptation constrains. There are knowledge gaps especially on the implications of the localized adaptation initiatives on climate change, that is, What are the major adaptation constrains and are they able to meet the needs of the community?, What are the implications of the adaptation measures on the environment and on efforts to mitigate climate change?, And finally what are the possible measures to strengthen adaptation and mitigate climate change? The forthcoming methodology section further articulates how these questions were addressed.

17.4 Materials and Methods

The aim of the study was to examine the emerging adaptation constrains emanating from Mount Darwin district, Mashonaland central Zimbabwe. Therefore, the main question for the research was What are the major adaptation constrains affecting agriculture-based rural communities in Mount Darwin district? The data for the Mount Darwin district case study was collected between November 2022 and May 2023. Eighteen key informants were purposively sampled based on their knowledge of the district especially the environmental dynamics in the region since resettlement in the commercial farming communities of the district in 1982. In-depth interviews were used to collect data from the participants about the climate change-driven adaptation trends, emerging constrains, and their environmental consequences (Allmark et al., 2009). This approach was ideal since in-depth interviews last between 2 and 3 h per session. Each interview session lasted an average of 1 h 30 min to 2 h. This time frame is most appropriate for the survey tool since the in-depth interviews require an average of 1 h to 5 h per session (Deterding & Waters, 2021).

The informants for the study were identified through snowball sampling technique, whereby the traditional leaders hand-picked knowledgeable inhabitants for the interviews and Focus Group Discussions (FGD) (Moretti et al., 2011). The age representation FDG participants was between 18 and 73 years. There was a total of 18 participants consisting of 11 women and 7 males since the population of females is generally higher than that of males. There were three FDGs each made up of six participants and the researchers who moderated the discussion. The first FGD was held on 29 November 2022 and the other two on 20 April 2023. Each FGD lasted an average of 2 h. To ensure objectivity in participant selection, each subgroup of six participants was selected from different parts of Mount Darwin district (town, rural, and resettlement regions). These informants provided a balanced and representative perspective about the region since adaptation patterns in the district were determined by region. To ensure objectivity, each FDG had a mixture of both men and women and also at least one community leader. Each FGD had six participants to ensure full participation of informants. Initially, the researchers had four key informants who included two village heads, the chief and a retired colonel. They helped in identifying and introducing the researchers to the other key informants. These four individuals like most of the informants were also knowledgeable about agriculture since they own farms in the resettlement area of the district.

Apart from the in-depth interviews and FGDs, the study also adopted participant observation. During the research period, the researcher(s) participated in environmental rehabilitation exercises in the district, including tree planting and pit filling held on the 9 and 12 of May 2023. These two activities provided the researchers with ample evidence of the negative effects of climate change and adaptation constrains on the environment especially deforestation for tobacco curing and illegal artisanal gold mining on arable, grazing, and residential land. Furthermore, systematic literature review was adopted to identify articles and related literature from various data bases including Scopus, PubMed, Web of Science, ScienceDirect, and google scholar. This data was used to analyze the climate landscape of rural Southern Africa and the adaptation measures and related constrains within the region (Bramer et al., 2018). The study adhered to all ethical considerations and all the informants voluntarily agreed to participate in the study.

17.5 Presentation and Discussion of Findings

Results from the study indicated that there are several adaptation limits/constraints in Mount Darwin district which have wider implications for other parts of rural Southern Africa. These adaptation constrains are intensifying the effects of climate change in the district. In Mount Darwin district, chief Chareka noted that overpopulation in the farming resettlement regions was a serious constrain. When resettlement occurred in 1982 in the former commercial farming regions there were only 129 families resettled in the district the environment was rich with fertile soils, lush vegetation, and several dams to irrigate the crops. Currently (April 2023), all the dams are silted, water levels are below average, the population has increased exponentially, and most of the vegetation has been cut down for tobacco curing and cooking purposes.

The district has also been experiencing climate shocks especially in terms of rainfall which has been erratic for the past decade (2013–2023). This has greatly impacted on the major segment of the population which still depends on rain-fed agriculture. The problem of inadequate rainfall has affected both farmers in the rural and resettlement regions of the district. Although there are several dams for irrigating crops in the resettlement farms, siltation and poor rains have made it difficult to effectively utilize the dam water for irrigation. Furthermore, widespread deforestation has led to far-flung erosion of fertile soils contributing to further siltation of dams which militate against efforts to irrigate surrounding tobacco farms. Consequently, farmers in Mount Darwin now rely on natural rainfall which is threatened by climate change evidenced by shorter and more erratic rainfall episodes and increased dry spells observed in the last 40 years in the region pushing smallholder tobacco farmers out of business.

Evidence from the FDGs also indicated that ignorance, culture of impunity, and low technical support were also drivers of acute climate change effects. In all the three FGDs with members of the Mount Darwin community, none of the participants including the local leaders knew or understood what climate change is or its effects on their environment despite being its victims for several decades. This lack of knowledge about climate change and what it entails poses a great setback on adaptation and the 1998/1999 World Development Report (WDR) also argued that knowledge/the basic component of any region's knowledge system, not financial capital, is more critical for adaptive capability against climate shocks (Nyong et al., 2007: 792).

The district traditional leaders also pointed on the culture of impunity especially on deforestation for tobacco curing, cooking, and ASM. Although there are environmental protection laws upheld by the traditional leaders, police, and the environmental management agency (EMA), local authorities stated that they were powerless to enforce the environmental laws since their main duty was only to report the offenders to either the police or EMA agents. However, the police and EMA authorities were accused of accepting bribes from offenders leading to a culture of impunity regarding environmental laws. Furthermore, lack of knowledge regarding energy efficient tobacco curing was also another constrain. All the participants in the FGDs including the local leaders involved in tobacco farming stated that they used firewood for curing purposes instead of using coal, charcoal, or electric heaters. Researchers identified that tobacco curing was the main cause of deforestation in the district and lack of technical knowledge on other environmentally friendly curing options was the main driver.

Furthermore, commercial tobacco farming in Mount Darwin has been practised in the area for more than 50 years initially by the white commercial farmers and then African resettled farmers. The effects of climate change in the district have, however, undermined livelihoods resulting in the introduction of other adaptation strategies to compliment tobacco farming. Artisanal and small-scale gold mining (ASGM) or *Chikorokoza* has become an alternative local adaptation initiative especially for Tsakare resettlement area where gold deposits were first discovered in 2016. Out of the 700 households identified in Tsakare, at least 250 were subsisting on ASGM. Other inhabitants also opened small businesses (grocery stores, hardware shops, and liquor houses) as an alternative to agriculture following the discovery of gold which led to high demand in goods and services. Also, the government and several non-governmental organizations (NGOs) also provided food aid to rural inhabitants.

However, these adaptation measures have been inhibited by various constrains especially in terms of mitigating climate shocks, improving livelihoods, and halting climate change in the region. For example, ASGM though a popular adaptation measure was widely considered as a liability to climate change adaptation and community peace. ASGM like tobacco farming has also further intensified the climate change effects in the resettlement due to the rampant and unsustainable way the mining activities are conducted. Evidence from the local farmers in the district drawn from the FDGs revealed that artisanal mining was encroaching on agricultural land, residential, and public infrastructure. Consequently, some of the farms in the area were no longer suitable for farming due to rampant digging by artisanal miners. Furthermore, the discovery of gold was also labelled as a harbinger of local price fluctuations, whereby the price of basic goods and services in the gold-rich regions of the district were at least twice more expensive compared to other areas. Moreover, several roads, bridges, and buildings in the district were damaged by underground tunnels criss-crossing the gold-rich communal farming regions. Reports of arable land being converted to mine dumps, pollution of open water bodies and the draining of underground water alongside the widespread collapsing of roads, buildings, and bridges during the rainfall season because of rampant ASGM were also made by the informants. In the past 7 years since ASGM started in Tsakare resettlement, there has been significant land degradation due to the increase in the number of gold miners and extraction wells occupying more and more space.

Likewise, climate resilience assistance provided by the state and NGOs to aid rural inhabitants was reported be too little too late for it to render meaningful contribution in aiding climate adaptation. The state and other stakeholders provided seeds and fertilizers to rural populations to aid in agriculture. However, this aid was said to be provided much later usually in the middle of the farming season such that its contribution towards combating food security was minimal. More so, due to climate change shocks much of the crops were usually destroyed by poor rains. As a result, there were reports that much of the agricultural aid provided was sold by the beneficiaries or bartered in exchange for basic goods especially maize meal, cooking oil, sugar, and also livestock. Furthermore, the aid did not cover the whole district. Residents of Mount Darwin town and surrounding resettlement farms were not part of the beneficiaries. The state and other stakeholders argued that resettled farmers were food secure because of the farms they were allocated despite evidence of climate shocks whilst Mount Darwin town dwellers were said to be financially well-off since most of them were employed in the surrounding business area. However, climate shocks and micro-economic fluctuations from AGM undermined the well-being of both farmers and town dwellers, respectively.

These findings from Mount Darwin district are a microcosm of emerging climate shocks on rural adaptation. Challenges of deforestation emanating from unsustainable agricultural activities are not limited to Mount Darwin district alone. An estimated 20% of Zimbabwe's annual forest loss is attributed to tobacco curing (Bafana, 2022). This is also evident across the Southern Africa region whose land area and population only accounts for 18% and 13% of the global land area and population, respectively, but the region accounts for 26% of the global total annual cost of land degradation (WFP, 2021).

Furthermore, unsustainable tobacco farming which militates against climate change mitigation measures is not unique to Mount Darwin district alone but is a common feature in most parts of rural Southern Africa where tobacco farming is practised. Studies that have been conducted regarding the environmental impact of tobacco farming also concur that there are a plethora of negative environmental effects of tobacco farming (World Health Organization, 2017). The general agricultural practices associated with tobacco farming, particularly in low-income and middle-income states, result in deforestation and soil degradation. Moreover, agrochemical pollution and deforestation consequently lead to ecological disruptions that trigger losses in ecosystem services, including biodiversity, land resources, and food sources, which pose an existential crisis considering the emerging threat of climate change (Lecours et al., 2012; Lecours, 2014). Also, policies and practices of multinational tobacco companies further impact on environmental challenges connected to tobacco production (Ruckert et al., 2023).

Likewise, the shift from agriculture to artisanal mining as an adaptation strategy poses greater danger to climate mitigation efforts as it damages the environment and poisons it, which increases the threat of climate change in parts of rural Southern Africa where ASM is emerging as an adaptation measure. Fritz et al. (2018) also stated that increasing population, poverty, climate change, and the desire to escape poverty are the main drivers for the increase in global artisanal miners. In Zimbabwe, the ASGM sector is one of the country's key economic livelihoods especially for the rural population such that it accounts for 26.4% or 3.5 million Zimbabweans (Mkodzongi & Spiegel, 2019). Land degradation activities through ASM activities decrease the availability of land for agriculture which has negative consequences for food security, macroeconomic effects through food price fluctuation; in the long term, this has adverse implications for exports of cash crops and efforts to eradicate hunger (Bafana, 2022). Consequently, reports of conflict over resources especially water for mineral processing and irrigation purposes have been rife in some parts of Mount Darwin. The informal nature of ASM activities in rural Southern Africa has multiple negative consequences which impact on other pre-existing rural livelihood activities especially agriculture through shifting of labour towards the ASM sector, water and mercury pollution, land degradation, and farm invasions (African Centre for Economic Transformation, 2017; Ofosu et al., 2020).

Hence, the livelihoods activities in rural Southern Africa are unsustainable due to climate change and the alternative adaptation strategies meant to compliment them are no different. The environmental impact of these activities augments climate change effects in the region such that measures such as proscribing them are currently being implemented at a global scale against Southern African countries (Marufu, 2022, 2023). In February 2022 and July 2023, the World Health Organization (WHO) and European Union (EU), respectively, reported intentions of banning tobacco production for health and environmental reasons (Ruckert et al., 2023). Nevertheless, these measures of proscribing tobacco farming are currently being resisted by the rural population especially in rural Southern African countries including Zimbabwe, Zambia and Malawi which continue to embark on unsustainable tobacco farming activities due to lack of alternative livelihood activities (Ruckert et al., 2023). In Mount Darwin, the culture of impunity against environmental laws further encourages community resistance against environmental laws. Furthermore, the provision of tobacco inputs on credit by foreign companies especially from China, which is Zimbabwe's major tobacco importer also nullifies the WHO and EU threat to proscribe tobacco farming. Also, Zimbabwe is the sixth largest producer of tobacco globally and tobacco is the country's fourth largest foreign currency earner after gold, platinum, and diaspora remittances with annual exports receipts close to US$1 billion (about $3 per person) annually. The sector also directly supports 200,000 rural households in Zimbabwe alone whilst accounting for 50% of agriculture export receipts and contributes 25% of the agriculture gross domestic product (GDP) (Marufu, 2022).

Drawing from these results, what is currently transpiring based on the findings from rural Southern Africa in general and Mount Darwin district in particular is that domestic climate change adaptation mechanisms are failing to alleviate the effects of climate change but are instead posing a greater threat to the environment and rural livelihoods. The current proposition to proscribe tobacco farming to address environmental and health issues is currently being resisted partly because of inadequate capital investment in adaptation and lack of alternative livelihood options, further deepening the climate crisis. This partly explains why Southern Africa is one of the world's most vulnerable regions to climate change effects which is partly exacerbated by low capital distribution towards climate change resilience. Therefore, in order to minimize climate change effects robust domestic adaptation mechanisms urgently need to be mainstreamed in rural Southern Africa in order for the region to realize the targets for the green development agenda. Hence, there is urgent need to channel funding towards climate adaptation. The United Nations (UN) Secretary-General, António Guterres, also shared similar sentiments that

> Donor countries and multilateral, regional and national development banks need to significantly increase the volume and predictability of their finance for adaptation and resilience…50% of the total share of climate finance provided by all developed countries and multilateral development banks to be allocated to adaptation and resilience in developing countries (Guterres, 2021).

Another local measure that is critical in addressing climate change effects is to lobby rural populations through awareness campaigns in the affected regions. Findings from the study indicated that climate change is a phenomenon that is not known let alone understood by the inhabitants of Mount Darwin district. This is a likely scenario across other parts of rural Southern Africa. Therefore, educating rural populations about climate change will not only empower the vulnerable populations to embark on efforts to minimize its effects but will also provide the much-needed local buy-in of external adaptation measures which are at present being resisted.

Regionally, there is urgent need to diffuse domestically produced climate change mitigatory measures that have been scientifically tested and proven effective towards minimizing climate change effects especially in regions affected by similar climate change patterns. For example, in 2011, the Zimbabwean government's agriculture research unit developed and successfully tested an efficient version of tobacco curing barns. These curing barns were able to reduce firewood consumption by at least 50% whilst also minimizing the curing cycle to 5–6 days from 7–8 days, without compromising the quality of the tobacco (Marufu, 2022). This technology, if adopted by small-scale farmers, will drastically reduce deforestation which will minimize climate change effects in parts of rural Southern Africa where large-scale tobacco farming is conducted including in Zambia and Malawi.

Another climate change mitigatory measure is re-afforestation especially in regions affected by deforestation. Re-afforestation, if done properly can also become a livelihood activity since trees are a raw material for domestic and commercial purposes. However, in Mount Darwin no local measures are currently being made to restore trees lost annually in the area mainly from tobacco curing. The state can also play a critical role by addressing the issue of impunity against environmental offenders through upholding regulations and enforce compliance through robust governmental and institutional policies.

Likewise, there is need for diversification of rural livelihood activities especially in climate change-prone regions. Since tobacco farming and ASGM are systematically destroying the environment thereby speeding up climate change effects, diversifying agricultural activities can contribute towards addressing some of these issues. These alternative options can include the production of cannabis or industrial hemp which was legalized in Zimbabwe, citrus, stevia, macadamia nuts, cut flowers, saffron crocus, soya beans among other environmentally friendly crops can mitigate climate change effects whilst safeguarding livelihoods. Since Mount Darwin district is located within Zimbabwe's major agricultural region, most of these crops are suitable to its environment (Musemwa et al., 2013). However, diversification has its challenges which may include high initial costs; unidentified domestic and export markets, and a lack of improved seed varieties (Chanza et al., 2019).

In terms of addressing adaptation constrains created by ASM on the environment and agriculture, there is urgent need for governmental policy to first consider the impact of mineral-extraction pollution on the agrarian economy and local income especially in areas where mining takes place close to agriculturally productive regions like Mount Darwin. Also, the revenue from emerging adaptation interventions must be cautiously evaluated against the environmental costs. Therefore, policy makers need to distinguish the socio-economic value of emerging adaptation strategies and properly align them with other existing livelihood patterns so that each adaptation measure will have a positive impact on livelihoods, the environment and climate change.

17.6 Conclusion

Climate adaptation is a challenge which is beset by numerous setbacks. Evidence from the Mount Darwin district shows that existing agricultural dependant livelihood activities can no longer sustain the growing population and this is worsened by climate change compounded by unsustainable farming practices. Adaptation efforts to compliment agriculture especially AGM have brought about new challenges that threaten both the environment and agriculture whilst undermining efforts to mitigate climate change. Current measures instituted by the government and propositions by global institutions to bolster climate resilience are also not yielding positive results. Since agriculture is the main rural livelihood source for most parts of rural Southern Africa, findings from Mount Darwin district have a regional implication. Most countries in Southern Africa are predominantly vulnerable to climate change because of restricted adaptive capacity, due to widespread poverty, persistent droughts, unfair land distribution, and over dependence on rain-fed agriculture. Therefore, a risk-sharing approach among Southern African states can go a long way in overcoming current climate change adaptation constrains. Thus, a robust adaptive capacity for rural Southern Africa requires a concerted global, regional, national, and domestic approach which embraces and incorporates all stakeholders.

References

African Center for Economic Transformation. (2017). *The impact of expanding artisanal and small-scale mining on small holder agriculture in West Africa. A Case Study of Burkina Faso, Ghana and Sierra Leone*. ACET, Ford Foundation.

African Development Bank Group. (2023). *2023 Southern Africa Economic Outlook: Southern Africa's economic prospects subdued, yet abounds with investment opportunity in climate change initiatives*. https://www.afdb.org/en/news-and-events/press-releases/2023-southern-africa-economic-outlook-southern-africas-economic-prospects-subdued-yet-abounds-investment-opportunity-climate-change-initiatives-63343 Accessed 25 October 2023.

Aitsi-Selmi, A., Egawa, S., Sasaki, H., Wannous, C., & Murray, V. (2015). The Sendai framework for disaster risk reduction: Renewing the global commitment to people's resilience, health, and well-being. *International Journal of Disaster Risk Science, 6*, 164–176. https://doi.org/10.1007/s13753-015-0050-9

Allan, R. P., Hawkins, E., Bellouin, N., & Collins, B. (2021). *IPCC*, 2021: Summary for policymakers.

Allmark, P., Boote, J., Chambers, E., Clarke, A., McDonnell, A., Thompson, A., & Tod, A. M. (2009). Ethical issues in the use of in-depth interviews: Literature review and discussion. *Research Ethics, 5*(2), 48–54. https://doi.org/10.1177/174701610900500203

Bafana, B. (2022). *Climate change is smoking Zimbabwe's valuable tobacco crop*. https://allianceforscience.org/blog/2022/01/climate-change-is-smoking-zimbabwes-valuable-tobacco-crop/. Accessed 20 August 2023.

Bansah, K. J., Arthur-Holmes, F., & Assan, E. (2023). Climate induced transformation of agriculture to artisanal mining economy in dry regions. *Journal of Rural Studies, 99*, 11–19. https://doi.org/10.1016/j.jrurstud.2023.02.005

Barnett, J., Evans, L. S., Gross, C., Kiem, A. S., Kingsford, R. T., Palutikof, J. P., Pickering, C. M., & Smithers, S. G. (2015). From barriers to limits to climate change adaptation: Path dependency and the speed of change. *Ecology and Society, 20*(3), 5. https://doi.org/10.5751/ES-07698-200305

Bechtoldt, M. N., Götmann, A., Moslener, U., & Pauw, W. P. (2021). Addressing the climate change adaptation puzzle: A psychological science perspective. *Climate Policy, 21*(2), 186–202. https://doi.org/10.1080/14693062.2020.1807897

Bramer, W. M., De Jonge, G. B., Rethlefsen, M. L., Mast, F., & Kleijnen, J. (2018). A systematic approach to searching: An efficient and complete method to develop literature searches. *Journal of the Medical Library Association: JMLA, 106*(4), 531. https://doi.org/10.5195/jmla.2018.283

Cartier, L. E., & Bürge, M. (2011). Agriculture and artisanal gold mining in Sierra Leone: Alternatives or complements? *Journal of International Development, 23*(8), 1080–1099. https://doi.org/10.1002/jid.1833

Chagutah, T. (2010). *Climate change vulnerability and adaptation preparedness in Southern Africa Zimbabwe country report*. Heinrich Böll Stiftung.

Chanza, N. (2018). Limits to climate change adaptation in Zimbabwe: Insights, experiences and lessons. In W. Leal Filho & J. Nalau (Eds.), *Limits to climate change adaptation. Climate change management*. Springer. https://doi.org/10.1007/978-3-319-64599-5_6

Chanza, N., Chigona, A., Nyahuye, A., Mataera-Chanza, L., Mundoga, T., & Nondo, N. (2019). Diagnosing barriers to climate change adaptation at community level: Reflections from Silobela, Zimbabwe. *GeoJournal, 84*, 771–783. https://doi.org/10.1007/s10708-018-9890-3

Chenzi, V., & Ndamba, A. (2023). Surviving the Covid-19 lockdown: Zimbabwe's informal sector, 2020–2021. *Review of African Political Economy*. https://doi.org/10.1080/03056244.2023.2246276

Deloitte. (2012). *The rewards of a green economy: Cultivating operational value through sustainability partnerships*. Deloitte.

Department of Environmental Affairs (DEA). (2010). *Green economy summit report*. DEA.

Deterding, N. M., & Waters, M. C. (2021). Flexible coding of in-depth interviews: A twenty-first-century approach. *Sociological Methods & Research, 50*(2), 708–739.

Dicker, S., Unsworth, S., Byrnes, R., Ward, B., Bhatt, M., & Paul, A. (2021). *Saving lives and livelihoods: The benefits of investments in climate change adaptation and resilience*. Leeds, UK.

Epule, T. E., Chehbouni, A., Dhiba, D., & Moto, M. W. (2021). The readiness index for climate change adaptation in Africa: The role of climate and adaptive capacity proxies. *Applied Sciences, 11*(20), 9413. https://doi.org/10.3390/app11209413

Esdaile, L. J., & Chalker, J. M. (2018). The mercury problem in artisanal and small-scale gold mining. *Chemistry–A European Journal, 24*(27), 6905–6916. https://doi.org/10.1002/chem.201704840

Fritz, M., McQuilken, J., Collins, N., & Weldegiorgis, F. (2018). *Global trends in artisanal and small-scale mining (ASM): A review of key numbers and issues* (pp. 81–81). International Institute for Sustainable Development.

Global Climate Risk Index. (2021). *Who suffers Most from Extreme Weather Events? Weather-related Loss Events in 2019 and 2000 to 2019*. https://www.germanwatch.org/en/19777. Accessed 20 August 2023.

Guterres, A. (2021). *Remarks to the Climate Adaptation Summit, 25 January*. https://www.un.org/sg/en/content/sg/speeches/2021-01-25/remarks-climate-adaptation-summit Accessed 29 October 2023.

Helen Suzman Foundation. (2024). *The Southern African Development Community I – Population*. https://hsf.org.za/publications/hsf-briefs/the-southern-african-development-community-i-population

Kamlongera, P. J. (2011). Making the poor 'poorer'or alleviating poverty? Artisanal mining livelihoods in rural Malawi. *Journal of International Development, 23*(8), 1128–1139. https://doi.org/10.1002/jid.1836

Kelman, I. (2015). Climate change and the Sendai framework for disaster risk reduction. *International Journal of Disaster Risk Science, 6*, 117–127. https://doi.org/10.1007/s13753-015-0046-5

Khanal, U., Wilson, C., Rahman, S., Lee, B. L., & Hoang, V. N. (2021). Smallholder farmers' adaptation to climate change and its potential contribution to UN's sustainable development goals of zero hunger and no poverty. *Journal of Cleaner Production, 281*, 124999. https://doi.org/10.1016/j.jclepro.2020.124999

Klein, R. J. T., Midgley, G. F., Preston, B. L., Alam, M., Berkhout, F. G. H., Dow, K., et al. (2014). Adaptation opportunities, constraints, and limits. In C. B. Field,

V. R. Barros, D. J. Dokken, K. J. Mach, M. D. Mastrandrea, T. E. Bilir, M. Chatterjee, K. L. Ebi, Y. O. Estrada, R. C. Genova, B. Girma, E. S. Kissel, A. N. Levy, S. MacCracken, P. R. Mastrandrea, & L. L. White (Eds.), *Climate change 2014: Impacts, adaptation, and vulnerability. Part A: Global and sectoral aspects. Contribution of Working Group II to the fifth assessment report of the Intergovernmental Panel on Climate Change* (pp. 899–943). Cambridge University Press.

Kumah, C., Hilson, G., & Quaicoe, I. (2020). Poverty, adaptation and vulnerability: An assessment of women's work in Ghana's artisanal gold mining sector. *Area, 52*(3), 617–625. https://doi.org/10.1111/area.12639

Kumar, P., Sahu, N. C., Kumar, S., & Ansari, M. A. (2021). Impact of climate change on cereal production: Evidence from lower-middle-income countries. *Environmental Science and Pollution Research, 28*(37), 51597–51611. https://doi.org/10.1007/s11356-021-14373-9

Lecours, N. (2014). The harsh realities of tobacco farming: A review of socioeconomic, health and environmental impacts. In *Tobacco control and tobacco farming: separating myth from reality* (pp. 99–137).

Lecours, N., Almeida, G. E., Abdallah, J. M., & Novotny, T. E. (2012). Environmental health impacts of tobacco farming: A review of the literature. *Tobacco Control, 21*(2), 191–196. https://doi.org/10.1136/tobaccocontrol-2011-050318

Marufu, L. (2022). *Zim panics over tobacco ban.* https://businesstimes.co.zw/zim-panics-over-tobacco-ban/ Accessed 17 August 2023.

Marufu, L. (2023). *Zim faces new tobacco ban threat.* https://businesstimes.co.zw/zim-faces-new-tobacco-ban-threat/. Accessed 17 August 2023.

Menghistu, H. T., Abraha, A. Z., Tesfay, G., & Mawcha, G. T. (2020). Determinant factors of climate change adaptation by pastoral/agro-pastoral communities and smallholder farmers in sub-Saharan Africa: A systematic review. *International Journal of Climate Change Strategies and Management, 12*(3), 305–321. https://www.emerald.com/insight/content/doi/10.1108/IJCCSM-07-2019-0049/full/html

Mkodzongi, G., & Spiegel, S. (2019). Artisanal gold mining and farming: Livelihood linkages and labour dynamics after land reforms in Zimbabwe. *Journal of Development Studies, 55*(10), 2145–2161. https://doi.org/10.1080/00220388.2018.1516867

Mkodzongi, G., & Spiegel, S. J. (2020). Mobility, temporary migration and changing livelihoods in Zimbabwe's artisanal mining sector. *The Extractive Industries and Society, 7*(3), 994–1001. https://doi.org/10.1016/j.exis.2020.05.001

Moretti, F., van Vliet, L., Bensing, J., Deledda, G., Mazzi, M., Rimondini, M., Zimmermann, C., & Fletcher, I. (2011). A standardized approach to qualitative content analysis of focus group discussions from different countries. *Patient Education and Counseling, 82*(3), 420–428. https://doi.org/10.1016/j.pec.2011.01.005

Multi-hazard Mitigation Council. (2018). Natural hazard mitigation saves: 2017 interim report.

Musemwa, L., Mushunje, A., Muchenje, V., Aghdasi, F., & Zhou, L. (2013). Efficiency of resettled farmers in Mashonaland Central Province of Zimbabwe in crop production: A DEA approach. *African Journal of Agricultural Research, 8*(22), 2722–2729.

Nhamo, G. (2013). Green economy readiness in South Africa: A focus on the national sphere of government. *International Journal of African Renaissance Studies - Multi-, Inter- and Transdisciplinarity, 8*(1), 115–142. https://doi.org/10.1080/18186874.2013.834628

Nhemachena, C., Nhamo, L., Matchaya, G., Nhemachena, C. R., Muchara, B., Karuaihe, S. T., & Mpandeli, S. (2020). Climate change impacts on water and agriculture sectors in Southern Africa: Threats and opportunities for sustainable development. *Water, 12*(10), 2673. https://doi.org/10.3390/w12102673

Nyong, A., Adesina, F., & Osman Elasha, B. (2007). The value of indigenous knowledge in climate change mitigation and adaptation strategies in the African Sahel. *Mitigation and Adaptation Strategies for Global Change, 12*, 787–797.

Ofosu, G., Dittmann, A., Sarpong, D., & Botchie, D. (2020). Socio-economic and environmental implications of Artisanal and Small-scale Mining (ASM) on agriculture and livelihoods. *Environmental Science and Policy, 106*, 210–220. https://doi.org/10.1016/j.envsci.2020.02.005

Pauw, W. P. (2021). The adaptation finance gap can only be closed by limiting the adaptation costs. *One Earth, 4*(10), 1352–1355. https://doi.org/10.1016/j.oneear.2021.09.002

Pauw, P., Weikmans, R., Watson, C., Jahns, H., Prowse, M., Quevedo, A., & Puri, J. (2021). Global progress on financing for adaptation. In *Adaptation Gap Report 2020* (pp. 23–31).

Persaud, A. W., Telmer, K. H., Costa, M., & Moore, M. L. (2017). Artisanal and small-scale gold mining in Senegal: Livelihoods, customary authority, and formalization. *Society & Natural Resources, 30*(8), 980–993. https://doi.org/10.1080/08941920.2016.1273417

Rabezanahary Tanteliniaina, M. F., & Andrianarimanana, M. H. (2023). Projection of future drought characteristics in the Great South of Madagascar using CMIP6 and bias-correction spatial disaggregation method. *Theoretical and Applied Climatology*, 1–13. https://doi.org/10.1007/s00704-023-04727-3

Ruckert, A., Labonté, R., Lencucha, R., Goma, F., & Drope, J. (2023). Exploring the political economy nexus of tobacco production and control: A case study from Zambia. *Critical Public Health, 33*(1), 25–36. https://doi.org/10.1080/09581596.2021.1981540

Savo, V., Lepofsky, D., Benner, J. P., Kohfeld, K. E., Bailey, J., & Lertzman, K. (2016). Observations of climate change among subsistence-oriented communities around the world. *Nature Climate Change, 6*(5), 462–473. https://doi.org/10.1038/nclimate2958

Seyuba K., & Garcia T. F. (2022). *Climate-related security risks in the SADC region*. https://www.sipri.org/commentary/topical-backgrounder/2022/climate-related-security-risks-sadcregion#:~:text=Southern%20Africa%20is%20among%20the,majority%20of%20the%20region's%20population. Accessed 25 October 2023.

Statista. (2023). *Unemployment rate in Southern Africa from 2010 to 2023*. https://www.statista.com/statistics/1323958/unemployment-rate-in-southern-africa/ Accessed 5 February 2024.

The World Bank. (2019). *Agriculture and Food: Overview – Context*. https://www.worldbank.org/en/topic/agriculture/overview Accessed 19 October 2023.

Thinda, K. T., Ogundeji, A. A., Belle, J. A., & Ojo, T. O. (2020). Understanding the adoption of climate change adaptation strategies among smallholder farmers: Evidence from land reform beneficiaries in South Africa. *Land Use Policy, 99*, 104858. https://doi.org/10.1016/j.landusepol.2020.104858

Thomas, A., Theokritoff, E., Lesnikowski, A., Reckien, D., Jagannathan, K., Cremades, R., Campbell, D., Joe, E. T., Sitati, A., Singh, C., & Segnon, A. C. (2021). Global evidence of constraints and limits to human adaptation. *Regional Environmental Change, 21*(3), 1–15. https://doi.org/10.1007/s10113-021-01808-9

Tschakert, P. (2009). Recognizing and nurturing artisanal mining as a viable livelihood. *Resources Policy, 34*(1-2), 24–31. https://doi.org/10.1016/j.resourpol.2008.05.007

United Nations Climate Change Secretariat (UNCCS). (2017). *Opportunities and options for integrating climate change adaptation with the Sustainable Development Goals and the Sendai Framework for Disaster Risk Reduction 2015–2030*.

United Nations Development Programme. (2020). *Zimbabwe launches new climate change adaptation project to reach 2.3 million vulnerable people*. https://www.adaptation-undp.org/zimbabwe-launches-new-climate-change-adaptation-project-reach-23-million-vulnerable-people

United Nations Office for Disaster Risk Reduction (UNDRR). (2019). *Global Assessment Report on Disaster Risk Reduction 2019 (GAR19)*. https://www.undrr.org/publication/global-assessment-report-disaster-risk-reduction-2019. Accessed 25 October 2023.

World Food Programme (WFP). (2021). *Climate Change in Southern Africa*. https://executiveboard.wfp.org/document_download/WFP0000129015#:~:text=Impacts%20of%20changes%20in%20temperatures%20and%20rainfall%20patterns%20are%20already,intensity%20of%20droughts%20and%20floods

World Health Organization. (2017). *Tobacco and its environmental impact: An overview*. World Health Organization.

18 Conclusions and Policy Recommendations: Dynamics, Prospects, and Challenges of Climate Resilience in Rural Southern Africa

Lazarus Chapungu, Godwell Nhamo, and Mark Matsa

18.1 Introduction

In examining the dynamics, prospects, and challenges of climate change resilience in rural Southern Africa, this book has illuminated the multifaceted and interconnected nature of climate vulnerability and adaptation. Rural communities in this region face an array of climate-related challenges, including prolonged droughts, unpredictable rainfall patterns, and increased frequency of extreme weather events. These challenges are exacerbated by socio-economic factors such as poverty, limited access to resources, and inadequate infrastructure. This concluding chapter provides a summary of main findings and the major conclusions and recommendations emerging from the whole book.

L. Chapungu (✉) · G. Nhamo
Institute for Corporate Citizenship, University of South Africa, Pretoria, South Africa
e-mail: tchapul@unisa.ac.za; nhamog@unisa.ac.za

M. Matsa
Department of Geography, Environmental Sustainability and Resilience Building, Midlands State University, Gweru, Zimbabwe
e-mail: matsam@staff.msu.ac.zw

18.2 Emerging Key Findings

It emerged that the climate crisis in Southern Africa poses significant challenges for vulnerable rural populations, confirming earlier observations by various scholars (Nhamo et al., 2019; Shoko Kori et al., 2024; Ziti et al., 2024). Varied temperature extremes and erratic rainfall patterns contribute to droughts, floods, and crop failures, impacting agricultural activities across the region. Without proactive intervention efforts, these challenges are expected to worsen, leading to heightened vulnerability. Limited adaptation support and institutional responses exacerbate the situation, risking setbacks in sustainable development efforts, including poverty alleviation and nutrition initiatives. The resulting vulnerabilities may prompt rural-to-urban migration, magnifying social strains and economic pressures. These and other emerging key findings are now summarized herein.

18.2.1 Increasing Trends of Climate Variability and Vulnerability

Results show that the increasing occurrences of temperature extremes and droughts disproportionately affect poorer communities, leading to income loss, agricultural failures, and food short-

M. Matsa et al. (eds.), *Climate Change Resilience in Rural Southern Africa*, Sustainable Development Goals Series, https://doi.org/10.1007/978-3-031-73600-1_18

ages. These vulnerabilities are compounded by limited adaptive capacity within rural communities. Earlier studies (e.g. Jiri & Mafongoya, 2017; Shoko Kori et al., 2024; Thierfelder et al., 2017; Thierfelder & Steward, 2022) confirm this observation. The book also highlights a significant rise in temporal climate vulnerability, with climate change amplifying existing inequalities and intersectional vulnerabilities. One notable consequence is the increased displacement resulting from climate-induced events, as also noted by Graham (2020) and Guo et al. (2024). The impacts of climate change span various sectors and are closely intertwined with social, economic, political, and geographical factors. Despite high overall vulnerability, disparities exist both between and within countries due to divergent adaptation strategies, economic conditions, and socio-political dynamics.

Several drivers of vulnerability to climate change are identified, including the efficacy of disaster management systems, dependence on agriculture, poverty levels, access to social protection mechanisms, exposure to disasters, economic fragility, and water scarcity. Moreover, vulnerabilities are deeply entrenched in social structures and intersectional marginalization, such as gender disparities, underscoring the imperative for structural reforms to foster gender-inclusive and climate-resilient development.

18.2.2 Matters Pertaining to Adaptive Capacity

The book underscores the effectiveness of integrating socio-ecological and economic systems management to bolster adaptive capacity. Strong governance frameworks and sufficient financial backing for institutions and research are highlighted as pivotal in charting clear adaptation pathways and fostering resilience-building initiatives. This finding is also supported by earlier studies (e.g. Cassim et al., 2021; Gilder & Rumble, 2021). Financing, including support for community-based adaptation activities, emerges as a critical component of successful adaptation efforts, particularly given the urgency imposed by worsening climate change impacts. However, adaptation encounters various barriers, necessitating prompt and comprehensive responses, considering the rapid manifestation of climate change outpacing mitigation initiatives. Among these barriers, social obstacles pose considerable challenges, significantly influencing the success or failure of adaptation activities. Mutambisi et al. (2021) and Shoko Kori et al. (2024) also observed some of these barriers. Societal changes stemming from climate variations may exacerbate discriminatory practices and widen societal disparities and marginalization, thus warranting holistic and integrated approaches to address adaptation barriers.

18.2.3 Disaster Risk Reduction Strategies to Address Climate Hazards

Rural Southern Africa is highly susceptible to various climate-related hazards, including droughts, floods, cyclones, and tornadoes, with a rising trend in their occurrence across the region. These disasters often result in the loss of productive assets, particularly as rural livelihoods are predominantly dependent on agriculture. To address these challenges, numerous policies, organizations, and institutions have been established for disaster risk reduction (DRR) purposes, emphasizing community involvement and decentralization of DRR structures to the grassroots level. Local institutions, including traditional leadership, play a crucial role as the primary responders to climate-related disasters, particularly rapid-onset events. However, they are often overlooked in DRR initiatives, despite their significant influence within local communities, leading to notable gaps in the process. The financial burden of DRR activities predominantly falls on governments and donor agencies, posing challenges, especially in financially constrained regions. Effective communication emerges as a cornerstone in preparing at-risk communities for disasters, with localized sources proving most effective for issuing early warnings. While anticipatory action to DRR is partially practiced across many countries in rural Southern Africa, there is a need for its upscale to realize desired impacts. Key obstacles to DRR initiatives in the region include limited funding, technical expertise, and logistical challenges.

18.2.4 Status of Climate Governance Structure and Mechanisms

It emerged that governance mechanisms aimed at fostering climate resilience in rural Southern Africa exhibit fragmentation, leading to incomplete support for communities adapting to climate change effects, confirming observations by Addaney et al. (2023), Garland (2014), and Stringer et al. (2009). Although theoretical structures for climate-responsive planning exist, their practical implementation often fails to yield more resilient communities. Inadequate communication and integration between the identification of floodplains or other disaster-prone areas, and critical functions like building permitting and land use planning, hinder the translation of sound policy directives into resilient communities.

Furthermore, mismatches between organizational mandates and capacities become evident during crises, such as delays in responsible institutions in responding when individuals are missing during disasters. The scope of the problem posed by climate change exceeds the capacities of national crisis response organizations, as it is primarily driven by global forces beyond their control. Consequently, coordination of central activities for integrated planning and prevention is challenging and often falls under the purview of transboundary and multilateral organizations, which may not align with disaster timelines effectively.

The experiences in Chimanimani and Matopos in Zimbabwe highlight the need for greater alignment of incentives among various stakeholders, ranging from individual household-level climate adaptation efforts to government support and response. Addressing this requires strengthening integrated planning processes.

18.2.5 Communities in Rural Arid Regions Are Extremely Susceptible to Climate Change

Using case studies from Zimbabwe, it was established that rural communities in the arid regions experience frequent and severe droughts, significantly undermining their resilience to climate-related adversities. The resilience of rural communities is notably influenced by the prevalence of crop diseases or pests. Moreover, the resilience of these communities is substantially impacted by farming activities, encompassing both crop and livestock production in terms of variety and quantity. Ownership of cattle emerges as a particularly beneficial factor in enhancing rural community resilience. Furthermore, communities reliant on indigenous food crops such as sorghum and pearl millet demonstrate higher levels of resilience compared to those solely dependent on maize cultivation. Key determinants of resilience include education level, access to information, land tenure, agro-processing training, market proximity, and income diversification. Consequently, the study advocates for the improvement of climate information systems and the mobilization of resources to address the underlying causes of climate vulnerability from a policy and strategic perspective.

18.2.6 Climate-Smart Opportunities Are Available for Resilience Building

The transition to climate-smart communities and the augmentation of resilience to extreme weather conditions appear to be progressing at a sluggish pace. Several barriers impede the swift adoption of climate-smart initiatives and transition. These barriers encompass inadequately developed rural infrastructure and limited access to modern technology. Additionally, rural inhabitants often lack sufficient knowledge and awareness of climate change adaptation strategies, while public extension service providers exhibit limited capacity to adequately support rural communities in adopting climate-smart strategies. Moreover, inadequate institutional support emerges as a significant challenge in the region, leading to a lack of policy implementation even when policies are in place. Furthermore, social and cultural values entrenched within certain communities pose barriers to the adoption of adaptation strategies and the transition to climate-smart livelihoods. Although opportunities and challenges are shared among countries in the region, variations may occur due to differing contextual factors. Several opportunities for building climate-resilient communities have been noted. Southern Africa boasts abundant renewable energy resources, including solar, wind, and hydroelectric power. Expanding the use of renewable energy technologies presents an opportunity to reduce greenhouse gas emissions, enhance energy security, and mitigate climate change impacts. More so, the region's diverse agroecological zones and fertile soils provide significant potential for agricultural production. Adopting climate-smart agricultural practices, such as conservation agriculture, agroforestry, and crop diversification, can increase resilience to climate variability and improve food security.

Despite water scarcity challenges in some areas, Southern Africa has access to abundant freshwater resources, including rivers, lakes, and aquifers. Sustainable water management practices, such as rainwater harvesting, water recycling, and efficient irrigation techniques, can enhance water security and resilience to droughts and water scarcity. In addition, the region is rich in biodiversity and ecosystem services, which provide critical support for human well-being and livelihoods. Protecting and restoring natural ecosystems, such as forests, wetlands, and mangroves, can enhance resilience to climate change impacts, regulate local climates, and provide essential ecosystem services.

Southern Africa also has seen significant advancements in technological innovation and ICT infrastructure in recent years. Leveraging technology, such as climate information systems, remote sensing, and mobile applications, can improve early warning systems, facilitate climate-smart decision-making, and enhance community resilience. Moreover, indigenous knowledge systems and traditional practices embedded within local communities offer valuable insights into adapting to climate change. Integrating indigenous knowledge with modern scientific approaches can inform context-specific climate-smart solutions and enhance community resilience. Lastly, Southern Africa benefits from international collaboration and support through partnerships, funding mechanisms, and capacity-building initiatives. Engaging with international organizations, donor agencies, and research institutions can facilitate knowledge sharing, technology transfer, and financial support for climate-smart initiatives.

18.2.7 Climate-Pandemics Interface Amplify Existing Vulnerabilities

Southern Africa's healthcare infrastructure plays a crucial role in determining its resilience to pandemics. In many parts of the region, healthcare systems are already strained due to limited resources, inadequate facilities, and workforce shortages. Climate change has exacerbated these challenges by increasing the frequency and severity of extreme weather events, such as heatwaves and floods, which can damage healthcare facilities and disrupt service delivery. Moreover, changing climate patterns facilitate the spread of vector-borne diseases, such as malaria and dengue fever, further burdening healthcare systems.

Southern Africa is characterized by high levels of poverty, inequality, and food insecurity, which can exacerbate vulnerability to pandemics. Climate change can exacerbate these socio-economic vulnerabilities by affecting agricultural productivity, water availability, and livelihoods, leading to increased food insecurity and economic hardship. Vulnerable populations, such as informal settlers and rural communities, face heightened risks during pandemics due to inadequate access to healthcare, sanitation, and social protection measures.

Climate change-induced environmental degradation, such as deforestation, land degradation, and loss of biodiversity, can compromise ecosystem resilience and increase the risk of zoonotic disease transmission. Deforestation, for example, can bring humans into closer contact with wildlife habitats, increasing the likelihood of emerging infectious diseases, such as Ebola and COVID-19. Additionally, changes in temperature and precipitation patterns can alter the distribution and abundance of disease vectors, such as mosquitoes and ticks, further increasing the risk of disease transmission.

Access to clean water and sanitation is essential for preventing the spread of infectious diseases, including pandemics. In Southern Africa, climate change can exacerbate water scarcity and water quality issues, making it challenging to maintain adequate hygiene and sanitation practices. Droughts and floods can disrupt water supply systems, contaminate water sources, and increase the risk of waterborne diseases, such as cholera and typhoid fever. Southern Africa's resilience to pandemics is also influenced by its adaptive capacity, including the ability to anticipate, respond to, and recover from health crises. Strengthening healthcare systems, investing in disease surveillance and early warning systems, and promoting community engagement and resilience-building initiatives are essential for enhancing adaptive capacity. Additionally, integrating climate change considerations into pandemic preparedness and response strategies can help address the interconnected challenges of climate change and public health.

18.2.8 Mainstreaming Climate Resilience Strategies and Policies

It emerged that climate resilience strategies in Southern Africa often employ integrated approaches that consider multiple sectors and stakeholders. These approaches recognize the interconnectedness of social, economic, and environmental systems and aim to address climate risks holistically. Integrated strategies may involve collaboration between government agencies, civil society organizations, the private sector, and local communities to develop and implement climate resilience measures.

18.2.9 Financing Gap for Climate Resilience

The financial deficit confronting countries in Southern Africa amounts to approximately US$1 trillion, with approximately 70% allocated to adaptation efforts and 21.7% earmarked for mitigation endeavours. While these estimations of requisite climate finance for adaptation and mitigation have sparked considerable debate among scholars, there exists a consensus that the climate finance deficit facing Southern African countries is of considerable magnitude, rendering it insufficient to address the pressing adaptation imperatives of the region.

18.2.10 Climate Mitigation Policies Not in Sync with Mitigation Efforts

Climate change mitigation policies in Southern Africa encompass a range of strategies aimed at reducing greenhouse gas emissions and promoting low-carbon development pathways. These policies are designed to address the drivers of climate change, while supporting sustainable economic growth and development. They reflect a multifaceted approach that integrates sector-specific measures with broader strategies for sustainable development and emissions reduction. By implementing and scaling up these policies, Southern African countries aim to contribute to global efforts to combat climate change while advancing their own development goals.

However, this book underscores the reality that climate change mitigation interventions remain weak, while policy frameworks remain high on a political agenda. The policy implementation gap is marred by fragmented policies, limited resources, a lack of expertise, and heavy reliance on consultancy when formulating policies.

18.2.11 Women Remain at the Coalface of Climate Vulnerability

It emerged that women in rural areas of Southern Africa are disproportionately affected by climate change due to their roles as primary caregivers, food producers, and water managers, as well as their limited access to resources and decision making power within households and communities. Climate change-induced changes in precipitation patterns, water availability, and ecosystem degradation can exacerbate women's workload, increase their exposure to environmental hazards, and undermine their livelihoods. Limited access to education, training, and formal employment opportunities further constrains women's economic resilience and adaptive capacity. Gender inequalities, patriarchal norms, and cultural practices often limit women's decision-making power, mobility, and voice in household and community affairs. These social barriers can hinder women's ability to participate in climate change adaptation and mitigation initiatives, advocate for their needs, and access support services and resources.

The impact of climate change on women is exacerbated by the existing climate policies and strategies, which are largely not engendered. The resilient strategies at national and local levels are also not contextual, are externally defined and driven. Women are not active participants in resilience because of poverty and gender inequality barriers. Resilient projects and programmes do not transform unequal power relations of access, participation, decision-making and control. Invisible resources and the knowledge of communities is not tapped. There is an assumption that issues of power and equality are invalid, yet they are core. A positive correlation between women empowerment and climate resilience was also noted, and empowering women is crucial for enhancing climate resilience and sustainable development in the region.

18.2.12 Digital Technologies Play a Pivotal Role

Digital technologies and artificial intelligence (AI) tools are extensively employed to bolster farmers' resilience, both directly and indirectly, in the face of challenges induced by climate change. This finding confirms earlier studies which have shown that there is an opportunity for smallholder farmers to leverage on the availability of smart innovations to bolster their resilience to climate change (Makate et al., 2019; Nhamo et al., 2019; Nkhata, 2022). Digitalization has streamlined the dissemination of climate information among stakeholders and rural populations, facilitating weather prediction, conducting risk assessments, and facilitating evacuations in the aftermath of climate-related disasters. Several conducive factors contribute to this phenomenon, with potential avenues for enhancement. Factors such as the expansion of network coverage, proliferation of mobile phone ownership among households, initiatives for rural electrification, rising literacy rates, extension efforts by agricultural departments, and heightened awareness regarding digital applications collectively present significant opportunities for leveraging digital platforms and AI tools. The policy landscape has provided a framework for farmers to enhance their digital proficiency and augment the utilization of information and communication technologies (ICT) to surmount the challenges posed by climate change.

However, several drawbacks associated with the utilization of digital and AI tools have been identified. These include limited access to energy, insufficient digital literacy within certain communities, a lack of proficiency in ICT, limited access to smart devices, geographic inaccessibility of certain regions, inadequate energy availability to power AI and digital technologies, and unreliable network services provided by network operators. Furthermore, the emergence of digital technologies has been recognized as a potential threat to citizens' security, with concerns regarding misinformation being closely linked to the process of digitalization. Rather than bolstering communities' resilience, digitalization may inadvertently exacerbate vulnerability through the dissemination of misinformation and susceptibility to cyberattacks. Moreover, these technologies heavily rely on consistent power supply and are subject to the interdependencies among their operators. Additionally, there exists a palpable tension in human–machine interaction during decision-making processes, stemming from a perceived lack of assurance that these technologies prioritize users' best interests. This scepticism often arises from challenges associated with explaining and interpreting digital data into actionable and comprehensible terms for human users.

Despite the prevailing challenges within the digital landscape, there is a noticeable upward trend in the adoption of digital technologies across rural Southern Africa, with promising prospects for further expansion of ICT usage to bolster the capacity of rural farmers in confronting the challenges posed by climate change. Continued governmental support through infrastructure development is imperative to sustain this momentum. The pivotal role played by non-governmental organizations cannot be overstated, as they have significantly contributed to the digitalization of rural areas, particularly in supporting agricultural initiatives aimed at enhancing resilience to climate change. It is essential for governments in Southern Africa to foster collaboration among stakeholders by enacting supportive sector-wide policies and facilitating the provision of necessary infrastructure to enhance the resilience of rural populations against climate-induced phenomena and mitigate the associated damage and loss attributable to climate change.

18.2.13 Rural Dwellers Have Climate Change Knowledge and Are Adapting

Findings reveal that local farmers have diversified their livelihood strategies through engagement in both on-farm and off-farm activities. Several studies confirm the effectiveness of indigenous knowledge systems in climate adaptation in rural southern Africa (e.g. (Chapungu & Sibanda, 2015; Leal Filho et al., 2022; Makondo & Thomas, 2018; Mutambisi et al., 2021). On-farm activities encompass market gardening and cultivation of traditional small grains known for their drought-resistance properties. Off-farm pursuits include beer brewing and vending for monetary gain, vending of fruits and vegetables, trading of clothing items, as well as the establishment of poultry and goat rearing projects. However, in their endeavours to adapt to climate change, local rural dwellers face a plethora of challenges including poor soils, increased drought frequency and severity, dwindling river flow rates, and diminished crop yields.

18.3 Policy Recommendations

Given the complex dynamics, prospects, and challenges of climate change resilience in rural Southern Africa, it is imperative to adopt comprehensive and multifaceted policy and practical approaches. Such selected approaches are presented herein.

Investment in climate-resilient infrastructure: Investment in climate-resilient infrastructure, such as water storage facilities, irrigation systems, and sustainable energy sources, should be given top priority by governments and international organizations. This would lessen the effects of problems brought on by climate change, like droughts and irregular rainfall patterns.

Promotion of sustainable Agricultural practices: Policies should promote the adoption of sustainable agricultural practices such as agroforestry, conservation agriculture, and the use of drought-resistant crop varieties. Providing incentives and support for farmers to adopt these practices can enhance their resilience to climate change.

Enhanced access to climate information and early warning systems: Efforts should be made to improve access to climate information and early warning systems for rural communities. This could involve the development of mobile-based platforms, community radio programs, and other communication channels to disseminate timely information about weather patterns and potential climate-related risks.

Capacity building and training programmes: Governments and other key stakeholders should invest in capacity building and training programmes to enhance the climate resilience of rural populations. This could include training in sustainable land management practices, disaster preparedness and response, and the use of climate-smart technologies.

Integration of climate resilience into development policies: Climate resilience considerations should be integrated into broader development policies and strategies at the national and regional levels. This could involve mainstreaming climate adaptation and mitigation measures into agricultural development plans, land use policies, and rural development initiatives. In addition, DRR must be incorporated into all sectors and policies, from local to national governance. It is imperative that a certain amount of the overall money allocated to DRR be allocated toward this integration. The mainstreaming of DRR into development planning at all societal levels must take place concurrently with integration. Building capacity is necessary for DRR initiatives in order to support the extension of knowledge across various spatial coverages. Such capacity building must employ a cross-sectoral strategy. More local players must be included into DRR institutions in rural southern Africa in order to enhance them.

Community-based adaptation and resilience building: Policies should support community-based adaptation and resilience-building initiatives, recognizing the unique knowledge and resources that local communities possess. This could involve providing funding and technical support for community-led projects such as reforestation efforts, water harvesting schemes, and sustainable livelihood diversification strategies.

Strengthening of institutional frameworks and governance mechanisms: There is a need to strengthen institutional frameworks and governance mechanisms for climate resilience at the local, national, and regional levels. This could involve establishing multi-stakeholder platforms for climate adaptation planning and coordination, enhancing the capacity of government agencies responsible for climate-related issues, and fostering partnerships between government, civil society, and the private sector.

Financial support mechanisms: Governments, development partners, and the private sector should provide financial support mechanisms, such as climate adaptation funds and insurance schemes, to help rural communities cope with and recover from climate-related shocks and disasters. In addition, it is essential to put in place innovative financial and investment instruments as well as financial mechanisms that can provide regional-tailored and all-inclusive solutions to the climate finance challenges confronting Southern Africa. For instance, it is crucial to strengthen domestic resource mobilisation to better meet the climate finance needs of the region particularly rural community's adaptation which remains structurally underfunded (Nhamo & Chapungu, 2024). In addition, enhancing equity in the allocation of climate finance resources between rural communities and urban areas within Southern Africa is imperative to cushion the most vulnerable groups of people and sectors. This must involve gender considerations more comprehensively and explicitly. Equity financing aligns the interests of investors and farmers, fostering a sense of shared responsibility. By sharing the risk, investors will be motivated to support farmers adopting more climate-resilient agricultural practices.

International donor financial support is essential for building climate resilience in rural Southern Africa. Although philanthropic contributions in the region primarily target rural communities, much of this support focuses on mitigation and is typically programme-based, rendering it unsustainable in the long term. Therefore, it is crucial for regional governments to collaborate with and guide the efforts of the international donor community towards adaptation solutions, where the funding gap is significant. International donor funding, anchored in a multi-stakeholder approach and informed by data-driven evidence, has the potential to stimulate investments in rural climate adaptation projects and bridge the adaptation finance gap. For example, international financial institutions like the World Bank and the African Development Bank can offer low-cost loans and innovative financial mechanisms to mitigate risks forecasted by the private sector. Additionally, donors can advance climate adaptation in rural Southern Africa through research and development, infrastructure expansion and management, and by supporting governments in harnessing indigenous knowledge systems to build climate resilience.

Promoting private sector investments in climate action can help address the substantial gap in climate finance needed for adaptation, thereby fostering climate-resilient rural communities in Southern Africa. The private sector's effective role is crucial for promoting green investments and enhancing modern climate-smart technologies essential for managing and predicting future climate shocks. Policy-makers must create a favourable and conducive environment for private sector investments in climate action in rural Southern Africa by simplifying business processes and offering incentives such as tax holidays, free land, and ensuring the availability of supporting infrastructure and utilities. Although public infrastructure for building climate resilience in many Southern African countries is financed by public institutions, fostering public–private partnerships through guarantees, co-financing, or other risk-reduction strategies is vital for attracting private capital.

Beyond providing climate finance, rolling out capacity-building and knowledge-transfer programmes is vital for enhancing rural climate resilience in Southern Africa. For instance, institutional capacity-strengthening programmes can be directed at international and local organizations that implement climate resilience projects in rural areas to improve their project and financial management skills. Good governance of climate finance requires strengthening the capacity of community-based organizations in Southern Africa to engage effectively in the design and implementation of fundable climate resilience projects. Moreover, governments, with support from other stakeholders, can provide mentorship, technical expertise, and skills, as well as networking and empowerment programmes for youths and women to adapt to and effectively respond to climate change challenges. Investors may also train local people on how to use new climate-smart technologies to boost agricultural productivity, enhance income diversification, and increase climate resilience.

Promotion of research and innovation: Policies should promote research and innovation in climate resilience technologies and practices, with a focus on locally appropriate solutions. This could involve supporting research institutions, universities, and innovation hubs to develop and disseminate climate-smart technologies and practices tailored to the needs of rural Southern Africa. While existing indices provide insight into rural exposure to climate hazards, they may not fully capture the diversity of rural livelihoods, such as ecotourism and subsistence crop farming. Future studies could expand spatial data coverage to include alternative "village indices" that encompass different rural livelihoods. Additionally, analyzing combined impacts of temperature, humidity, wind, and rainfall extremes can provide a more comprehensive understanding of climate change impacts, highlighting areas where interventions are most urgently needed. While it is evident that Southern Africa is increasingly vulnerable to climate change, more localized research is necessary to elucidate the nuanced dynamics of this vulnerability, considering the region's diverse contexts and localised impacts. Hence, future research endeavours should encompass dryland areas across all southern African countries to provide a more comprehensive understanding of resilience dynamics in the region.

Promotion of cross-border cooperation: Given the transboundary nature of climate change impacts, there is a need for increased cooperation and collaboration among countries in the Southern African region. Policies should promote cross-border initiatives for sharing knowledge, resources, and best practices in climate resilience and adaptation.

18.4 Conclusions

A complexity of dynamics, prospects, and barriers influencing climate resilience activities in Southern Africa has been presented from an interdisciplinary and multidisciplinary lens. The narratives emerging depict a plethora of adaptation tactics based on local knowledge and resourcefulness. These narratives, which range from smallholder farmers' communal resilience plans to grassroots groups' innovative responses, highlight the agency and ingenuity of local actors in dealing with climate challenges. However, with these narratives of resilience, there are persistent systemic constraints that prevent the manifestation of adaptive capacities. Structural inequalities, which are defined by unequal access to resources and opportunity, create vulnerabilities, especially among excluded groups. Moreover, the interplay of socio-economic factors, environmental degradation, and political instability amplifies the impacts of climate change, further compromising resilience.

The book underscores the imperative of advancing climate resilience through a multifaceted approach that integrates local perspectives, scientific insights, and policy interventions. Central to this endeavour is the empowerment of local communities as agents of change, through participatory decision-making processes and capacity-building initiatives that leverage indigenous knowledge systems. Interdisciplinary and multidisciplinary collaboration emerges as the linchpin for effective climate resilience action, bridging the divide between scientific research, policy formulation, and community engagement. By fostering dialogue and knowledge exchange across sectors, disciplines, and stakeholders, the region can catalyze innovative solutions that address the complex and interconnected challenges of climate change.

Going forward, achieving climate resilience in rural Southern Africa requires a long-term commitment to transformative change. This needs not only investments in adaptable infrastructure and technology, but also systemic reforms that address the underlying causes of vulnerability, such as poverty, inequality, and unsustainable development practices.

References

Addaney, M., Kwadwo Dumenu, W., & Jarbandhan, D. B. (2023). Introduction to climate change, adaptation policy and resilient development in Africa. In *Climate change in Africa: Adaptation, resilience, and policy innovations* (pp. 1–9). https://doi.org/10.1007/978-3-031-30050-9_1

Cassim, A., Radmore, J.-V., Dinham, N., & Mccallum, S. (2021, January). South African climate finance landscape 2020.

Chapungu, L., & Sibanda, F. (2015). Effectiveness of conventional indigenous practices in climate change adaptation and mitigation in Masvingo District, Zimbabwe. In *Indigenous knowledge in Zimbabwe* (pp. 22–39). Lay Found.

Garland, R. M. (2014). National policy response to climate change in South Africa. *South African Medical Journal, 104*(8), 584. https://doi.org/10.7196/SAMJ.8605

Gilder, A., & Rumble, O. (2021). *Implications of COVID-19 for climate finance in Africa.* https://www.africaportal.org/publications/implications-covid-19-climate-finance-africa/

Graham, C. (2020). Climate-induced population displacement in sub-Saharan Africa: A review of resilience-building strategies. *Geoforum, 117*, 300–303. https://doi.org/10.1016/j.geoforum.2020.07.004

Guo, X., Zhang, Z., Zhang, X., & Feng, S. (2024). Spatiotemporal changes of climatic disasters and population displacements in Africa. *Climate and Development*. https://doi.org/10.1080/17565529.2024.2323523

Jiri, O., & Mafongoya, P. (2017). A synthesis of smallholder farmers' adaptation to climate change in southern Africa: Averting adaptation vacuum. In *Climate change management* (pp. 247–265). https://doi.org/10.1007/978-3-319-49520-0_16

Leal Filho, W., Barbir, J., Gwenzi, J., Ayal, D., Simpson, N. P., Adeleke, L., Tilahun, B., Chirisa, I., Gbedemah,

S. F., Nzengya, D. M., Sharifi, A., Theodory, T., & Yaffa, S. (2022). The role of indigenous knowledge in climate change adaptation in Africa. *Environmental Science & Policy, 136*, 250–260. https://doi.org/10.1016/j.envsci.2022.06.004

Makate, C., Makate, M., Mango, N., & Siziba, S. (2019). Increasing resilience of smallholder farmers to climate change through multiple adoption of proven climate-smart agriculture innovations. Lessons from Southern Africa. *Journal of Environmental Management, 231*, 858–868. https://doi.org/10.1016/j.jenvman.2018.10.069

Makondo, C. C., & Thomas, D. S. G. (2018). Climate change adaptation: Linking indigenous knowledge with western science for effective adaptation. *Environmental Science & Policy, 88*, 83–91. https://doi.org/10.1016/j.envsci.2018.06.014

Mutambisi, T., Chanza, N., Matamanda, A. R., Ncube, R., & Chirisa, I. (2021). Climate change adaptation in southern Africa: Universalistic science or indigenous knowledge or hybrid. In *African handbook of climate change adaptation: With 610 figures and 361 tables* (pp. 1751–1766). https://doi.org/10.1007/978-3-030-45106-6_8

Nhamo, G., & Chapungu, L. (2024). Prospects for a sustainable and climate-resilient African economy post-COVID-19. *Global Environmental Change, 86*, 102836. https://doi.org/10.1016/j.gloenvcha.2024.102836

Nhamo, L., Matchaya, G., Mabhaudhi, T., Nhlengethwa, S., Nhemachena, C., & Mpandeli, S. (2019). Cereal production trends under climate change: Impacts and adaptation strategies in Southern Africa. *Agriculture (Switzerland), 9*(2). https://doi.org/10.3390/agriculture9020030

Nkhata, B. (2022). Climate change and water resources in southern Africa: A resilience perspective. In *The Palgrave handbook of climate resilient societies: Volumes 1-2* (Vol. 1, pp. 413–434). https://doi.org/10.1007/978-3-030-42462-6_98

Shoko Kori, D., Kelso, C., & Musakwa, W. (2024). Climate change adaptation by smallholder farmers in Southern Africa: A bibliometric analysis and systematic review. *Environmental Research Communications, 6*(3). https://doi.org/10.1088/2515-7620/ad3127

Stringer, L. C., Dyer, J. C., Reed, M. S., Dougill, A. J., Twyman, C., & Mkwambisi, D. (2009). Adaptations to climate change, drought and desertification: Local insights to enhance policy in southern Africa. *Environmental Science and Policy, 12*(7), 748–765. https://doi.org/10.1016/j.envsci.2009.04.002

Thierfelder, C., & Steward, P. (2022). Increasing adaptation to climate stress by applying conservation agriculture in Southern Africa. In *Conservation agriculture in Africa: Climate smart agricultural development* (pp. 270–283). https://doi.org/10.1079/9781789245745.0016

Thierfelder, C., Chivenge, P., Mupangwa, W., Rosenstock, T. S., Lamanna, C., & Eyre, J. X. (2017). How climate-smart is conservation agriculture (CA)? – Its potential to deliver on adaptation, mitigation and productivity on smallholder farms in southern Africa. *Food Security, 9*, 537. https://doi.org/10.1007/s12571-017-0665-3

Ziti, C., Chapungu, L., & Nhamo, G. (2024). Climate change response strategies and implications on sustainable development goals in Mutirikwi River Sub-Catchment of Zimbabwe. *Journal of Environmental Geography, 17*(1–4), 1–14. https://doi.org/10.14232/jengeo-2024-44881

Index

M. Matsa et al. (eds.), *Climate Change Resilience in Rural Southern Africa*, Sustainable Development Goals Series, https://doi.org/10.1007/978-3-031-73600-1

The manufacturer's authorised representative in the EU is Springer Nature Customer Service Centre GmbH, Europaplatz 3, 69115 Heidelberg, Germany. If you have any concerns regarding our products, please contact ProductSafety@springernature.com

Printed and bound by CPI Group (UK) Ltd, Croydon, CR0 4YY
01/10/2026
02225937-0020